Multiscale Modelling and Optimisation of Materials and Structures

Multiscale Modelling and Optimisation of Materials and Structures

Tadeusz Burczyński
Institute of Fundamental Technological Research
Polish Academy of Sciences
Warszawa, Poland

Maciej Pietrzyk
AGH University of Science and Technology
Kraków, Poland

Wacław Kuś
Silesian University of Technology
Gliwice, Poland

Łukasz Madej
AGH University of Science and Technology
Kraków, Poland

Adam Mrozek
AGH University of Science and Technology
Kraków, Poland

Łukasz Rauch
AGH University of Science and Technology
Kraków, Poland

This edition first published 2022

Registered Offices
John Wiley & Sons, Inc., 111 River Street, Hoboken, NJ 07030, USA
John Wiley & Sons Ltd, The Atrium, Southern Gate, Chichester, West Sussex, PO19 8SQ, UK

Editorial Office
111 River Street, Hoboken, NJ 07030, USA

For details of our global editorial offices, customer services, and more information about Wiley products visit us at www.wiley.com.

Wiley also publishes its books in a variety of electronic formats and by print-on-demand. Some content that appears in standard print versions of this book may not be available in other formats.

Library of Congress Cataloging-in-Publication data applied for

Hardback ISBN: 9781119975922

Cover Design: Wiley
Cover Image: © Dr Adam Mrozek

Set in 9.5/12.5pt STIXTwoText by Straive, Pondicherry, India
Printed and bound by CPI Group (UK) Ltd, Croydon, CR0 4YY

C9781119975922_170522

Contents

Preface

Multiscale Modelling and Optimisation of Materials and Structures presents an important and challenging area of research that enables the design of new materials and structures with better quality, strength, and performance parameters. It also provides a possibility for the creation of reliable models that take into account structural, material, and topological properties at different scales. The book addresses four major areas: (i) the basic principles of macroscale, microscale, and nanoscale modeling techniques; (ii) the connection of microscale and/or nanoscale models with macrosimulation software; (iii) optimization development in the framework of multiscale engineering and the solution of identification problems; (iv) the computer science techniques used in this class of models and advice for scientists interested in developing their own models and software for multiscale analysis and optimization.

Therefore, the book presents several approaches to multiscale modelling, such as the bridging and homogenization methods, as well as the general formulation of complex optimization and identification problems in multiscale simulations. It also presents the application of global optimization algorithms based on robust bioinspired algorithms, proposes parallel and multi-subpopulation approaches in order to speed-up computations, and discusses several numerical examples of multiscale modeling, optimization, and identification of composite and functionally graded engineering materials.

Multiscale Modelling and Optimisation of Materials and Structures is thereby a valuable source of information for young scientists and students looking to develop their own models, write their own computer programs, and implement them into simulation systems.

Biography

Tadeusz Burczyński: His expertise is in computational sciences, including computational intelligence, computational mechanics, and computational materials science, especially in optimization and multiscale engineering.

Maciej Pietrzyk: His research focuses on numerical modelling including multiscale approach and the application of optimization techniques in materials science.

Wacław Kuś: His scientific interests are related to the applications of parallel and HPC methods in the optimization of multiscale problems in mechanics and biomechanics.

Łukasz Madej: His expertise is in computational materials science and process engineering. The main area of interest is full-field multiscale modelling of industrial processes and phenomena.

Adam Mrozek: His research focuses on molecular dynamic simulations, optimization of mechanical properties of the 2D materials, and multiscale modelling.

Łukasz Rauch: The main interest of his research is focused on computer science applied in industry including conventional way of modelling as well as application of surrogate models.

1

Introduction to Multiscale Modelling and Optimization

A wide selection of materials exhibits unusual in-use properties gained by control of phenomena occurring in mesoscale, microscale, and nanoscale during manufacturing. Examples of such materials range from constructional steels (e.g. AHSS – Advanced High Strength Steels for automotive industry [8] and titanium alloys for aerospace industry [10]) through various materials for energy applications [6] to new biocompatible materials for ventricular assist devices [9] or other biomedical applications [11]. Due to potential advances in materials science that could dramatically affect the most innovative technologies, further development in this field is expected. For this to happen, materials science has to be supported by new tools and methodologies, among which numerical modelling plays a crucial role.

On the other hand, to predict the correlation between processing parameters and product properties properly, one needs to investigate macroscopic material behaviour and phenomena occurring at lower dimensional scales, at grain level or even at atomistic levels. Thus, multiscale modelling with the digital materials representation (DMR) concept [7] is a research field that potentially can support the design of new products with unique in-use properties. The development of new materials modelling techniques that tackle various length scale phenomena is observed in many leading research institutes and universities worldwide. Multiscale analysis of length and temporal scales has already found a wide range of applications in many areas of science. Advantages provided by a combination of numerical approaches: finite element (FEM), crystal plasticity finite element (CPFEM), extended finite element (XFEM), finite volume (FVM), boundary element (BEM), meshfree, multigrid methods, Monte Carlo (MC), cellular automata (CA), molecular dynamics (MD), molecular statics (MS), phase field and level set methods, fast Fourier transformation, etc. are already being successfully applied in practical applications [14].

This book's main feature, which distinguishes it from other publications, is that it is focused on modelling of processing of materials and that it combines the problem of multiscale modelling with the optimization tasks providing a wide range of possibilities for practical industrial applications.

The first part of the book contains a presentation of basic principles of the microscale and nanoscale modelling techniques. The second part supplies information about applications of optimization and identification techniques in multiscale modelling. The book is recapitulated by presenting information on the computer science techniques used in multiscale modelling, and it is focused on computer implementation issues, advising scientists interested in developing their multiscale models.

Multiscale Modelling and Optimisation of Materials and Structures, First Edition. Tadeusz Burczyński, Maciej Pietrzyk, Wacław Kuś, Łukasz Madej, Adam Mrozek, and Łukasz Rauch.

1.1 Multiscale Modelling

1.1.1 Basic Information on Multiscale Modelling

During the last years, numerical modelling became a widely used tool that successfully supports and comprehends experimental research on various modern materials. Basic principles of this modelling technique, as well as the classification of models, can be found in fundamental works of Allix [1] or Fish [4], and Horstemeyer and Wang [5] which discussed possible scopes of applications of multiscale modelling in the industry. However, within the last several years, various papers on multiscale modelling have been published in the scientific literature concerning both theoretical issues and practical applications. A variety of problems was raised in these papers, and several ambiguities connected with nomenclature and definitions can be found. The problem of naming various scales in multiscale modelling is an example of a lack of consistency in definitions. Commonly, 'macro', 'meso', 'micro', and 'nano' scales are mentioned (Figure 1.1).

Nevertheless, this nomenclature can be misleading and often not sufficient. The term 'macroscale' is used for different scales, from large structures to single crystals. Others describe the single-crystal level as 'microscale'. In many cases, the characteristic entities in the particular scale are important, not the characteristic length. For example the characteristic length of the scale, in which grains are distinguishable, can vary from nanometres to centimetres. Moreover, the same grain-level scale can be applied as macroscale in connection with an atomistic scale model or as microscale with a macroscopic model of a large structure deformation. Therefore, this book uses the term 'coarse' scale for the scale with greater characteristic length and 'fine' scale for the scale with smaller characteristic length. It allows using the terms 'coarser' and 'finer' to describe relations between more than two scales. To keep the concept clear, the fine or coarse scales are usually related to the governing structures in particular cases (e.g. grains, atoms). Despite the length scale problem, there is also an issue with the time scale as many of the phenomena occur in significantly different time regimes. Usually, the temporal scale is unified across the different length scales to provide physically relevant results, which sometimes makes the computational model expensive.

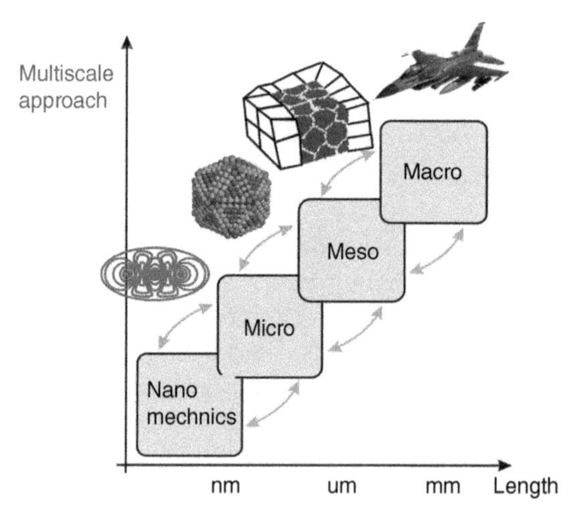

Figure 1.1 Multiscale concept diagram which illustrates 'macro', 'meso', 'micro', and 'nano' scales.

The lack of a definition for the term multiscale in the numerical modelling community is another problem. The broad understanding of multiscale modelling includes all problems with important features in multiple scales (temporal and spatial) [15]. Stricter definitions are also used. In the narrow sense of the term, modelling is multiscale when the models at different scales are simultaneously considered [13]. There is a fundamental difference between these two definitions. Important features at more than one scale are present in almost all problems of material science. A microstructural base of work hardening is one of the most common examples. Hardening is a process at a clearly different scale than plastic deformation modelled with a continuity assumption. Hence we can describe the simulation of plastic deformation and work hardening as the multiscale model. One can also define the multiscale model as a system of submodels at two or more scales but not necessarily simultaneously computed. Some examples are discussed in the book. Two submodels must work simultaneously when both are dependent on results from the second one. Such models are two-way coupled, contrary to one-way coupling where one of the models is independent.

The range of possible applications of multiscale modelling is extensive. It includes such different topics as social issues, economy, ecosystems, weather forecast, and physical and chemical processes occurring in gases, fluids, and solids. Material models are an essential branch of multiscale modelling, combining physical and chemical processes, mainly in solids. Such a wide variety of topics makes revising all multiscale modelling applications impossible. Even in material science, several families of applications can be distinguished. The main areas are modelling of polymers, composites, solidification, thin layers, or polycrystalline materials. Each of these areas has its own characteristic features, distinguishing them from the others. While the bases are common for all areas, possible solutions to particular problems are different.

In this book, applications of multiscale modelling to processes occurring in polycrystalline materials, mainly metals and their alloys, are reviewed and addressed. The most distinguishing feature of metals and their alloys is their unrepeatable microstructure. However, the atomistic structure of crystals is regular, while properties of these materials depend mostly on imperfections of atomic structures (dislocations, solid solutions of atoms, grain boundaries, etc.). The mesoscale picture of the material structures includes disordered (usually) distribution of grain shapes, sizes and orientations, as well as strong anisotropy of single crystals and grain boundaries. It is a significant difference in comparison with composites and polymers (with its repeatable microstructure). Then, the difference between the modelling of crystallization and thin layers deposition lies within lack or a very small amount of fluids. It must be remembered that modelling of fluids flow is governed by different types of equations. Moreover, numerical techniques for microscale models and coupling of scales are different in the presence of fluids. Due to this diversity in materials and different techniques used for their description, the book focuses mostly on mentioned metals and their alloys to provide in-depth information on the practical application of dedicated multiscale modelling and optimization techniques.

1.1.2 Review of Problems Connected with Multiscale Modelling Techniques

As mentioned, in the field of metallic materials, several classifications of multiscale models exist. They depend on such issues as the character of coupling between the scales or methodology of data transformation between the scales. Coupling between the scales can be either strong or weak. Strong coupling combines a description of both (or more) scales into one equations system. In weak coupling, only some data are transferred between the subsequent scales. In the fine scale, the boundary or initial conditions can be developed based on coarse scale data (top-down approach/downscaling), while data from the fine scale can be used as material properties or a material state in the coarse scale (bottom-up approach/upscaling). Weakly coupled models have a relatively flexible structure and can be classified into top-down and bottom-up approaches.

The strength of coupling has some significant consequences. The strongly coupled models are usually faster and have a better mathematical and theoretical background. Usually, such models are solved with a single numerical method. However, phenomena in all scales must be described with consistent mathematical formulations (usually partial differential equations), which is rather hard or even not possible for some phenomena characterized, e.g. by stochastic behaviour. Moreover, the strongly coupled submodels cannot be separated, and their parts cannot be replaced with other submodels, which make their adaptation difficult. The weakly coupled models are more flexible, from both mathematical and numerical points of view, which makes their development and adaptation much easier. The coupling strength is linked with a methodology of data transfer between the scales.

Various classifications of the multiscale modelling methods were proposed having in mind all the aforementioned features of these techniques. Since this book is dedicated to simulations of materials processing methods, the multiscale models are classified into two groups: upscaling and concurrent approaches. In the upscaling class of methods, constitutive models at higher scales are constructed from observations and models at lower, more elementary scales [2]. By a sophisticated interaction between experimental observations at different scales and numerical solutions of constitutive models at increasingly larger scales, physically based models and their parameters can be derived at the macroscale. In this approach, it is natural that the microscale problem has to be solved at several locations in the macro-model. For example, in the application of the concept of computational homogenization for each integration point of finite elements in the macroscale, the representative volume element (RVE) is considered in the microscale (Figure 1.2).

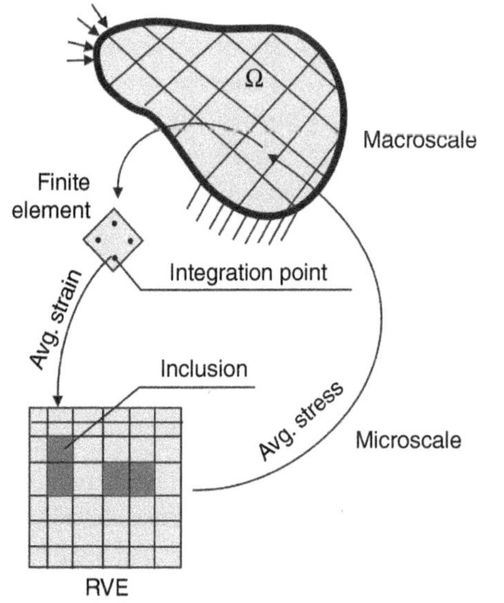

Figure 1.2 Illustration of computational homogenization concept.

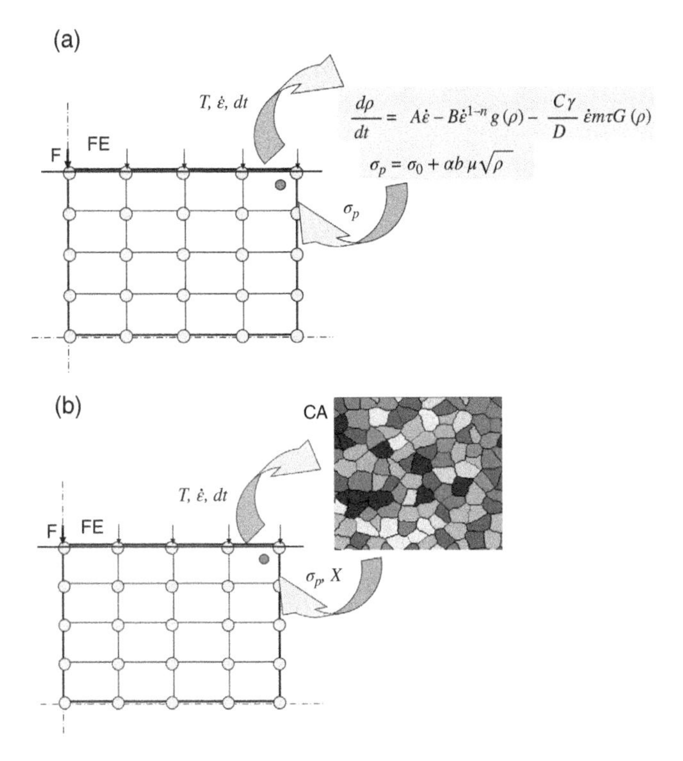

Figure 1.3 (a) Internal variable method as a precursor of upscaling and (b) the idea of the CAFE model.

Thus, classification of the multiscale problems with respect to the upscaling goals and the upscaling costs is needed, which is one of the book's objectives. The idea of upscaling has, in an intuitive manner, been used in engineering for decades. A commonly used approach based on a calculation of the flow stress in the FE model by a solution of the differential equation for the evolution of dislocation populations (Figure 1.3a) can be considered the precursor of upscaling.

Significant progress was made when more advanced discrete methods were used in the microscale; see Figure 1.3b, where the CAFE method (cellular automata finite element) is shown as an example. Now we witness further development of multiscale methods in computational science, which couple fine and coarse scales more systematically. A variety of models and methods is used in both microscale/nanoscale and macroscale. The models are generally different for the two scales, but solutions using the same model in the coarse and fine scales are common. FE^2 method is a typical example of such an approach [3].

In concurrent multiscale computing, one strives to solve the problem simultaneously at several scales (in practice two-scales) by an a priori decomposition of the computational domain. Large-scale problems are solved, and local data (e.g., displacements) are used as boundary conditions for a more detailed part of the problem. The question of how the fine scale is coupled to the coarse scale is essential in this approach. The major difficulty in coupling occurs when different models describe fine scales and coarse scales, for example coupling FE to MD. In other words, the objective is to find a computationally inexpensive but still accurate approach to the decomposition problem. The terms 'homogenization' and 'localization' are commonly used for operations during data transfer. Again, this terminology is not well defined. Homogenization, in multiscale modelling, has two meanings. In the context of strongly coupled models, based on differential equations, it is understood as the asymptotic homogenization, the method of studying partial differential equations with rapidly oscillating coefficients. The in-depth description of the homogenization and the averaging based on the perturbation theory, as well as their application to multiscale modelling, can be found e.g. in [12]. However, most authors use this term differently, describing it as 'any method of transforming a heterogeneous field into a homogeneous one'. Weighted averaging is usually used; however, more advanced techniques are also present. In this book, we use the term homogenization in common understanding and asymptotic homogenization in the sense of the element of perturbation theory. Details on the multiscale classification and examples of its applications can be found in Chapter 5 of the book.

1.1.3 Prospective Applications of the Multiscale Modelling

In recent years, a gradual paradigm shift has been taking place in the selection of materials to suit particular engineering requirements, especially in high-performance applications. The empirical approach adopted by materials scientists and engineers in choosing materials parameters from a database is being replaced by design based on the DMR concept. Features that span across a large spectrum of length scales are altered and controlled to achieve the desired properties and performance at the macroscale. Research efforts, in this aspect, include the development of engineering materials by changing the composition, morphology, and topology of their constituents at the microscopic/mesoscopic level. The objective of this book is to show multiscale methods and their applications in computational materials design. From one side, we present computational multiscale material modelling based on the bottom-up/top-down, one-way coupled description of the material structure in different representative scales. On the other side, our intention is to show possibilities of a combination of multiscale methods with optimization techniques.

1.2 Optimization

Solution of optimization problems in multiscale modelling allows finding structures with better performance or strength in one scale with respect to design variables in another scale. In this case, the typical situation is to find a vector of material or geometrical parameters on the micro-level, which minimizes an objective function dependent on state fields on a macro-level of the structure.

A special case of optimization problems associated with the multiscale approach is the optimization of atomic clusters for the minimization of the system's potential energy. This case has an important consequence, especially in the design of new 2D nanomaterials and nanostructures.

The identification problem is formulated as the evaluation of some geometrical or material parameters of structures in one scale having measured information in another scale. The important case of identification in multiscale modelling is to find material properties, the shape of the inclusions/fibres or voids in the microstructure having measurements of state fields made on the macro-object. The identification problem is formulated and considered as a special optimization task.

The analysis methods of multiscale models based on computational homogenization are adopted for these classes of problems. To solve optimization and identification problems, global optimization methods based on bioinspired algorithms are used.

1.3 Contents of the Book

The aspects of multiscale modelling mentioned previously have already been discussed in numerous publications, including several books [4, 12, 13, 15]. However, multiscale modelling still remains a difficult task, and its valid and reliable application is quite difficult. Moreover, the lack of an unambiguous definition leads to misunderstandings and mistakes. The book supplies some practical information concerning the development and application of multiscale material models, in particular in combination with optimization techniques.

As mentioned, the book is divided into three main sections. The first section is composed of Chapters 2 and 3, and it is focused on discussion of phenomena occurring in materials in processing and on models used to describe these phenomena. Such phenomena as recrystallization, phase transformations, cracking, fatigue, and creep are discussed very briefly. The modelling methods are divided into two groups. The first includes computational methods for continuum such as FEM, XFEM, and BEM. Discrete methods describing microscale and nanoscale phenomena include MS as well as MD, CA, and MC approaches. Computational homogenization methods, which are used in the coupled multiscale models, are also discussed in this part of the book. Basic principles of methods of optimization are also described in this part of the book, with a particular emphasis on the methods inspired by nature.

The second section of the book is composed of Chapters 4 and 5 and focuses on DMR. Case studies based on DMR are presented. Applications of methods described in Chapters 2 and 3 to modelling various processes and phenomena are shown.

The third section of the book is connected with applications of multiscale optimization methods. Such issues as optimization of atomic clusters, material parameters optimization, shape optimization, and topological optimization are discussed. The problem of identification in multiscale modelling is also presented.

Computer implementation issues for multiscale models are recapitulated in the book. Implementation of selected algorithms concerning visualization as well as scales coupling with use of commercial software are described in Chapter 7. The part will highlight the possibilities of increasing the computational efficiency, which is especially important when optimization based on a direct problem model of multiscale nature is considered.

References

1 Allix, O. (2006). Multiscale strategy for solving industrial problems. In: *III European Conference on Computational Mechanics* (ed. C.A. Motasoares, J.A.C. Martins, H.C. Rodrigues, et al.), 107–126. Dordrecht: Springer.

2 De Borst, R. (2008). Challenges in computational materials science, multiple scales, multiphysics and evolving discontinuities. *Computational Material Science* 43: 1–15.

3 Feyel, F. (1999). Multiscale FE2 elastoviscoplastic analysis of composite structures. *Computational Materials Science* 16: 344–354.

4 Fish, J. (2014). *Practical Multiscaling*. Wiley.

5 Horstemeyer, M.F. and Wang, P. (2003). Cradle-to-grave simulation-based design incorporating multiscale microstructure-property modeling: reinvigorating design with science. *Journal of Computer-Aided Materials Design* 10: 13–34.

6 Kilner, J.A., Skinner, S.J., Irvine, S.J.C., and Edwards, P.P. (ed.) (2012). *Functional Materials for Sustainable Energy Applications*. Oxford, Cambridge, Philadelphia, New Delhi: Woodhead Publishing Ltd.

7 Madej, L., Wang, J., Perzynski, K., and Hodgson, P.D. (2014). Numerical modeling of dual phase microstructure behavior under deformation conditions on the basis of digital material representation. *Computational Materials Science* 95: 651–662.

8 Matlock, D.K., Krauss, G., and Speer, J.G. (2005). New microalloyed steel applications for the automotive sector. *Materials Science Forum* 500–501: 87–96.

9 Milenin, A. and Kopernik, M. (2011). Microscale analysis of strain-stress state for TiN nanocoating of POLVAD and POLVAD_EXT. *Acta of Bioengineering and Biomechanics* 13: 11–19.

10 Muszka, K., Lopez-Pedrosa, M., Raszka, K. et al. (2014). The impact of strain reversal on microstructure evolution and orientation relationships in Ti-6Al-4V with an initial alpha colony microstructure. *Metallurgical and Materials Transactions A* 45: 5997–6007.

11 Niinomi, M. (2002). Recent metallic materials for biomedical applications. *Metallurgical and Materials Transactions A* 33: 477–486.

12 Pavliotis, G. and Stuart, A. (2008). *Multiscale Methods: Averaging and Homogenisation*. Springer.

13 Weinan, E. (2011). *Principles of Multiscale Modeling*. Cambridge University Press.

14 Yanagimoto, Y., Banabic, D., Banu, M., and Madej, L. (2022). *Simulation of Metal Forming – Visualization of Invisible Phenomena in the Digital Era*. CIRP Annals Manufacturing Technology in press.

15 Yip, S. (ed.) (2005). *Handbook of Materials Modeling*. Dordrecht: Springer.

2

Modelling of Phenomena

The major physical phenomena responsible for material behaviour under manufacturing and exploitations stages are described in this chapter. It highlights the limitations of conventional modelling approaches and introduces possibilities provided by the multiscale solutions. Both nanoscale and microscale phenomena, including quantum scale, are discussed, and nanostructures and microstructures are analyzed.

2.1 Physical Phenomena in Nanoscale

The laws of quantum mechanics underlie the basis of the description of physical phenomena, as well as simulation methods in the nanoscale. However, the solution of the Schrödinger equation – the fundamental equation which describes the behaviour of the particles in quantum scale – even the numerical one, is restricted to the relatively simple cases of molecular systems. The high complexity of the Schrödinger equation and the high dimension of the space in which it is defined exclude the pure *ab initio* methods [70] from the mechanical engineering computations presented in this book. Therefore, a series of subsequent simplifications and approximations procedures are used to overcome this problem and make the molecular simulations possible on the desired level of accuracy. The quantum mechanics computations, such as methods based on the density functional theory (DFT) [23, 94], are also beyond the scope of this book.

This chapter describes the route of the derivation of classical molecular dynamics (MD) [33, 38, 103], based on the Newtonian equations of motion, from the Schrödinger wave equation. Presented methodology and hierarchy of introduced simplifications also reveal the basis of the formulation of atomic potentials – the simplified models of the interatomic interactions. The problems with a proper approximation of the interactions, which determine behaviour of the atoms in classical MD, underlie the development of the mixed, quantum-classical approaches (like Car-Parinello MD [12, 93]), where the motion of the nuclei is determined by the Newtonian equation of motion, while the elements of the electronic structure are still treated as quantum objects. Such approaches allow estimation of the potential energy and interatomic forces on the fly in each MD integration step without engaging any predefined potentials. However, in the case of large mechanical computations at the nanoscale (up to billions of atoms), the classical MD method still plays an important role [101]; thus, the most popular atomic potential models are presented in the second section of this chapter.

Multiscale Modelling and Optimisation of Materials and Structures, First Edition. Tadeusz Burczyński, Maciej Pietrzyk, Wacław Kuś, Łukasz Madej, Adam Mrozek, and Łukasz Rauch.

2.1.1 The Linkage Between Quantum and Classical Molecular Mechanics

In quantum mechanics, behaviour of the particles (nuclei, electrons, etc.) is described by the Schrödinger equation. Unlike the classical equation of motion, the solution of the Schrödinger equation provides probabilistic information about the quantum mechanical system.

The time-dependent wave function Ψ of a quantum system with i electrons and I nuclei has a general form:

$$\Psi = \Psi(\mathbf{r}_1, \ldots, \mathbf{r}_i, \mathbf{R}_1, \ldots, \mathbf{R}_I, t), \tag{2.1}$$

where: vectors $\mathbf{r}_i$ and $\mathbf{R}_i$ denote, respectively, the electronic and nuclear coordinates in the three-dimensional space. In this chapter, the abbreviated notation ($\mathbf{r}$ and $\mathbf{R}$) is used. The probability of finding the quantum system in the elementary volume element dV of the considered configuration space centred at the point ($\mathbf{r}$, $\mathbf{R}$) at time t is expressed as:

$$\Psi^*(\mathbf{r}, \mathbf{R}, t)\Psi(\mathbf{r}, \mathbf{R}, t)dV. \tag{2.2}$$

The full wave function (2.2) can be obtained as the solution of the time-dependent Schrödinger equation:

$$i\hbar\frac{\partial\Psi(\mathbf{r}, \mathbf{R}, t)}{\partial t} = H\Psi(\mathbf{r}, \mathbf{R}, t). \tag{2.3}$$

The standard Hamiltonian H of the quantum system is given as a sum over potential and kinetic energy operators:

$$H = T_e + V_{ee} + V_{eN} + V_{NN} + T_N, \tag{2.4}$$

where: T_e and T_N denote kinetic energies of the electrons and the nuclei, respectively:

$$T_e = -\frac{\hbar^2}{2m_e}\sum_i \nabla_i^2, \tag{2.5}$$

$$T_N = -\frac{\hbar^2}{2M_I}\sum_I \nabla_I^2. \tag{2.6}$$

The operators: V_{ee}, $V_{eN,}$, and V_{NN} are the potential energies of the interactions (Coulomb potentials) between the electrons, the nuclei, and between the electrons and the nuclei, respectively:

$$V_{ee} = \sum_{i<j}\frac{e^2}{|\mathbf{r}_i - \mathbf{r}_j|}, \tag{2.7}$$

$$V_{eN} = \sum_{i<j}\frac{e^2 Z_I}{|\mathbf{R}_I - \mathbf{r}_j|}, \tag{2.8}$$

$$V_{ee} = \sum_{i<j}\frac{e^2 Z_I Z_J}{|\mathbf{R}_i - \mathbf{R}_j|}, \tag{2.9}$$

where: Z_I and M_I refer to the atomic number and mass of the I-th nucleus, and m_e and e denote mass and charge of the electron, respectively.

One of the possible approaches of the derivation of classical MD, according to [33, 71, 128], utilizes the equations of the *time-dependent self-consistent field* (TDSCF) method. These equations can be obtained by decomposition of the Hamiltonian (2.4) and subsequent separation of the nuclear and electronic contributions to the wave function (2.1).

The full Hamiltonian operator (2.4) can be rearranged to a simpler form:

$$H = H_e + T_N, \tag{2.10}$$

where electronic Hamiltonian H_e is defined as follows:

$$H_e = T_e + V_{ee} + V_{eN} + V_{NN}, \tag{2.11}$$

and may be written as:

$$H_e = T_e + V_e. \tag{2.12}$$

The term V_e corresponds to the total potential energy of the considered system:

$$V_e = V_{ee} + V_{eN} + V_{NN}. \tag{2.13}$$

The full wave function, given in the form (2.1), depends on the electronic as well as nuclear coordinates (also known as *fast* and *slow* coordinates, respectively). These two contributions have to be separated. The traditional way is the approximation of the full wave function with the following product form:

$$\Psi(\mathbf{r}, \mathbf{R}, t) \approx \phi(\mathbf{r}, t)\chi(\mathbf{R}, t) \exp\left[\frac{i}{\hbar}\int_{t_0}^{t} \widetilde{E}_e(t')dt'\right]. \tag{2.14}$$

This kind of simplification refers to the one-determinant or single-configuration version of the full wave function. The electronic $\phi(\mathbf{r},t)$ and nuclear $\chi(\mathbf{R},t)$ wave functions should be normalized to unity for any time point t, so the following conditions should be satisfied:

$$\langle \phi, t \mid \phi, t\rangle = 1, \quad \langle \chi, t \mid \chi, t\rangle = 1. \tag{2.15}$$

The $\widetilde{E}_e$ term in the formulation (2.14) is the phase factor, specified as:

$$\widetilde{E}_e(t) = \int \phi^*(\mathbf{r}, t)\chi^*(\mathbf{R}, t)H_e\phi(\mathbf{r}, t)\chi(\mathbf{R}, t)d\mathbf{R}d\mathbf{r}. \tag{2.16}$$

The Schrödinger equation for the electronic wave function $\phi(\mathbf{r},t)$ might be obtained by substituting approximation (2.14) into the Schrödinger Eq. (2.3) with the full Hamilton operator (2.4). After the multiplication of (2.3) from the left by $\langle\varphi|$, $\langle\chi|$, and assuming energy conservation:

$$\frac{d\langle H\rangle}{dt} = 0, \tag{2.17}$$

the following coupled system of equations can be formulated:

$$i\hbar\frac{\partial\phi}{\partial t} = -\sum_i \frac{\hbar^2}{2m_e}\nabla_i^2\phi + \left\{\int \chi^*(\mathbf{R}, t)V_e(\mathbf{R}, \mathbf{r})\chi(\mathbf{R}, t)d\mathbf{R}\right\}\phi, \tag{2.18}$$

$$i\hbar\frac{\partial\chi}{\partial t} = -\sum_I \frac{\hbar^2}{2M_I}\nabla_I^2\chi + \left\{\int \phi^*(\mathbf{r}, t)H_e(\mathbf{R}, \mathbf{r})\phi(\mathbf{r}, t)d\mathbf{r}\right\}\chi. \tag{2.19}$$

Equations (2.18) and (2.19) are the fundamentals of the TDSCF method, developed by Dirac [21]. This is a mean-field method, i.e. the nuclei ('slow' degrees of freedom) move in the average field of the electrons (the 'fast' ones) and vice versa. Both types of degrees of freedom are coupled together, and the proper feedback in the two directions is assured. It can be noted that mean-field description is a price to pay for the simplest separation of electronic and nuclear coordinates [71].

Since the TDSCF method is still based on quantum-mechanical equations, other approximations and simplifications must be introduced to reveal the linkage between quantum and classical MD. In the subsequent step, the nuclei should be treated as classical point particles. It can be done by rewriting the nuclear wave function as:

$$\chi(\mathbf{R},t) = A(\mathbf{R},t)\exp\left[\frac{i}{\hbar}S(\mathbf{R},t)\right]. \tag{2.20}$$

The amplitude and phase factors, denoted by A and S, respectively, should be both real and positive in the polar representation. In the next stage, the nuclear wave function (2.20) is substituted into Eq. (2.19), and after splitting the real and imaginary parts, the coupled equations for the nuclei expressed in the variables A and S can be obtained:

$$\frac{\partial S}{\partial t} + \sum_I \frac{1}{2M_I}(\nabla_I S)^2 + \int \phi^* H_e \phi d\mathbf{r} = \hbar^2 \sum_I \frac{1}{2M_I}\frac{\nabla_I^2 A}{A}, \tag{2.21}$$

$$\frac{\partial A}{\partial t} + \sum_I \frac{1}{M_I}(\nabla_I A)(\nabla_I S) + \sum_I \frac{1}{2M_I} A\left(\nabla_I^2 S\right) = 0. \tag{2.22}$$

Equations (2.21) and (2.22) formulate quantum fluid dynamics representation [71] and can be used to solve the time-dependent Schrödinger equation. In the further process of derivation of the classical MD, Equation (2.21) plays an even more important role. The right side of this equation directly depends on $\hbar$, but this term can be neglected when the classical limit ($\hbar \to 0$) is applied:

$$\frac{\partial S}{\partial t} + \sum_I \frac{1}{2M_I}(\nabla_I S)^2 + \int \phi^* H_e \phi \, d\mathbf{r} = 0. \tag{2.23}$$

After neglecting the terms proportional to $\hbar$, the newly obtained equation is isomorphic to the Hamilton-Jacobi formulation of equations of motion – a well-known form of the classical mechanics:

$$\frac{\partial S}{\partial t} + H(\mathbf{R}, \nabla_I S) = 0. \tag{2.24}$$

The Hamiltonian operator in Eq. (2.24) is now defined in terms of generalized coordinates $\mathbf{R}=(\mathbf{R}_1 ... \mathbf{R}_I)$ and their conjugated momentum $\mathbf{P}=(\mathbf{P}_1 ... \mathbf{P}_I)$:

$$H(\mathbf{R},\mathbf{P}) = T(\mathbf{P}) + V(\mathbf{R}), \tag{2.25}$$

where: $T(\mathbf{P})$ and $V(\mathbf{R})$ are appropriate kinetic and potential energy operators, respectively. The momentum $\mathbf{P}_I$ of the I-th nuclei and associated force $\dot{\mathbf{P}}_I$ is defined as follows:

$$\mathbf{P}_I \equiv \nabla_I S, \tag{2.26}$$

$$\dot{\mathbf{P}}_I = -\nabla_I V(\mathbf{R}_I). \tag{2.27}$$

Now, according to formulation (2.23), the Newtonian equations of motion can be formulated in the following way:

$$\frac{d\mathbf{P}_I}{dt} = -\nabla_I \int \phi^* H_e \phi = -\nabla_I V_e^E(\mathbf{R}(t)), \tag{2.28}$$

or

$$M_I \ddot{\mathbf{R}}_I(t) = -\nabla_I V_e^E(\mathbf{R}(t)). \tag{2.29}$$

At this stage, the motion of the nuclei is described by the familiar law of classical Newtonian mechanics (2.29) in the field of the effective potential V_e^E (the *Ehrenfest potential*) generated by the electronic system. This kind of potential is a function of only 'slow' (nuclear) coordinates $\mathbf{R}$ at time point t and can be interpreted as a result of averaging H_e over the 'fast' (electronic) degrees of freedom, i.e. computing the quantum expectation value of the Hamiltonian H_e at fixed, current positions of the nuclei.

It can be noted that the electronic Eq. (2.18) of the TDSCF method is still expressed in terms of the nuclear wave function $\chi(\mathbf{R},t)$. The wave function has to be substituted by the positions of the nuclei. One of the convenient ways to do this is to replace the probability density of the nuclei $|\chi(\mathbf{R}, t)|^2$ with a product of delta functions $\prod_I \delta(\mathbf{R}_I - \mathbf{R}_I(t))$ and apply again the classical limit ($\hbar \rightarrow 0$) to (2.18). The delta functions should be centred at the current positions $\mathbf{R}(t)$ of the nuclei, obtained by solving Eq. (2.29). This approach leads to the time-dependent electronic Schrödinger equation:

$$i\hbar \frac{\partial \phi}{\partial t} = H_e(\mathbf{r}, \mathbf{R}(t))\phi(\mathbf{r}, \mathbf{R}(t), t). \tag{2.30}$$

Equations (2.29) and (2.30) are the base of the Ehrenfest MD – one of the hybrid quantum-classical approaches, like Born-Oppenheimer or Car-Parinello MD [12, 93, 128]. The nuclei are treated now as slow classical particles which move according to the Newtonian equation of motion (2.29), while the fast electrons are still quantum objects. The Hamiltonian operator H_e and electronic wave function ϕ now depend on the classical nuclear positions $\mathbf{R}(t)$, so the feedback between classical and quantum parts is guaranteed in both directions. Keeping in mind conventional MD, the last problem which has to be solved is the proper approximation of the effective potential function V_e^E in the classical equation of motion (2.28).

As mentioned earlier, Ehrenfest MD is a mean-field method derived from the TDSCF approach. However, transitions between electronic states are still possible. The electronic wave function should be written as a linear combination of the base functions ϕ_j of the electronic states and complex coefficients $c_j(t)$ for given time point t:

$$\phi(\mathbf{r}, \mathbf{R}, t) = \sum_{j=0}^{\infty} c_j(t)\phi_j(\mathbf{r}, \mathbf{R}), \tag{2.31}$$

where one assumes: $\sum_j |c_j(t)|^2 \equiv 1$. The convenient choice for the base functions ϕ_j is a set of adiabatic base functions, which can be obtained as solution of the time-independent Schrödinger equation, treated as the eigenvalue problem:

$$H_e(\mathbf{r}, \mathbf{R})\phi_j(\mathbf{r}, \mathbf{R}) = E_j(\mathbf{R})\phi_j(\mathbf{r}, \mathbf{R}). \tag{2.32}$$

It can be noted that vector $\mathbf{R}$ refers now to the instantaneous nuclear coordinates at time t, and E_j are the energy eigenvalues of the electronic Hamiltonian H_e.

Deriving an appropriate approximation model of the potential V_e^E requires further simplification of the total electronic wave function to the single state function. Typically, only the ground state function ϕ_0 of the H_e is taken under consideration [71], and the expansion (2.31) is truncated after the first term. This approximation is admissible only if the difference in energy between the ground and the first excited state is large compared to the thermal energy.

Restriction to the ground state means that the nuclei move according to the classical Newtonian Eq. (2.29) on a single potential energy surface (PES):

$$V_e^E = \int \phi_0^* H_e \phi_0 \equiv E_0, \tag{2.33}$$

which can be determined by solving the independent Schrödinger equation for ground state only:

$$H_e \phi_0 = E_0 \phi_0. \tag{2.34}$$

Now, the PES computations may be separated from the computations of the motion of the nuclei. The general scheme for deriving pure classical MD follows:

- Compute the ground state energy $E_0(\mathbf{R})$ for as many nuclear configurations $\mathbf{R}$ as possible, solving eigenvalue problem (2.34).
- Render the PES V_e^E from obtained data points $(\mathbf{R}, E_0(\mathbf{R}))$ using numerical interpolation methods; it can be noted that numerical computation of the gradients of V_e^E should be possible.
- Solve the system of the Newtonian equations of motion for given initial conditions to obtain the trajectories of the nuclei (molecules). This can be done using, e.g. the velocity Verlet integration algorithm described in the Section 3.2.1.

Reconstructing the PES for any large atomic system is, generally, a non-trivial task. With an increasing number of nuclei, the number of possible nuclear configurations rises extremely fast, making it impossible to determine ground state energies for each configuration (see the optimization of the atomic clusters). It is unclear how many and which particular configurations must be considered to compute the PES with an admissible error. The standard way to overcome this problem is to approximate the PES V_e^E by the *truncated* expansion of many-body potentials:

$$V_e^E \approx V_e^{approx}(\mathbf{R}) = \sum_{I=1}^{N}(\mathbf{R}_I) + \sum_{I<J}^{N}(\mathbf{R}_I, \mathbf{R}_J) + \sum_{I<J<K}^{N}(\mathbf{R}_I, \mathbf{R}_J, \mathbf{R}_K) + \ldots. \tag{2.35}$$

As a consequence of this formulation, the electronic coordinates are replaced by interaction potentials V_N and no longer exist as explicit degrees of freedom in the equations of motion. This means that hybrid quantum-classical problems (2.29) and (2.33) turn into pure classical mechanics and classical MD:

$$M_I \ddot{\mathbf{R}}_I(t) = -\nabla_I V_e^{\text{approx}}(\mathbf{R}_I(t)). \tag{2.36}$$

At this stage, only the proper model of atomic potential has to be determined. In practical MD calculations, two basic methods are used. First, mentioned earlier, is based on the ab initio method: solving the time-independent Schrödinger equation for a given set of nuclear configurations and further approximating the obtained results. This approach has fundamental meaning for the embedded atom method (EAM) [19, 20, 28] used in the MD simulations of metallic materials. The second method is the empirical approach: fitting a set of parameters of the potential function, given in the analytic form. In this approach, expansion (2.35) is usually truncated after the second or third term, i.e. only interactions between two or three particles are considered. This is a radical simplification and introduces an error in the total potential energy estimation. Typical representatives of the analytic potential functions are pair-wise Lennard-Jones or Stillinger-Weber (three-body interactions) potential. The next section will discuss all the popular potential models with the physical meaning of their parameters.

2.1.2 Atomic Potentials

The models of interatomic interactions, used in the methods of computational nanomechanics to describe the potential energy and atomic forces, determine the reality, efficiency, and precision of the performed simulations. The right type of atomic potential should be chosen before starting the non-quantum molecular simulation. The modern implementations of MD [33, 38, 56, 103] or molecular statics (MS) [10] methods have many built-in classical atomic potential models. Most of the types of atomic potentials can be divided into two main classes. The first class consists of the simple two-body (such as Coulomb, Morse, and Lennard-Jones) or three-body (Stillinger-Weber) interactions, given in the parametric analytical form. The second group is the more sophisticated many-body approaches, like EAM, and whole family of bond order (BO) potentials: from several generations of reactive empirical bond order (REBO) potentials to the ab initio based approaches, like reactive force fields (ReaxFF).

These models have become popular and are widely used in the atomic-level modelling of metallic materials (EAM), hydrocarbons, 2D materials (Stillinger-Weber, REBO, and ReaxFF), and complex molecules. Such potentials can describe the short- and long-range interactions, as well as different types of the hybridization states of the atom. The values of the potential's parameters can be obtained experimentally or during ab initio computations. The simple pair-wise models are commonly used in large-scale molecular simulations especially due to the low cost of the computations. The popular mechanical engineering computations, pair-wise and many-body approaches, will be further discussed in detail.

2.1.2.1 Lennard-Jones Potential

One of the most popular and commonly used pair-wise potentials was derived from the van der Waals interaction forces by the Lennard-Jones in 1924 [61, 62]. This potential has a well-known '12-6' form:

$$V^{LJ}(r_{ij}) = 4\varepsilon\left[\left(\frac{\sigma}{r_{ij}}\right)^{12} - \left(\frac{\sigma}{r_{ij}}\right)^{6}\right], \tag{2.37}$$

and is parametrized by the two factors σ and ε; r_{ij} denotes the distance between two atoms i and j. The parameter ε refers to the bonding energy and describes the depth (minimum) of the potential (Figure 2.1). Increasing the value of ε leads to stronger interatomic bonds, and therefore harder materials can be modelled. The factor σ defines the zero crossing of the potential, which is conjugated with one of the most important parameters of the potential – the equilibrium distance r_0:

$$r_0 = \sqrt[6]{2}\sigma. \tag{2.38}$$

The derivative of the atomic potential, with respect to the interatomic distance vector $\mathbf{r}_{ij}$, defines the forces acting between two atoms i and j:

$$\mathbf{f}_{ij} = -\frac{\partial V(r_{ij})}{\partial r_{ij}}\frac{\mathbf{r}_{ij}}{r_{ij}} \qquad \mathbf{f}_{ji} = -\mathbf{f}_{ij}. \tag{2.39}$$

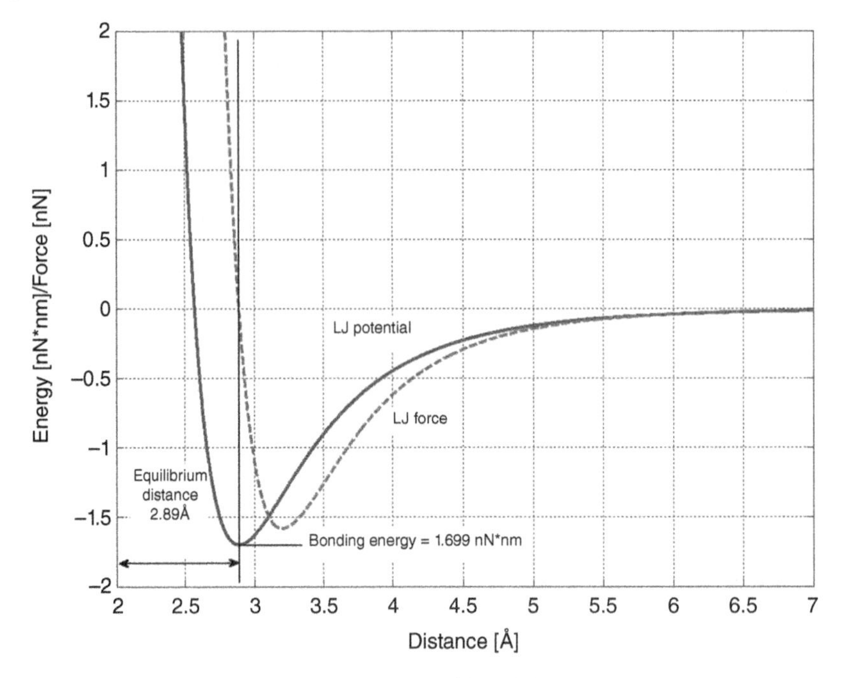

Figure 2.1 The Lennard-Jones atomic potential and force.

As a consequence of relation (2.39), the equilibrium distance corresponds to the zero crossing of the atomic interaction force (Figure 2.1), which defines the behaviour of the considered particles. Two elements attract each other on distances greater than r_0 and repulse on smaller distances to avoid a possible collision.

The Lennard-Jones potential presented in the form given by Eq. (2.37) was basically developed to model weak interaction in the noble gases (such as argon or xenon); however, it is commonly used for modelling general effects at the nanoscale in solids and liquids due to simplicity and low computation costs.

2.1.2.2 Morse Potential

Another popular two-body interaction model is the formula introduced by Philip Morse:

$$V^{\text{Morse}}\left(r_{ij}\right) = D_e\left[e^{-2\alpha\left(r_{ij}-r_0\right)} - 2e^{-\alpha\left(r_{ij}-r_0\right)}\right], \tag{2.40}$$

as a potential energy representation in the Schrödinger equation for diatomic molecules [78]. The Morse potential has three fitting parameters: the equilibrium distance r_0, the dissociation energy D_e which, similar to the ε in Lennard-Jones formulation, refers to the minimum energy level at r_0. The last parameter α defines the curvature of overall potential, which is related to the frequency of the bond vibrations. It can be noted that for $\alpha = 2.2$; the Morse potential has a similar shape to the Lennard-Jones one but slightly different interaction at large distances. The visual comparison of the Lennard-Jones and Morse potentials with different curvatures is shown in Figure 2.2.

The native applications of the Morse potential are interactions in the diatomic molecules (such as N_2, F_2) or modelling the covalent bondings, but this formula is also widely used in

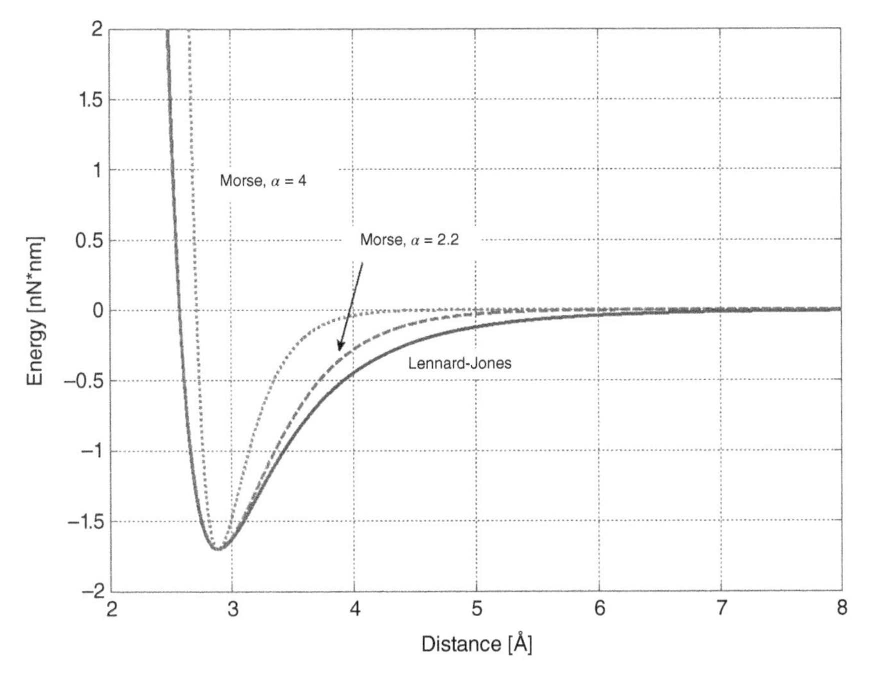

Figure 2.2 The Morse potential.

the simplified or large-scale simulations of metallic solids. Application of this kind of potential to the monoatomic metals is discussed in [32], while sets of parameters are also given by [115]. Determination of the values of the Morse potential's parameters on the basis of the EAM (see Section 2.1.2.7) for face centred cubic (FCC) crystals is presented in [1]. Although the Eq. (2.40) looks simple, the computational cost is higher than in the case of Lennard-Jones potential due to exponential terms existing in the considered formulation [27].

2.1.2.3 Stillinger-Weber Potential

In the pair-wise potentials, only the distance between two atoms is taken into account. The introduction of the higher-order terms of (2.35) provides more information about the local environment and spatial configurations of the particles because, in addition to the linear distances, interatomic angles and coordination numbers (i.e. local densities of atoms) may be taken into account. Thus, this kind of potential is usually fitted to certain, specific types of atomic lattices, such as hexagonal close packed (HCP) cell or face centered cubic (FCC) cell. The many-body interaction models play an important role, especially in the simulations of complex molecular structures, metallic solids, and 2D materials like graphene, its derivatives, and single-layered MoS_2.

One of the typical many-body potentials that provides a good description of the covalent bondings in the amorphous and the diamond crystal silicon is the Stillinger-Weber potential [118], constructed from two- and three-body terms:

$$V^{SW} = \sum_{i,j>i} V_2^{SW}\left(r_{ij}\right) + \sum_{i,j>i,k>j} V_3^{SW}\left(\mathbf{r}_{ij}, \mathbf{r}_{ik}\right), \tag{2.41}$$

where the pair-wise and many-body terms (respectively, V_2^{SW} and V_3^{SW}) have the following forms:

$$V_2^{SW}(r_{ij}) = \varepsilon A\left(B\sigma^p r_{ij}^{-p} - \sigma^q r_{ij}^{-q}\right)e^{\left[\sigma\left(r_{ij} - a\sigma\right)^{-1}\right]}, \tag{2.42}$$

$$V_3^{SW}(\mathbf{r}_{ij}, \mathbf{r}_{ik}) = \varepsilon\lambda e^{\left[\gamma\sigma\left(r_{ij} - a\sigma\right)^{-1} + \gamma\sigma\left(r_{jk} - a\sigma\right)^{-1}\right]}\left(\cos\theta_{jik} - \cos\theta_0\right)^2. \tag{2.43}$$

As in previous cases, the parameters σ and ε are related to the equilibrium bond length and bonding energy. However, the many-body term V_3^{SW} depends on the angular configuration between a particular triad of atoms j-i-k, and θ_{jik} refers to the angle between vectors $\mathbf{r}_{ij}$ and $\mathbf{r}_{ik}$. The values of the scaling parameters existing in the formulas (2.41)–(2.43) can be fitted using one of the classical or bio-inspired [47] optimization algorithms on the base of the data obtained experimentally or during ab initio computations [131].

The practical example, simulation of the tensile test of the single-layered MoS_2, showing the difference in the results obtained using two different parameterizations [44, 47] of the Stillinger-Weber potential, is presented in Section 4.1.1.6.

In the case of simulations of silicon-based or other covalently bonded materials, the Tersoff potential [126], its multi-component modifications [127], the modified embedded atom method (MEAM), and the whole class of the BO potentials are frequently used. Both of them are usually implemented in many MD simulation packages such as large-scale atomic/molecular massively parallel simulator (LAMMPS) [56]. The BO models and the MEAM are presented in the subsequent sections of this chapter.

2.1.2.4 Reactive Empirical Bond Order (REBO) Potential

REBO potential is another, more sophisticated than the Stillinger-Weber, many-body interaction model, which combines the common pair-wise interactions with the BO function, which depends on angular distances and coordination numbers of atoms. The original formulation comes from an extension of the Abell-Tersoff formalism [8]:

$$V^{\text{REBO}} = \frac{1}{2}\sum_{i \neq j} f_{ij}^C(r_{ij})\left[V^R(r_{ij}) - b_{ij}V^A(r_{ij})\right]. \tag{2.44}$$

The b_{ij} and f_{ij}^C denote, respectively, the BO and the cutoff functions. The V^R and V^A terms in formulation (2.44) are responsible for a pair-wise repulsion between atomic cores and attraction of valence electrons.

The idea of the REBO model has been intensively developed over the years. The second generation of this potential [8], originally fitted for hydrocarbons, is given as:

$$V^{\text{REBO}} = \frac{1}{2}\sum_{i \neq j} f_{ij}^C(r_{ij})\left[\left(1 + \frac{Q}{r_{ij}}\right)A{\cdot}e^{-\alpha \cdot r_{ij}} - b_{ij}B{\cdot}e^{-\beta \cdot r_{ij}}\right], \tag{2.45}$$

where the set of parameters: A, B, Q, α, and β is responsible for determining the bonding energies.

Another version, called adaptive intermolecular reactive bond order (AIREBO) potential, extends the basic model with an additional torsion and Lennard-Jones-like long-range terms [119]. Such a formulation and proper parameterization allow to simulate the behaviour of various allotropes of carbon and 2D graphene-like materials with high accuracy [81, 82].

The previously mentioned BO function b_{ij} has the general complex form:

$$b_{ij} = \left[1 + \sum_{k \neq i,j} f_{ij}^{C}(r_{ij}) \cdot G\left[\cos\left(\theta_{ijk}\right)\right] + P(N_i)\right]^{-1/2}, \tag{2.46}$$

and depends on the coordination term P, bond angle function G, and the number of neighbours of i-th atom, denoted by N_i. Explicit formulations of these functions depend on the type of modelled atoms or molecules. For example, for $SLMoS_2$ atomic systems (please refer to Section 4.1.1.6. for details), BO, neighbour, and cutoff functions are expressed as [63, 64]:

$$\begin{aligned} P(N_i) &= -a_0 \cdot (N_i - 1) - a_1 \cdot e^{-a_2 \cdot N_i} + a_3 \\ N_i &= N_i^{Mo} + N_i^{S} = \sum_{k \neq i,j}^{Mo} f_{ik}^{C}(r_{ik}) + \sum_{l \neq i,j}^{S} f_{il}^{C}(r_{il}) \\ f_{lk}^{C}(r_{ij}) &= \begin{cases} 1 & r_{ij} < R_{lk}^{\min} \\ \dfrac{\left\{1 + \cos\left[\left(r_{ij} - R_{lk}^{\min}\right)\pi / \left(R_{lk}^{\max} - R_{lk}^{\min}\right)\right]\right\}}{2} & R_{lk}^{\min} \leq r_{ij} \leq R_{lk}^{\max} \\ 0 & r_{ij} > R_{lk}^{\max} \end{cases} \end{aligned} \tag{2.47}$$

The cutoff function restricts the range of the interactions only to the nearest neighbours, placed in the area between cutoff radii $R^{\min}$ and $R^{\max}$. The many-body parameters, denoted by a_0–a_3, have to be fitted on the base of the particular type and spatial arrangement of atoms. Similar to the original formulation of REBO for hydrocarbons [8], the bond angle function G is a sixth-order polynomial with seven additional parameters.

2.1.2.5 Reactive Force Fields (ReaxFF)

One of the most versatile and complex approaches to modelling atomic interactions has been developed by van Duin et al. [14, 24]. In the case of ReaxFF, the overall potential energy is expressed as a general sum of partial energies, which characterize the structure and state of the atomic system:

$$\begin{aligned} V^{\text{REAX}} = &E_{bond} + E_{over} + E_{under} + E_{lp} + \\ &+ E_{val} + E_{tor} + E_{vdW} + E_{coul} + E_{ep} \end{aligned}. \tag{2.48}$$

Individual terms in Eq. (2.48) correspond to the energy contributions from bonds (E_{bond}), over- and under-coordinated atoms (E_{over}, E_{under}), and lone pairs (E_{lp}). The E_{val} and E_{tor} denote, respectively, energy dependent from valence and torsion angles. Terms E_{vdW} and E_{coul} are energy contributions introduced by the non-bonded van der Waals and Coulomb interactions, respectively. The last term, E_{ep}, is penalty energy which helps control the stability of the modelled system. Depending on the type of atoms or molecules, certain additional terms, like triple-bond energy correction (in the case of the carbon monoxide), can be introduced [14] to the total potential energy formulation.

The sets of fitting data for the ReaxFF potentials are usually obtained on the basis of the ab initio, or DFT, computations and published in tabularized form. Due to complex formulations, the need of interpolation of discrete data, and the involvement of additional procedures like conjugated gradient-based charge equilibration [104], the high accuracy of the ReaxFF approach costs the increased time of computations, compared to the other many-body potentials.

Computational times, expressed as the number of the time steps per second, of the three MD simulations with, respectively, Stillinger-Weber [47], REBO [63, 64], and ReaxFF [79, 91]

Table 2.1 Computational times of the $SLMoS_2$ MD simulation with different atomic potentials, expressed as the number of the time steps per second.

	SW	REBO	ReaxFF
Time steps per second	113.7	4.4	2.9

MoS_2 potentials, are presented in Table 2.1. [80]. In all cases, the fully relaxed, 100 Å × 100 Å $SLMoS_2$ lattices with 3648 atoms were subjected to 10 000 steps of the NVT MD run (canonical ensemble, see Section 3.2.1.3 for details) at a constant temperature of 30 K and time step equal to 1 fs. Simulations were performed using the LAMMPS software [56] on the single core of the Xeon E5-2630 CPU, eliminating the communication overhead. The constant computational effort on building neighbour lists has been neglected.

2.1.2.6 Murrell-Mottram Potential

This potential often finds an application in modelling the metallic solids and the atomic clusters [67, 83, 84]. A standard two-plus-three body formulation is used:

$$V^{MM}\left(r_{ij}, r_{ik}, r_{jk}\right) = V_2^{MM}\left(r_{ij}\right) + V_3^{MM}\left(r_{ij}, r_{ik}, r_{jk}\right). \tag{2.49}$$

The two-body term has exponential form, similar to the Morse potential (see Eq. 2.40):

$$V_2^{MM}\left(r_{ij}\right) = -D_e\left(1 + a_2\rho_{ij}\right)\exp\left(-a_2\rho_{ij}\right), \tag{2.50}$$

where: D_e is the dissociation energy and $\rho_{ij} = (r_{ij} - r_0)/r_0$ means the reduced interatomic distance; r_0 is the equilibrium bond length. One requires that the value of the three-body contribution has to be unchanged upon interchanging identical atoms. This can be done by expressing the three-body potential in terms of the symmetry coordinates Q_1, Q_2, Q_3:

$$V_3^{MM} = D_e P_s(Q_1, Q_2, Q_3) F_D(a_3, Q_1), \tag{2.51}$$

which are defined as follows:

$$\begin{Bmatrix} Q_1 \\ Q_2 \\ Q_3 \end{Bmatrix} = \begin{bmatrix} \sqrt{\frac{1}{3}} & \sqrt{\frac{1}{3}} & \sqrt{\frac{1}{3}} \\ 0 & \sqrt{\frac{1}{2}} & -\sqrt{\frac{1}{2}} \\ \sqrt{\frac{2}{3}} & -\sqrt{\frac{1}{6}} & -\sqrt{\frac{1}{6}} \end{bmatrix} \begin{Bmatrix} \rho_{ij} \\ \rho_{jk} \\ \rho_{ki} \end{Bmatrix}. \tag{2.52}$$

For monoatomic metals, the second-order polynomial is applied [17]:

$$\begin{aligned} P_s(Q_1, Q_2, Q_3) = c_0 + c_1 Q_1 + c_2 Q_1^2 + c_2\left(Q_2^2 + Q_3^2\right) + \\ + c_4 Q_1^3 + c_5 Q_1\left(Q_2^2 + Q_3^2\right) + c_6\left(Q_3^3 - 3Q_3Q_2^2\right) \end{aligned}. \tag{2.53}$$

The damping function F_D in Eq. (2.51) should make the three-body term vanish exponentially as Q_1 goes to infinity. This can be achieved in many different ways:

$$F_D(a_3, Q_1) = \operatorname{sech}(a_3 Q_1), \tag{2.54}$$

or

$$F_D(a_3, Q_1) = \exp\left(a_3 Q_1^2\right). \tag{2.55}$$

The parameters defining Murrell-Mottram potential for aluminium atoms are given by Cox et al. [17], while the values of the MM-related potential's parameters for silicon and monoatomic metals such as Cu, Au, and Ag are presented in the following papers: [65, 66].

2.1.2.7 Embedded Atom Method

A slightly different approach to overcome the limitations of the simple pair potentials was proposed by Baskes, Daw, and Foiles [19, 20, 28]. The information about the surrounding environment is provided by an additional term, which describes the amount of energy necessary to embed an atom in the background electron density of the neighbouring atoms. In the EAM, the total energy of the monoatomic system is typically represented as:

$$V^{\text{EAM}} = \frac{1}{2}\sum_{i,j>i} V_2\left(r_{ij}\right) + \sum_i F^{embed}(\overline{\rho}_i), \tag{2.56}$$

where: V_2 is the standard two-body potential, and F^{embed} denotes the embedding energy of the i-th atom in the surrounding with the spherically averaged electron density:

$$\overline{\rho}_i = \sum_{j \neq i} \rho\left(r_{ij}\right), \tag{2.57}$$

developed by the electrons of all other atoms in the considered system.

The values of V_2, F^{embed}, and ρ, as the functions of appropriate variables, are usually obtained from ab initio calculations. Thus they are usually given in a discrete, tabularized form. A deep study of the developing EAM potentials for aluminium and nickel atoms is presented by Mishin et al. [75].

The EAM is currently one of the most popular approaches and plays an important role in modelling metallic materials. Especially in the simulations of some physical phenomena like crack growth, the EAM performs particularly well. The main drawbacks of this model are the higher computational cost compared to the potentials given in the direct analytic formulas and the preclusion of the directional bonding due to the spherically averaged electron density. To overcome the bonding limitations, the angular effects have been incorporated into the original formulation (2.57). Such improved potential, known as modified embedded atom method (MEAM) [6, 7], is widely used in modelling metals, alloys, as well as covalently bonded systems based on carbon and silicon atoms.

Most of the MD solver implementations such as LAMMPS [56] have ready, built-in routines to compute potential energy and interatomic forces using complex formulations like ReaxFF and EAM. Unfortunately, due to many variants and modifications, there is no one standard form for tabularized data of such kind of potentials; thus, usually, the popularity of the software determines the file format to be used. However, in the case of EAM, one of the most popular is the DYNAMO format. In addition to the necessary, discrete values of the potential functions, this format contains some information such as optimal cutoff radius and the type and parameters of the atomic lattice.

Apart from scientific literature, several WWW services distribute the parameters and data for a given type of atomic potentials. An excellent archive of the atomic potential's files is located at [43].

2.2 Physical Phenomena in Microscale

As we proceed in the twenty-first century, advances in materials research and technology offer great promise. Materials science forms the foundation for engineers in product development because the structures, components, and devices that engineers design are limited by the properties of the materials that are available and the techniques that can be used for fabrication. Modelling of materials behaviour at microscale is an extremely broad subject. Due to the variety of structures of various materials and the variety of deformation mechanisms, it is impossible to consider all engineering materials in this book. Therefore, the analysis of the physical phenomena occurring at the microscale is limited to metallic materials with polycrystalline structures, and phenomena such as grain deformation and phase transformations are considered.

The two criteria that the company considers when selecting the materials are its performance and costs. The present book is focused on the former one, although the costs will also be addressed occasionally. The composition of the material is the main factor, which decides about its performance. The selection of a proper composition for a particular application is the task for chemists, and it will not be considered. Instead, the emphasis will be put on various methods of improvement of materials performance by specific thermo-mechanical treatment without changing the chemistry. Materials subjected to certain combinations of plastic deformation and temperature changes can obtain particularly useful properties. Grain deformation and phase transformations are the mechanisms which can be used to control the properties, and these mechanisms are discussed in this chapter.

Conventional methods of modelling these phenomena are presented, and limitations of these methods, which justify the application of multiscale modelling, are highlighted. Beyond this, the problem of the accuracy of simulations and selection of the relevant model for particular applications is discussed.

Three materials are considered in the case study section in this chapter. Two of them are dual-phase (DP) steel, which, due to their excellent combination of high strength and workability, is commonly used by the automotive industry. These steel contain different amounts of carbon, and the effect of this element is a part of the case studies. The third investigated steel is eutectoid steel 900A with a special microstructure providing high wear resistance and making this steel useful for the manufacturing of rails. The chemical compositions of all steels are given in Table 2.2.

2.2.1 Microstructural Aspects of Selection of a Microscale Model

Accuracy of numerical simulations of materials processing depends, to a large extent, on the correctness of the description of material properties, as well as mechanical and thermal boundary conditions. Both these aspects are discussed in the present book. One of the objectives of

Table 2.2 Chemical compositions of investigated steels.

Steel	C	Mn	Si	P	S	Cr	Mo	V	Ti
DP 1	0.11	1.45	0.19	0.014	0.009	0.27	0.03	0.005	0.013
DP 2	0.07	1.45	0.25	0.01	0.006	0.55	0.03	0.005	0.002
900A	0.71	1.05	0.31	0.016	0.018	0.03	–	–	–

materials processing is to create a product with relevant dimensions and required microstructure and properties. In typical practical applications of metallic materials, the properties of the product that are of primary importance are required strength, good ductility, good weldability, and formability. The process of controlled thermo-mechanical treatment, in which the process parameters are chosen to suit a particular material and manufacturing route, in order to attain those attributes, may realize some or all of these attributes. The parameters available are the temperature, strain, strain rate per pass, and cooling rates.

2.2.1.1 Plastometric Tests

As has been mentioned, obtaining the required shape and properties of products is the main task of thermomechanical processing. These objectives are obtained by precise control of dimensions and by control of changes occurring in the microstructure during processing. The former is based mainly on automatic control systems, and it is not considered in the present book. The latter include:

- Phase transformations during heating of the material before hot deformation. Reasonably uniform microstructure should be obtained after heating.
- Grain growth during heating before hot deformation.
- Recrystallization during hot deformation processes. Various types of recrystallization have to be considered, and the effect of recovery has to be accounted for.
- Grain growth after recrystallization during interpass times.
- Precipitation of carbon nitrides during hot forming and an effect of precipitates on kinetics of recrystallization.
- Phase transformations during cooling after hot deformation.

Having in mind strong correlation between microstructure of product and microstructure obtained during hot deformation, the latter has to be particularly emphasized in the present book. The macroscopic reaction of a material subjected to plastic deformation is an effect of competition between the two following phenomena:

- Work hardening, which involves an increase of the dislocation density.
- Restoration, which leads to a decrease of the dislocation density due to recovery and recrystallization.

Changes in the microstructure during materials processing are reflected by flow curves, which represent the relation of flow stress on a strain. Typical possible responses of materials are shown in Figure 2.3. The type of the response depends on the stacking fault energy (SFE) of the material. Curves 1 and 2 in Figure 2.3 show the response of materials with the low SFE (1) and high SFE (2). In the former case, low energy is accumulated in the material, and dynamic recrystallization (DRX) is not launched. Response of the material is a competition between hardening and recovery. In the latter case, high energy accumulated in the material leads to initiation of the DRX. Curves 3 and 4 represent deformation of materials at low values of the Zener-Hollomon coefficient:

$$Z = \dot{\varepsilon} \exp\left[\frac{Q_{def}}{R(T + 273)}\right], \tag{2.58}$$

where: $\dot{\varepsilon}$ – strain rate, Q_{def} – activation energy for deformation in J/(molK), R – gas constant, T – temperature in °C.

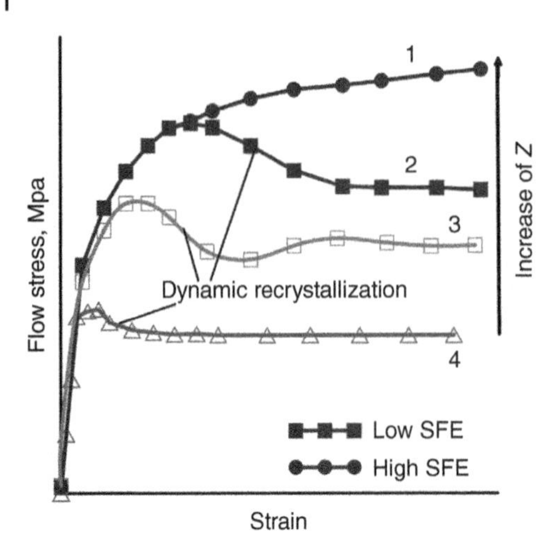

Figure 2.3 Typical material responses to plastic deformation.

In this book, activation energy is given in J/(molK). Deformation at low Z may involve oscillation of the material response (curve 3) or hardening followed by fast softening and saturation (curve 4).

If for materials characterized by curve 1 in Figure 2.3 a deformation is interrupted, a static recrystallization (SRX) will occur during an interpass time. The critical strain necessary to launch SRX is much lower than that needed for DRX.

The prime objective of the e.g. mill engineer is to design the manufacturing chain, referred to as the thermal-mechanical treatment. In doing so, the knowledge and understanding of certain critical temperatures is essential, as these affect the hardening and softening processes, as well as a final microstructure. The considered processes include precipitation hardening, recrystallization, and recovery, each of which may be static or dynamic, depending on whether loads are applied or not. Also, work hardening resulting from deformation below recrystallization stop temperature, which causes pan-caking of the grains, needs to be considered. The critical temperatures affecting microstructure evolution are [60]:

- precipitation start and stop temperature;
- recrystallization start and stop temperature; and
- transformation start and stop temperature.

All three temperatures are functions of the process and material parameters. Their importance in understanding the problems associated with the control of the material's microstructure and structure is best examined by studying the phase equilibrium diagrams. It should be mentioned that thermomechanical processing yields non-equilibrium microstructures. Therefore, modelling of these processes becomes particularly difficult. The objective of this chapter is a presentation and critical evaluation of models describing the previously mentioned phenomena, in particular their compatibility with the thermomechanical approach to the hot metal forming processes. Modelling of the thermal and mechanical phenomena during materials processing has improved significantly during recent years. Models of various complexities and abilities have been developed, and accurate predictions of the metal flow and temperature fields in two- and three-dimensional domains do not present particular difficulties. Similar to the development of thermal-mechanical models, problems of modelling of the recrystallization, grain growth, and phase transformations have been of interest for

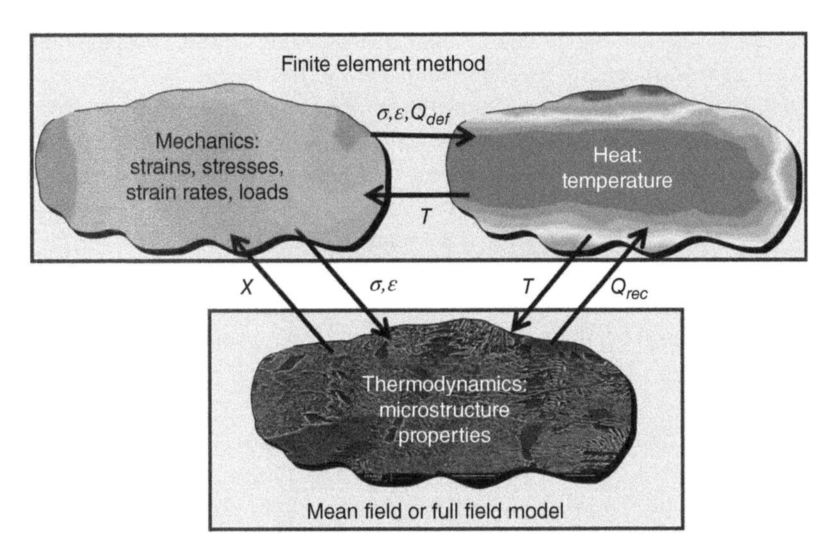

Figure 2.4 Schematic illustration of FE model application for a prediction of local microstructure evolution during hot forming of steels.

scientists for a long time, and a number of models, based mainly on closed-form equations, have been developed for various materials. Lack of information regarding strain and temperature distribution in the deformation zone for the more complex geometry of tooling forced scientists a few decades ago to assume uniform and isothermal deformation in the processes. Development of the finite element (FE) technique and its applications to the simulation of materials processing gave a new perspective for modelling microstructure evolution in the metal forming process, as shown in Figure 2.4. Microstructure evolution equations were solved using local, current values of strains, strain rates, stresses, and temperatures calculated by the FE method (FEM) [97]. In consequence, the calculation of distribution of microstructural parameters in the whole domain became possible.

The information supplied by the FE models regarding the evolution of strain rates, strains, stresses, and temperatures is a useful input for the modelling of microstructural phenomena which take place in the deformation zone during hot rolling. The importance of modelling of microstructure evolution is discussed by Sellars [114] and can be summarized as follows:

- For a given composition of the alloy, the high-temperature flow stress is influenced to a large extent by the microstructure. Proper prediction of metal flow and deformation force during processing is possible only if the relevant microstructure is known.
- The microstructure present at the end of the rolling and cooling operations controls the product properties.

Advantages resulting from using FEM or alternative methods in the microstructure evolution modelling can be twofold. First is the possibility of the investigation of local situations in the deformation zone using conventional closed-form microstructure evolution equations. Solving these equations using temperatures, strain rates, strains, and stresses calculated by the FEM allows the prediction of the distribution of the grain size, recrystallization, phase transformation kinetics, and other microstructural phenomena in the volume of the deformed body. Variations of the temperature in time can be accounted for using the additivity rule [112]. The second advantage is the possibility of using more advanced phenomenological microstructure evolution models, which are usually based on the internal variable method or on more advanced multiscale modelling techniques described in the following chapters of this book.

Modelling of microstructure evolution based on a combination of the FEM with the conventional closed-form equations describing recrystallization, grain growth, and phase transformations is well described in an earlier book [60]. In this book, the emphasis is on advanced multiscale modelling techniques, in which discrete methods are used to describe phenomena in lower-dimensional scales. Beyond this, numerical techniques alternative to FEM are presented. Since the general idea of the connection of dimensional scale remains the same, the basic information concerning phenomena at microscale and possibilities of their modelling is repeated in this chapter. The flow stress, which is a result of microstructure development and which has an important influence on materials behaviour during processing, is discussed first. The following sections deal with phenomena of recrystallization and phase transformations.

One of the challenges in simulations of thermomechanical processes is the evaluation of rheological parameters in various conditions of deformation, and this problem is considered in the next section. Modelling of changes occurring in the microstructure during processing is the next challenge, and it is presented in the following sections on phase transformations and recrystallization. Workability is one of the most important properties of metallic materials. Therefore, beyond the microstructure evolution, phenomena affecting workability are discussed as well. These include crack initiation and propagation, creep, and fatigue.

2.2.1.2 Inverse Analysis

The goal of many researchers was the development of the method that eliminates the influence of the disturbances occurring in the plastometric tests and that allows estimation of rheological parameters independently of these disturbances. The problem of parameters evaluation is defined as the inverse problem, and it is widely presented in the scientific literature [29, 49, 123]. The inverse algorithm developed by the authors is described in [123], and its basic principles are repeated briefly next. The rheological and friction parameters are determined by searching for a minimum of a goal function:

$$\phi = \sqrt{\frac{1}{Nt}\sum_{i=1}^{Nt}\left[\frac{1}{Nr}\sum_{j=1}^{Nr}\left(\frac{R_{ij}^{m}-R_{ij}^{c}}{R_{ij}^{m}}\right)^{2}+\frac{1}{Ns}\sum_{j=1}^{Ns}\left(\frac{F_{ij}^{m}-F_{ij}^{c}}{F_{ij}^{m}}\right)^{2}\right]}, \tag{2.59}$$

where: Nt – number of tests, Nr – number of radius measurements along the height, Ns – number of load measurement sampling points in one test, F_{ij}^{m}, F_{ij}^{c} measured and calculated load, R_{ij}^{m}, R_{ij}^{c} – measured and calculated radius of the sample after the test.

Goal function (2.59) is used for the uniaxial and ring compression tests. Both friction coefficient and flow stress model are determined from these tests. Loads only are measured in the plane strain compression, and the goal function (2.59) contains only the second term under the square root [123]. Calculated values of loads and the shape of the samples are obtained from the direct problem model. In this book, this model is based on the rigid-plastic thermomechanical FE solution proposed in [60]. Since it is a well-known solution, it is not repeated in this book.

2.2.2 Flow Stress

2.2.2.1 Procedure to Determine Flow Stress

Proper evaluation of rheological parameters in various conditions of deformation is one of the challenges in simulations of thermomechanical processes. The general procedure is composed of performing plastometric tests of compression, tension, or torsion and interpretation of the results of these tests using preferably the inverse analysis [123]. The main sources of errors, which further influence the accuracy of numerical simulations, are due to:

- Inaccuracies in the plastometric tests.
- Errors of the inverse analysis caused by errors in evaluation of the friction conditions in the tests.
- Lack of capability of the selected rheological model to describe properly the real behaviour of the material.

The general objective of research in various laboratories is the design of the procedure, which will guarantee the development of accurate and reliable rheological models of materials. This procedure combines critical analysis of experimental tests, application of the method of automatic interpretation of results of tests, and finally, mathematical description of measured properties. The particular objectives are twofold. The experimental part results in the evaluation of the plastometric tests and the development of good practice guides. The numerical analysis's objective is to eliminate the influence of inhomogeneities of strains, localization of strains, effect of friction, and heat generated due to deformation and friction in the tests.

Due to the complex structure of metals, their behaviour depends on many factors like grain size, grain boundaries, dislocation density, SFE, etc. In polycrystalline materials, all crystallographic and structural effects should be accounted for, which is difficult due to their complexity and scale. To overcome this problem, polycrystals are described by flow (stress–strain) curve, which represents statistically all mentioned phenomena. In this approach, the stress versus strain, strain rate, temperature, etc. (depending on conditions) relations are determined on the basis of the results of plastometric tests. The tests can have various forms (tension, compression, torsion) depending on further use of the flow curve. Torsion tests, which were investigated elsewhere, e.g. [37], are not discussed in this book, and the focus is on compression tests. Applications of plain strain compression (PSC), plain strain compression in channel die (PSCc), uniaxial compression of a cylinder (UC), as well as compression of a ring (RC) are discussed next (Figure 2.5).

The advantages and disadvantages of typical tests (PSC, RC, UC) are discussed in [60] and are not repeated here. These tests proved their applicability to all bulk forming processes. The main goal, but also the problem with mentioned tests for polycrystalline materials, is to properly interpret the results, which are affected by errors caused by inhomogeneous distribution of strains, strain rates, temperatures, stresses, etc. Inverse technique [123] is commonly used to eliminate the influence of the mentioned errors and to obtain flow curves independent of the type of the test. The majority of errors in yield stress due to differences in shape and size of the specimen can be eliminated by inverse technique [123]; nevertheless, the variations caused by using different equipment and procedures of performing experiments are not so

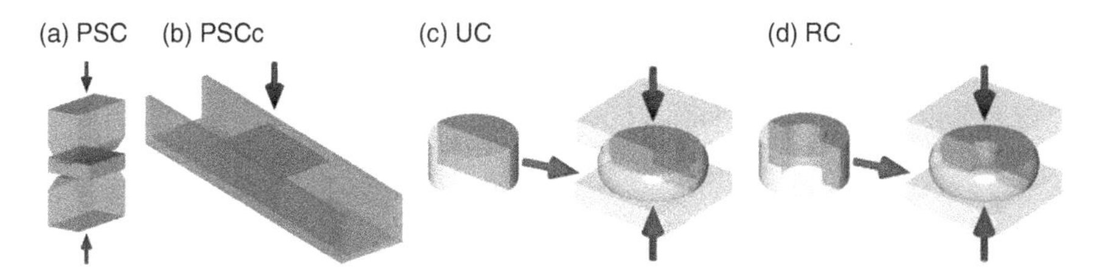

Figure 2.5 Schematic illustrations of selected plastometric tests.

easy to remove. In conventional tests, the preheating procedure is essential. Differences in the preheating conditions cannot be eliminated by the inverse analysis, see for example [124]. Thus, the results of the test have to include information about the preheating conditions, which should be selected according to the further application of the model (roughing or finishing rolling or forging, etc.)

2.2.2.2 Flow Stress Model

The objective of this part of the work is the evaluation of the capabilities of various mathematical models, which are used as rheological laws in the FE codes. Approximation errors are the main source of inaccuracies in this part of the model. It is difficult to select the function, which describes the flow stress properly in a wide range of temperatures and strain rates. Therefore, relatively complicated functions are often proposed [18], or more advanced models are developed, such as the internal variable model (IVM) [109]. The results of the evaluation of selected models regarding their capabilities to describe the flow stress properly, and to quantitatively determine errors due to approximation of the experimental data, are given next. Models in which flow stress is defined as a function of the three primary parameters, strain, strain rate, and temperatures, are the most commonly used models for practical simulations using FE codes. It is expected from the models that they properly describe typical material responses to plastic deformation, which are shown in Figure 2.3.

Three models of various complexity and various predictive capabilities can be distinguished. The models were proposed in [18, 31, 39], respectively:

$$\sigma_p = A\varepsilon^n \exp(-q\varepsilon)\dot{\varepsilon}^m \exp(-BT), \tag{2.60}$$

$$\sigma_p = \sqrt{3}\left[A\varepsilon_i^n \exp\left(\frac{B}{T+273}\right)\exp(-q\varepsilon_i) + [1-\exp(-q\varepsilon_i)]A_{sat}\exp\left(\frac{B_{sat}}{T+273}\right)\right]\left(\sqrt{3}\dot{\varepsilon}\right)^m. \tag{2.61}$$

$$\sigma_p = \sigma_0 + \left(\sigma_{ss(e)} - \sigma_0\right)\left[1-\exp\left(-\frac{\varepsilon}{\varepsilon_r}\right)\right]^{\frac{1}{2}} - R. \tag{2.62}$$

where:

$$R = \begin{cases} 0 & \varepsilon \le \varepsilon_c \\ \left(\sigma_{ss(e)} - \sigma_{ss}\right)\left[1-\exp\left(-\left[\frac{\varepsilon-\varepsilon_c}{\varepsilon_{xr}-\varepsilon_c}\right]^2\right)\right] & \varepsilon > \varepsilon_c \end{cases}.$$

$$\sigma_0 = \frac{1}{\alpha_0}\sinh^{-1}\left(\frac{Z}{A_0}\right)^{\frac{1}{n_0}}, \quad \sigma_{ss} = \frac{1}{\alpha_{ss}}\sinh^{-1}\left(\frac{Z}{A_{ss}}\right)^{\frac{1}{n_{ss}}}, \quad \sigma_{sse} = \frac{1}{\alpha_{sse}}\sinh^{-1}\left(\frac{Z}{A_{sse}}\right)^{\frac{1}{n_{sse}}}.$$

$$\varepsilon_r = \frac{1}{3.23}\left[q_1 + q_2\left(\sigma_{ss(e)}\right)^2\right], \quad \varepsilon_{xr} - \varepsilon_c = \frac{\varepsilon_{xs}-\varepsilon_c}{1.98}, \quad \varepsilon_c = C_c\left(\frac{Z}{\sigma_{ss(e)}^2}\right)^{N_c}, \quad \varepsilon_{xs} - \varepsilon_c = C_x\left(\frac{Z}{\sigma_{ss(e)}^2}\right)^{Nx}.$$

Equation (2.60) accounts for the softening during deformation. However, once the softening begins, it continues, and this model can give erroneous results (stress below zero) for large strains. The Eq. (2.61) is capable of describing softening followed by saturation. The Eq. (2.62) can predict the oscillatory response of the material.

All conventional rheological models do not account for the history of the process. After changing the conditions of deformation, the model's response moves immediately to the new equation of state, and the flow stress is a function of new values of external variables.

On the other hand, it was observed experimentally, see for example [129], that some metallic materials show a delay in the response to the change of conditions. This delay is due to microstructural phenomena, which require some time to proceed. Therefore, the rheological models which include internal variable as an independent parameter were developed. The internal variable in these models remembers the history of deformation and allows to account for this history in flow stress calculations. Thus, a better quality description of materials behaviour was obtained when an IVM with dislocation density introduced as an independent variable was used; see [89, 109] for the details of the IVM model developed by the authors. This model is also summarized next.

The dislocation density is the main internal variable in the IVM for metallic materials. The dislocation density can be treated either as average [89] or several kinds of dislocations can be considered, e.g. mobile and trapped [90]. Also, a probability function describing the distribution of dislocation density can be introduced [98]. It is obvious that the introduction of a more complicated treatment of the dislocation density leads to an increase in computation costs in the FEM. Thus, the model with the average dislocation density, which is adequate for the flow stress predictions, is discussed briefly next, and the identification of coefficients in this model for the investigated steels is presented. Since the stress during plastic deformation is governed by the evolution of dislocation populations, a competition of storage and annihilation of dislocations, which superimpose in an additive manner, controls a hardening. Thus, the yield stress accounting for softening is proportional to the square root of dislocation density ρ:

$$\sigma_p = \sigma_0 + \alpha b G \sqrt{\rho}, \tag{2.63}$$

where: σ_0 – stress due to elastic deformation, b – length of the Burgers vector, G – shear modulus, α – constant.

The evolution of dislocation populations accounting for restoration processes is given by:

$$\frac{d\rho}{dt} = \frac{\dot{\varepsilon}}{bl} - k_2 \dot{\varepsilon}^{-q} \rho - \frac{k_3}{D_\gamma} \rho^p \mathrm{R}[\rho - \rho_{cr}], \tag{2.64}$$

where: $\dot{\varepsilon}$– strain rate, ρ_{cr} – critical dislocation density, D_γ – grain size.

$$R[\rho(t) - \rho_{cr}] = 0 \qquad \text{for} \qquad \rho \le \rho_{cr}$$

$$R[\rho(t) - \rho_{cr})] = \rho\,(t - t_{cr}) \qquad \text{for} \qquad \rho > \rho_{cr}$$

t_{cr} – time at the beginning of DRX.

The average free path for dislocations l, the self-diffusion coefficient k_2, and the grain boundary mobility k_3 in the Eq. (2.64) are calculated as:

$$l = A_0 Z^{-A_1}, \tag{2.65}$$

$$k_2 = k_{20} \exp\left(\frac{Q_s}{RT}\right), \tag{2.66}$$

$$k_3 = k_{30} \rho \exp\left(\frac{Q_m}{RT}\right), \tag{2.67}$$

where: Q_m – activation energy for grain boundary mobility, Q_s – activation energy for self-diffusion.

The critical dislocation density for DRX is given by:

$$\rho_{cr} = -A_{11} + A_{12} Z^{A_{10}}. \tag{2.68}$$

Coefficients A_0, A_1, k_{20}, Q_s, k_{30}, Q_m, α, p, q, A_{10}, A_{11}, and A_{12} are determined using the inverse analysis of plastometric tests.

Further improvement of the quality of description of materials deformation can be obtained when multiscale modelling techniques are applied. Contrary to the previously mentioned conventional models, the multiscale modelling techniques allow to represent the real material microstructure explicitly and to account for stochastic and discontinuous phenomena occurring in the microstructure. The multiscale modelling techniques are the objective of this book, and they are thoroughly discussed in the following chapters.

2.2.2.3 Identification of the Flow Stress Model

As has been mentioned, three materials are considered in this chapter as a case study, two DP steels and eutectoid steel 900A with the chemical compositions given in Table 2.2. Plastometric tests were performed for these steels, and the selected results are presented next. Comparison of loads recorded for the two DP steels is in Figure 2.6a. A similar comparison for the flow stress is shown in Figure 2.6b. It is seen that as far as loads and flow stresses are considered, the difference between the two DP steels is small.

Full inverse analysis with the objective function (2.59) was used to determine the coefficients in the flow stress Eqs. (2.60)–(2.62), and (2.64). The obtained results are presented in Tables 2.3–2.6. Since differences between coefficients obtained for the two DP steels were negligible, the results for the DP 1 steel are only presented. In the last column of the tables the final value of the objective function (2.59) is given. This value represents the accuracy of the inverse analysis.

As it is seen in Tables 2.2–2.5, an increase of the number of coefficients in the flow stress equation allows obtaining better accuracy of the inverse analysis. Therefore, more complex Eqs. (2.61) and (2.62) will be used in the present work in a simulation of hot deformation of DP steels, which require better accuracy of loads prediction.

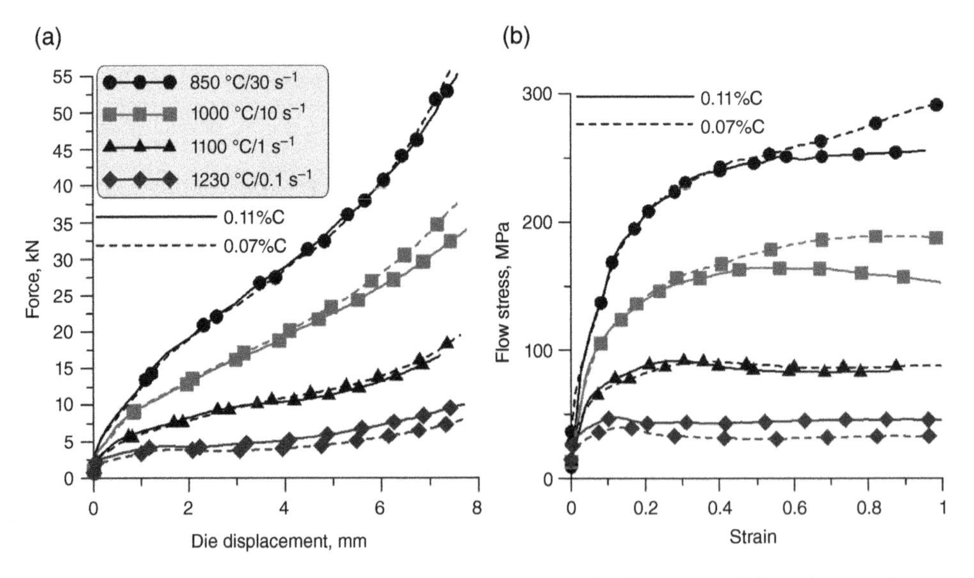

Figure 2.6 Selected examples of comparison of loads recorded for the two DP steels (a) and comparison of flow stress obtained for these steels from the inverse analysis (b).

Table 2.3 Coefficients in Eq. (2.60) obtained from the inverse analysis for the steel DP.

A	n	q	m	β	Φ
3255.3	0.196	0.283	0.119	0.003007	0.0868

Table 2.4 Coefficients in Eq. (2.61) obtained from the inverse analysis for the steel DP.

a	β	n	m	a_{sat}	β_{sat}	q	Φ
2.423	4517.1	0.208	0.122	0.0052	10 956	0.5335	0.0754

Table 2.5 Coefficients in Eq. (2.62) obtained from the inverse analysis for the steel DP.

A_0	n_0	α_0	A_{sse}	n_{sse}	α_{sse}	A_{ss}	n_{ss}	Φ
0.451×10^{12}	38.9	0.016	0.58×10^{12}	4.891	0.0105	0.156×10^{12}	8.609	0.0575
α_{ss}	q_1	q_2	C_c	N_c	C_x	N_x	Q_{def}	
0.00032	0.876	1×10^{-10}	0.00158	0.011	0.004	1.53	316 950	

Table 2.6 Coefficients in Eqs. (2.63)–(2.68) obtained from the inverse analysis for the steel DP.

A_0	A_1	k_{20}	Q_s	k_{30}	Q_m	α	Φ
0.785×10^{-4}	0.188	44.9	16 470	2.47	655 000	1.104	0.0628
p	q	A_{10}	A_{11}	A_{12}	Q_{def}		
2.46	0.027	0.412	0.873	0.0167	303 000		

Table 2.7 Coefficients in Eq. (2.60) obtained from the inverse analysis for the steel 900A.

A	n	q	m	β	Φ
4386.7	0.276	0.385	0.14	0.003256	0.0268

In contrast, eutectoid steel is used for the manufacturing of rails or rods, and requirements concerning the accuracy of calculations of loads are lower; therefore, only Eq. (2.60) was used in simulations of manufacturing of these steels. Coefficients obtained for the steel 900A from the inverse analysis are presented in Table 2.7.

IVM defined by Eqs. (2.63)–(2.68) requires a solution of the differential equation at each Gauss point in an FE mesh and leads to an increase of the computing time; therefore, it is used only when information about dislocation density is needed.

Coefficients in Tables 2.3–2.7 determined by the inverse method will be used in the analysis of case studies in this book. These coefficients for other materials, mainly metals, can be found in authors' other publications, e.g. [100, 124].

2.2.3 Recrystallization

Interpass softening is an important factor affecting the evolution of microstructure and load level in multipass hot working; both dynamic and static softening processes, recovery, and recrystallization can occur. In most investigations, SRX has been considered the dominant event that removes the effect of deformation, and a number of empirical equations have been proposed to describe its kinetics, particularly in such simple conditions as after single pass deformation at a constant temperature. Investigations of SRX reflected the significant roles of experimental process variables, such as a strain and a temperature, and the small contribution of a strain rate and a state of stress. This is acceptable for plate rolling in reversing mills, where the interpass times are relatively long.

Images of the microstructure showing various stages of the recrystallization are shown in Figure 2.7. Equiaxed austenite grains (A) are seen at the beginning of the deformation. If the strain is great enough to reach a certain critical value, DRX can begin. Elongated grains characteristic for the critical strain are seen in the microstructure B. Nuclei of new grains and small dynamically recrystallized grains are seen together with the previously deformed austenite grains in the microstructure C. DRX may be subsequently followed by post-dynamic softening processes, metadynamic recovery (MDRV), metadynamic recrystallization (MDRX), and possibly also by static processes of SRX and static recovery (SRV). Small equiaxed grains after MDRX are seen in the microstructure D. It is still unclear whether the softening after complete DRX takes place through a single mechanism or consists of two or even four separate competitive processes, and different opinions prevail, e.g. [107, 108]. MDRX is investigated quite seldom, and only limited data are available concerning C-Mn and certain microalloyed steels, but the results highlight the significant effect of strain rate and the insignificant roles of strain and temperature on its kinetics. The contribution of DRX and MDRX may become important in dictating the final microstructure, particularly

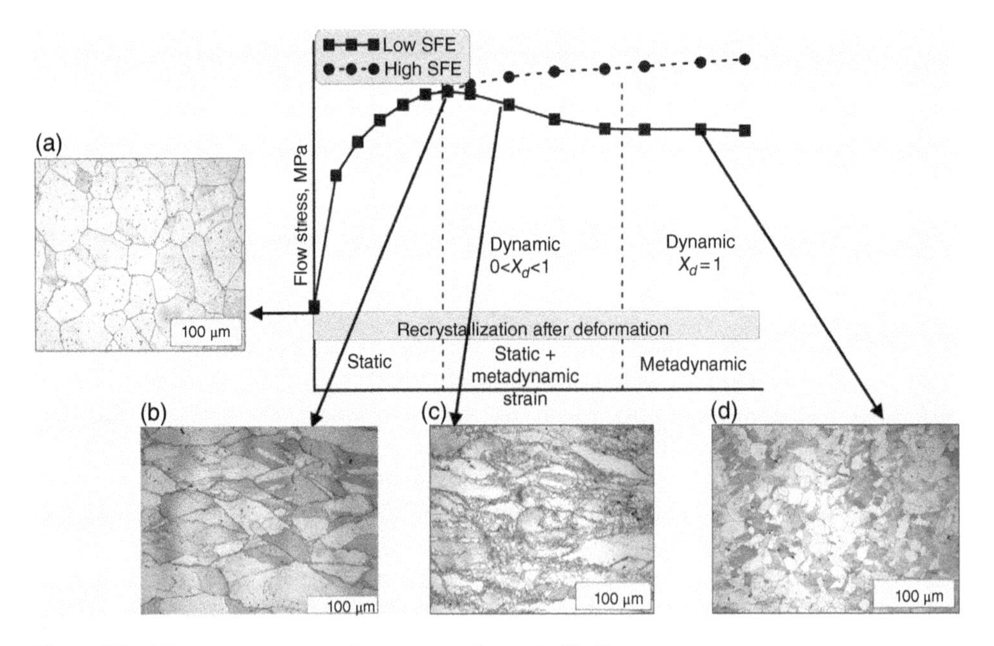

Figure 2.7 Microstructures at various stages of recrystallization.

during the finishing stage of strip, bar, and rod rolling in multistand mill trains, where, due to the brief interpass times, strain can accumulate and reach the critical value for DRX to occur, and on account of the high deformation rates, MDRX can be expected to proceed at an extremely fast rate.

Under certain instances, softening remains only partial due to kinetic factors at low deformation temperatures with short interpass times, or possibly also after the occurrence of DRX, which means that some strain, a residual strain, is retained for the subsequent pass. There are several methods for estimating the amount of residual strain or stored energy. Two models have been proposed to take into account the strain accumulation under partial SRX conditions, the uniform softening approach and the law of mixtures approach. The former assumes that all grains behave homogeneously and the residual strain can be calculated from the recrystallized fraction. In the latter approach, the recrystallized and non-recrystallized grains are thought to behave independently. Data on steels support the first model [40], but data also suggests that the law of mixtures model may apply even to steels.

The balance between dislocation generation and the removal of dislocations determines the rate of the hardening of alloys during the deformation by dynamic recovery. Austenite (steel) does not undergo dynamic recovery to the same degree as other metals. This ability to have extensive work hardening, without softening by recovery, may lead to DRX at higher strains. This kind of recrystallization is more probable at higher temperatures and at lower strain rates. In industrial hot deformation processes, however, usually, the strain rates are so high that there is not enough time to trigger dynamic softening of the work-hardened material. Therefore, it is much more common in hot rolling for the material to be deformed only in the work hardening regime, i.e. the strain per pass is not large enough (<0.4) to initiate DRX. Hence, concurrent SRV, accompanied by SRX, is usually occurring after deformation. There is a high driving force for static softening to take place between rolling passes and during cooling after the final pass prior to transformation. Both SRV and recrystallization have been observed in austenite, although the extent of the former is rather limited.

2.2.3.1 Static Microstructural Changes

SRV is defined as a softening process in which the decrease in density and change in the distribution of the dislocations after hot deformation or during annealing are the operating mechanisms. In a low temperature range, the acting mechanism involves vacancy motion; those operating in the intermediate temperature and those in the high temperature range (>0.5 Tm, where Tm is the melting temperature) involve dislocation climbing and cross slip. As mentioned previously, the extent of SRV in hot deformation processes is rather limited. There is a general consensus that the maximum amount of softening during holding, attributable to recovery, is approximately 20%.

The hot deformation of alloys at strains typically encountered in hot deformation processes leads to a significant work hardening, which is usually not removed by either dynamic softening processes or SRV. This hardening leaves a high driving force for static softening processes. Following SRV, there is usually partial or complete softening of microstructure by SRX. From the technological point of view, these processes are less important in hot forging. The strains in the majority of forging processes are reasonably large, and beyond cogging, the interpass times are long enough to allow static processes to be completed. In contrast, recrystallization plays a crucial role in hot rolling processes, and it is often used to control the microstructure of products, in particular in continuous hot rolling of microalloyed steel strips.

Recrystallization is usually described as taking place in two stages: nucleation of new grains and the growth of these grains at the expense of deformed ones. Some features of SRX are as follows:

- A minimum amount of deformation (critical strain) is necessary before SRX can take place.
- The lower the degree of deformation, the higher the temperature required to initiate SRX.
- The final grain size depends upon the degree of deformation and, to a lesser extent, upon the annealing temperature.
- The larger the original grain size, the slower the rate of recrystallization.

There is evidence that the nucleation of recrystallization occurs at the austenite grain boundaries by a bulge mechanism under the deformation conditions, which are encountered in hot rolling. As with most grain surface nucleation, grain corners, edges, and surfaces in that order of preference are the preferred sites for recrystallization. The bulge mechanism leads to a formation of critical size nuclei, which then grow, by grain boundary migration, due to a high driving force ahead of the growing point. This driving force is predominantly related to the dislocation density of the subgrain walls, the degree of misorientation across the subgrain boundaries, and the size of the subgrains. There appears to be a change in the recrystallization mechanism at a very large initial grain size (>100 μm), with the initial stage of recrystallization still occurring at the grain boundaries. At an early stage of transformation, however, the original grain boundaries are completely covered by new recrystallized grains, which have impinged upon one another, stopping further growth. Further softening by recrystallization would require intragranular nucleation of recrystallization, which can occur at subgrain boundaries or deformation bands. At high temperatures (>1200 °C), usually, only large grains are present, and it takes only a fraction of a second for the material to recrystallize completely. Therefore, there is little interest in the kinetics of recrystallization under these conditions. This type of recrystallization may become important, however, for lower preheating temperatures or in hot direct rolling.

In conventional preheating at high temperatures, incomplete recrystallization can take place at the early stage of rolling when small reductions are usually applied. If it happens, the accumulation of strains leads to full recrystallization in subsequent passes, and in consequence, the effect of the initial conditions on the downstream final microstructure is very small and is usually neglected. An exception is the thick plate rolling process when the total reduction ratio is low, insufficient to remove the reheated grain structure.

It has been shown in numerous experiments that for grain sizes below 100 μm recrystallization in carbon-manganese steels is very rapid above 1000 °C. Slowing down recrystallization to the extent important from the practical point of view takes place usually below 950 °C. The situation is different for microalloyed steels, in which the recrystallization is retarded by the precipitates. The modelling of SRX is an important part of rolling technology design.

Conventional modelling of recrystallization and modelling of phase transformations is formulated on a similar basis. Both these processes in metallic materials are governed by nucleation and a growth of a new phase. These phenomena can be explicitly taken into account when such discrete models as cellular automata or Monte Carlo are used to describe processes in microscale; see following chapters of this book. In conventional models, nucleation and growth are accounted for indirectly by a selection of a relevant function, which properly describes kinetics of transformation observed in experiments.

Modelling of recrystallization and modelling of phase transformations is a wide topic, which cannot be even briefly discussed in this chapter. Depending on the objective of modelling, different models and techniques are used, beginning from those based on the first principles and thermodynamic relations to experimental closed-form equations describing the kinetics of transformations. The present section is limited to the description of phenomenological models for engineering applications. The emphasis is put on the identification of parameters in these models and on the implementation of these models into the FE software.

Modelling of recrystallization and phase transformations for engineering applications has been for years based on the JMAK approach (from the names of Johnson and Mehl [45], Avrami [2–4], and Kolmogorov [52]). The general equation for an arbitrary transformation is:

$$X = 1 - \exp\left[k\left(\frac{t}{t_X}\right)^n\right], \tag{2.69}$$

where: X – volume fraction of a new phase, n – Avrami exponent, k – Avrami coefficient.

The value of the coefficient k depends on the time t_X, which is the basic time. It can be shown that:

$$k = \ln(1 - X_b), \tag{2.70}$$

where: X_b – volume fraction of a new phase for the basic time.

In modelling recrystallization, time for 50% recrystallization $t_{0.50}$ is usually used as basic time [113, 114], then $X_b = 0.5$ and $k = \ln(1-X_b) = 0.693$.

The problem of the correlation between the parameters of the hot deformation processes and the development of the resulting microstructure has been investigated extensively in the second half of the twentieth century, and a number of papers have been published in the scientific literature. Among them, the works of Sellars [113, 114], Hodgson and Gibbs [40], Roberts et al. [106], Laasraoui and Jonas [55], Choquet et al. [15], Yada [132], and Sakai [32] should be mentioned. In these publications, the authors present closed-form equations describing the processes of recrystallization and grain growth. The general form of equations describing time for 50% recrystallization, recrystallized grain size, and grain growth is similar in models proposed by various researchers, while the difference is in coefficients in these equations. Some of these equations concerning various chemistry of steels were reported in [60]. A review of these models is presented in this book, and new models developed for steels with chemical compositions given in Table 2.2. are added. The coefficients for these steels were obtained by approximation of experimental data. Double deformation tests and stress relaxation tests, which are compared in [130], are commonly used for the identification of recrystallization models. In the latter test, a relevant reheating cycle is applied, and after subsequent cooling to a specific deformation temperature, a first deformation pass is given. After this deformation, the stress is immediately relieved to a minimum value necessary to keep the sample in a position. After variable waiting times, a second deformation pass with the same strain and strain rate as the first deformation pass is applied. The softening fraction is determined by comparing the stress–strain curves for both deformation passes. The shape of the second deformation curve is strongly influenced by the time between the two deformation passes. If the interpass time is sufficiently high for full softening to occur, the second flow curve should be identical to the first flow curve. If there is no softening at all, the second flow curve should appear as an extrapolation of the first flow curve.

Stress relaxation tests are an alternative method of supplying data for the identification of recrystallization models. During a couple of recent years, the stress relaxation method has been widely investigated; see for example, research performed at the University of Oulu [48]. Stress relaxation tests are applied to measure the SRX kinetics in hot deformed metals.

The method has been employed for DP steels and eutectoid steel investigated in this book. It seems reliable and highly efficient because a single test is able to reveal the complete kinetics under given conditions, while 5–10 conventional, interrupted-deformation tests are needed to do the same.

In this book, stress relaxation tests have been used to investigate the characteristics of the static and post-dynamic softening processes and the accumulation of strain in DP steels and eutectoid steel. These tests were carried out on a Gleeble 3800 thermomechanical simulator using samples measuring ϕ10×12 mm. A graphite foil was inserted between the specimen and the tungsten carbide compression anvils to reduce the friction and a tantalum foil to prevent sticking. These foils, owing to their low electrical conductivity compared to that of steel, also decreased the harmful longitudinal temperature gradient present in the specimen. The specimens were heated at a rate of 10°C/s directly up to the deformation temperature in the range of 800–1100 °C, held for 300 s to stabilize the temperature and microstructure, and then compressed up to a prescribed strain at a constant true strain rate. Using the stroke control of the machine, the strain was held constant after the prior deformation and the compressive force relaxed was recorded as a function of time. The data acquisition rate was 20 Hz to provide sufficient data from the very beginning of the relaxation stage. A few specimens were quenched by water spray from the deformation temperature to determine the prior austenite grain size before the onset of the deformation stage.

Various empirical and semi-empirical equations describing time for 50% recrystallization ($t_{0.5X}$) have been published in the scientific literature. The most commonly used form of this equation is:

$$t_{0.5X} = b\varepsilon^p D^q \dot{\varepsilon}^s \exp\left[\frac{Q_{RX}}{R(T+273)}\right], \tag{2.71}$$

where: ε – strain, D – grain size prior to deformation, $\dot{\varepsilon}$ – strain rate, Q_{RX} – apparent activation energy for recrystallization, R – gas constant, T – temperature in °C.

The majority of published research on the identification of the coefficients b, p, q, s, and Q_{RX} in Eq. (2.14) and Avrami exponent n in Eq. (2.12) concerns various steels. Results for few selected steels are given in Table 2.8. Coefficients for DP steels and eutectoid steel in Table 2.2 were developed in the present work using stress relaxation tests. In the equation

Table 2.8 Coefficients in Eqs. (2.69) and (2.71) describing the kinetics of the static recrystallization.

Steel	n	b	p	q	s	Q_{RX}	References
C-Mn	1.7	2.5×10^{-19}	−4	2	0	300 000	[114]
C-Mn	1.7	5×10^{-21}	−4	2	0	330 000	[106]
C-Mn	1	2.3×10^{-15}	−2.5	2	0	230 000	[40]
C-Mn	1	1.14×10^{-13}	−3.8	0	−0.41	252 000	[55]
Ti-V	1.7	9.3×10^{-16}	−4	2	0	230 000	[54]
C-Mo	1.7	1.5×10^{-12}	−2.18	0.878	-0.28	232 000	[108]
DP1	1.7	$2.43\cdot10^{-13}$	−0.73	−0.421	1.8	215 880	
DP2	1.7	8.56×10^{-14}	−1.83	1.8	−0.54	218 630	
900A	1.7	2.4×10^{-8}	p	−0.2	−0.28	232 000	

developed for the pearlitic steel 900A, the strain sensitivity of time for 50% recrystallization depends on the austenite grain size prior to deformation, according to the formula $p = -1.006D^{0.22}$.

The rate of recrystallization is only one important aspect when modelling the microstructural evolution occurring during intervals between deformations and cooling from the last pass to the transformation temperature. It is also necessary to be able to predict the recrystallized grain size. Several authors investigated the influence of strain rate, strain, and temperature on the fully recrystallized grain size. It was observed that there is a power-law relationship between the recrystallized grain size and the applied strain. Using a constant value of this power for different temperatures seems possible to some extent. Furthermore, there appears to be no effect of the composition over the range valid for conventional carbon-manganese steels. This agrees with the lack of effect of composition on the SRX kinetics observed earlier. It also agrees with a previous study [120] where no effect of composition on the recrystallized grain size was observed for carbon contents from 0.1 to 0.8. Another feature concerning the prediction of the recrystallized grain size is its large sensitivity on temperature, observed by a majority of authors. There are, however, some models, like [114], which do not incorporate a temperature term. Various empirical and semi-empirical equations describing the grain size after recrystallization (D_r) have been published in the scientific literature. The most commonly used form of this equation is:

$$D_r = b_1 + b_2 \varepsilon^m \dot{\varepsilon}^n D^r \exp\left[\frac{-Q_d}{R(T+273)}\right]. \tag{2.72}$$

Not all components of Eq. (2.72) appear in all the models. The constants b_1, b_2, m, n, r, and Q_d in Eq. (2.72) for some steels are presented in Table 2.9. Since the coefficients in recrystallized grain size equation for the two DP steels were similar, only one steel is presented in Table 2.8.

There are several models, which contain equations with the structure different from that in Eqs. (2.71) and (2.72); see [60] to find some of these models. Situations when recrystallization is not completed between subsequent deformations and only partial recrystallization takes place during interpass times are possible in industrial processes. In these situations, a simple weighted average $D_p = XD_r + (1-X)D$ is usually used to calculate grain size at the entry to the next pass. D_p represents here the partially recrystallized grain size, D is the grain size prior to deformation, D_r is the recrystallized grain size, and X is the recrystallized volume fraction.

Table 2.9 Coefficients in Eq. (2.72) describing the recrystallized grain size.

Steel	b_1	b_2	m	n	R	Q_d	References
C-Mn	0	0.5	−1	0	0.67	0	[114]
C-Mn	6.2	55.7	−0.65	0	0.5	35 000	[106]
C-Mn	0	343	−0.5	0	0.4	45 000	[40]
C-Mn	0	0.5	−0.67	0	0.67	0	[55]
Ti-V	0	4.54	−0.53	−0.1	0.67	15 000	[54]
Ti-V	4.3	195.7	−0.57	0	0.15	35 000	[106]
DP	0	63.7	−0.74	−0.05	0.2	29 324	
900A	0	9.91	−0.65	−0.1	0.54	17 540	

2.2.3.2 Dynamic Softening

All softening processes that take place during plastic deformation are referred to as dynamic ones. These include dynamic recovery and DRX. Dynamic recovery involves the rearrangement of dislocations and consists of two processes. Dislocations of opposite signs annihilate each other or rearrange to form cells of relatively low dislocation density surrounded by boundaries of high dislocation density. At high temperatures, the mechanisms responsible for dynamic recovery are the cross slip of screw dislocations or climb of the edge ones [50]. Since in the conventional approach, the dynamic recovery has only an indirect influence on modelling the microstructure evolution by controlling the onset of DRX, it will not be discussed further here. More information about modelling dynamic recovery can be found in Section 2.2.1, where the internal variable method is described.

In metals of high SFE, dynamic recovery takes place rapidly, and a steady state of stress is reached. This is a result of a balance between work hardening and recovery (Figure 2.3). The steady state is characterized by a subgrain size, which in general depends on the Zener-Hollomon parameter Z (Eq. (2.58)). Deformation of materials with medium or low SFE is characterized by slow dynamic recovery. Thus, usually, dislocation density is permitted to increase to an appreciable level, and it causes an onset of DRX before the steady state is reached, as shown in Figure 2.3. It seems that DRX is a well-researched phenomenon now. A survey of research on hot deformation of steels shows, however, that researchers remain convinced that generally the material softening is caused by DRX and plastic instabilities become less likely or frequent as the temperature increases and/or the strain rate diminishes. For the objective of modelling of thermomechanical processing of steels, it is assumed in this work that when the dislocation density achieves its critical value, the DRX starts and becomes the dominant softening phenomenon. This critical value of the dislocation density corresponds to the critical strain ε_c, which corresponds to a critical dislocation density. For a given SFE, the critical strain is a function of temperature, strain rate, and austenite grain size.

The characteristic of microstructure evolution during and after DRX is connected with the mechanism of this recrystallization. Once the critical density (which depends on strain rate, temperature, and steel chemical composition) is reached, DRX is initiated by the bulging of pre-existing grain boundaries at low strain rates. At higher strain rates, DRX is initiated by the growth of the high-angle cell boundaries formed by dislocation accumulation. The driving force for the growth of the nuclei is the difference in dislocation density in front of and behind the boundary. However, the mechanism of nucleation differs for single peak and multiple peak behaviours. In the single peak case (grain refinement), nucleation occurs essentially along existing grain boundaries and is referred to as the necklace structure. The growth of each grain is stopped by concurrent deformation. When all the grain boundary sites are exhausted, further new grains are nucleated within the original grains at the interface of the recrystallized and unrecrystallized grains. In the multiple peak case, the growth of each new grain is terminated by boundary impingement and not by the concurrent deformation. In industrial hot rolling processes, the strain rate is relatively high, such that only single peak DRX is likely to occur, if any.

All information concerning DRX is essential for the development of the transition rules for the cellular automata model described in Chapters 3 and 5. On the other hand, for the purpose of the conventional modelling microstructural evolution, an analysis of research on DRX allows one to assume that in a given material, the characteristics of this recrystallization depend on three parameters only: the initial grain size D; temperature T; and strain rate $\dot{\varepsilon}$. The initial grain size affects the critical strain, ε_c; the peak strain, ε_p; and the kinetics of DRX.

The finer the initial grain size, the lower are the critical and peak strains. This is because dislocations accumulate more rapidly and the higher specific grain boundary area (per unit volume) leads to faster recrystallization kinetics. Peak stress is also found to be dependent on the initial grain size; however, the steady-state stress and final grain size are independent of it.

When DRX starts during deformation, then after deformation, other processes, leading to the decrease of dislocation density, take place. These processes include MDRX and MDRV. These processes are automatically accounted for by the models based on internal variables, see, e.g. [109], or by more advanced cellular automata or Monte Carlo models. In conventional modelling, however, separate equations are proposed to calculate kinetics of a MDRX and a grain size after a MDRX.

The conventional models of DRX involve equations describing the critical strain, kinetics of DRX, and grain size after DRX. Numerous models developed for various materials can be found in the scientific literature, and some of them are reviewed in [60]. Selected models are repeated next, and models for the DP steel and eutectoid steel are added. The general equation describing the critical strain for DRX is:

$$\varepsilon_c = AZ^r D^s. \tag{2.73}$$

Coefficients A, r, and s and activation energy Q_{def} in the Zener-Hollomon parameter (Eq. (2.1)) obtained by various researchers for various materials are given in Table 2.10.

Kinetics of DRX is usually described using strain as an independent variable instead of time. The general equation describing dynamically recrystallized volume fraction is:

$$X_{DRX} = 1 - \exp\left[B\left(\frac{\varepsilon - \varepsilon_c}{\varepsilon_p}\right)^k\right], \tag{2.74}$$

where: ε_p – strain at the peak stress, usually calculates as $\varepsilon_p = C\varepsilon_c$.

Coefficients in Eq. (2.74) obtained by various scientists for various materials are given in Table 2.11.

Experiments carried out by various scientists show that the fully dynamically recrystallized grain size is insensitive to the strain and grain size prior to deformation, but it depends on a joint effect of temperature and strain rate, expressed by the Zener-Hollomon parameter Z. The general equation describing the grain size after the DRX is:

$$D_{DRX} = BZ^q. \tag{2.75}$$

Coefficients in Eq. (2.75) activation energy Q_{def} in the Zener-Hollomon parameter (Eq. (2.58)) obtained by various authors for various steel compositions are presented in Table 2.12.

Table 2.10 Coefficients in Eq. (2.16) describing the critical strain for dynamic recrystallization.

Steel	A	r	s	Q_{def}	References
C-Mn	4.9×10^{-4}	0.15	0.5	312 000	[114]
C-Mn	6.82×10^{-4}	0.13	–	312 000	[108]
Ti-V	0	4.54	¯0.53	¯0.1	[54]
DP	0	63.7	¯0.74	¯0.05	
900A	4.3×10^{-4}	0.18	0.3	315 000	

Table 2.11 Coefficients in Eq. (2.74) describing kinetics of dynamic recrystallization.

Steel	B	k	C	References
C-Mn	−0.8	1.4	1.23	[40]
Ti-V	−0.4	1.5	1.12	[54]
DP	−0.176	1.813	0.8	
900A	−1.7	1.3	1.05	[54]

Table 2.12 Coefficients in Eq. (2.75) describing a grain size after dynamic recrystallization.

Steel	B	q	Q_{def}	References
C-Mn	1800	−0.15	312 000	[114]
C-Mn	16 000	−0.23	312 000	[40]
Ti-V	1400	−0.16	312 000	[54]
DP	500	−0.103	349 500	
900A	16 000	−0.2	315 000	[54]

Once the DRX is initiated during the deformation, the dynamically recrystallized nuclei continue to grow after the deformation is interrupted. This mechanism is identified as MDRX. Three distinct softening processes take place after dynamic recrystallization, and they are described as SRV, MDRX, and SRX. While the DRX nuclei are growing by MDRX, the rest of the material undergoes SRV and SRX. Unlike SRX, the metadynamic one apparently does not require an incubation time because it makes use of the nuclei formed by DRX. As a consequence, dynamically recrystallized microstructures are subject to rapid changes after unloading, and this results in coarser grain size.

The fact that the SRX laws do not apply once the strain exceeds the critical value was observed by several scientists. It was noticed that for strains above the critical strain, the time for 50% recrystallization becomes independent of strain, leading to much lower rates of recrystallization than predicted by the SRX models. The kinetics of microstructure restoration processes after the DRX is strongly dependent on strain rate. In contrast, during the SRX the transformation kinetics depends strongly on strain and temperature and weakly on strain rate. It is generally observed that at the early stage of SRX the grain size decreases, irrespective of whether the final fully recrystallized grains are finer or coarser than the original ones.

All the observations presented by various researchers [40, 106, 110, 114] form the bases for the MDRX models, which are quite different from the SRX ones. As it has been mentioned, metadynamic processes are automatically accounted for by IVMs, see, e.g. [109], or by such discrete methods as cellular automata or Monte Carlo. In conventional modelling, separate equations describing metadynamic phenomena are used. The softening curve for the MDRX can be adequately described by the Avrami Eq. (2.12) with the exponent n of approximately 1.5, which is approximately an average of those reported in Table 2.6 for the SRX. The general form of the equation describing time for 50% MDRX is:

$$t_{0.5MDRX} = A_1 Z^s \exp\left[\frac{Q_{MDRX}}{R(T + 273)}\right]. \tag{2.76}$$

Table 2.13 Coefficients in Eq. (2.76) describing kinetics of metadynamic recrystallization.

Steel	A_1	s	Q_{def}	Q_{MDRX}	References
C-Mn	1.12	−0.8	312 000	230 000	[21]
Ti-V	1.12	−0.8	312 000	230 000	[35]
DP	1.575×10^{-3}	−0.4522	349 500	230 000	
900A	1.12	−0.8	315 000	230 000	[35]

The coefficients in Eq. (2.76) suggested by various scientists for different steels are given in Table 2.13.

As mentioned earlier, the metadynamic grain size depends on the Zener-Hollomon parameter Z and is larger than that after DRX. The general equation describing the metadynamic grain size is:

$$D_{MDRX} = A_2 Z^u. \tag{2.77}$$

The coefficients in Eq. (2.77) obtained by various scientists for different steels are given in Table 2.14.

2.2.3.3 Grain Growth

Following complete static or MDRX, the equiaxed austenite microstructure coarsens by grain growth. Several grain growth algorithms based on the discrete methods are presented in the following chapters of the book. These advanced algorithms are capable of dealing with an anisotropic or abnormal grain growth. However, in most conventional research on this topic, an assumption is made that the growth is uniform and usually described by a power closed-form equation. These models for uniform growth are, in general, based on the isothermal law:

$$D(t)^n = D_{RX}^n + k_{GR} t \exp\left[\frac{-Q_{GR}}{R(T+273)}\right], \tag{2.78}$$

where: D_{RX} – the fully recrystallized grain size, t – time after complete recrystallization, Q_{GR} – apparent activation energy for grain growth, n, k_{GR} – coefficients.

Table 2.15 contains constants in Eq. (2.78) obtained by various researchers. A slightly different approach to the grain growth problem is proposed by Roberts et al. [106], and since it is used by many scientists, it is worth mentioning. Authors of [106] observed that a characteristic feature of the grain growth after the recrystallization is an abrupt decrease in the growth rate some time after the completion of recrystallization. A parabolic type equation with two

Table 2.14 Coefficients in Eq. (2.77) describing a grain size after metadynamic recrystallization.

Steel	A_2	u	Q_{def}	References
C-Mn	26 000	−0.23	312 000	[21]
Ti-V	23 000	−0.16	315 000	[35]
DP	600	−0.103	349 500	
900A	25 000	−0.23	312 000	[35]

Table 2.15 Coefficients in Eq. (2.78) describing a grain growth.

Steel	n	k_{GR}	Q_{GR}	References
C-Mn	2	4.27×10^{12}	66 600	[85]
C-Mn-V	7	1.45×10^{27}	400 000	[40]
DP	4.5	4.1×10^{23}	440 000	

Table 2.16 Coefficients in Eq. (2.79) describing a grain growth.

	Stage 1		Stage 2		References
Steel	a	b	a	b	
C-Mn	6.6	6200	8.1	9000	[106]
Ti-V	7.1	7180	9.5	10 920	[54]
900A	7.0	5900	8.4	8520	

sets of parameters for two stages of the growth is used to characterize this behaviour qualitatively:

$$D(t)^2 = D_{RX}^2 + t \times 10^{(a-(b/T))}, \tag{2.79}$$

The values of coefficients a and b for various steels are given in Table 2.16. The first stage of grain growth is assumed to last approximately 20 s.

2.2.3.4 Effect of Precipitation

The development of microalloyed steels in the second half of the twentieth century was one of the biggest breakthroughs in the progress of research on new steels. In these steels, small amounts of niobium, vanadium, or titanium were added. Significant grain refinement was obtained due to retarding recrystallization by precipitates of carbonitrides of these microelements. The precipitation process of carbonitride Nb(C,N) strongly influences the final microstructure and properties of products. Precipitates inhibit the recrystallization of austenite, which leads to a reduction of the size of the ferrite grains after the transformation of non-recrystallized austenite. A new generation of steels called high strength low alloyed (HSLA) steels was developed on this basis.

Modelling of HSLA steels is of particular importance, and it has to combine complex phenomena of thermodynamics of precipitates with microstructure modelling accounting for these precipitates. The models proposed for SRX of niobium steels in the absence of strain-induced precipitation are of the same character as those for the carbon-manganese steels, except that the coefficients in equations are different. The change in recrystallization kinetics caused by strain-induced precipitation due to the presence of niobium was handled by making the coefficient b and the activation energy Q_{RX} in Eq. (2.71) functions of temperature. However, the appearance of precipitation changes the character of recrystallization compared to that affected by the solute drag effect only. Prediction of the strain-induced precipitation start becomes the most important part of simulations. Dutta and Sellars [25] developed a model for strain-induced precipitation, which is a basis of several models suggested later by other scientists. The model describes the incubation period and the progress of precipitation [25, 26], and it is used to calculate the number and size of precipitates. Solubility product of Nb, C, and N in austenite is [25]:

$$k_s = \frac{[\mathrm{Nb}]([\mathrm{C}] + (12/14)[\mathrm{N}])}{10^q}, \qquad q = 2.26 - \frac{6770}{T}, \tag{2.80}$$

where: [Nb], [C], [N] – the contents of niobium, carbon, and nitrogen in austenite, expressed in percentage by mass.

Critical radius for nucleation, r_c, is determined by the driving force, and the equation is:

$$r_{cr} = -\frac{2\gamma}{\Delta G_v}, \tag{2.81}$$

where: γ – energy of the interface equal to 0.5 J/m^2, ΔG_v – the difference in free energy per unit volume, which is calculated from the formula:

$$\Delta G_v = \frac{-RT_K \ln k_s}{V_m}, \tag{2.82}$$

where: V_m – molar volume; for Nb(C,N), $V_m = 1.28 \times 10^{-5}$ m^3/mol.

At elevated temperatures, coagulation of the particles of precipitates occurs, which can be modelled by the Wagner equation:

$$r^3 - r_0^3 = \frac{8}{9}\frac{\gamma D_{Nb} C_0 V_m^2}{RT_K} t. \tag{2.83}$$

The model described previously calculates the number of precipitates per unit volume, volume fraction of precipitates, and average diameter of precipitates. It is a starting point for further modelling of recrystallization kinetics, accounting for the influence of precipitates on recrystallization kinetics during hot forming. In this model, the change in recrystallization behaviour is handled in a recursive manner. The amount of recrystallization during a time increment is modelled using the solute drag equation, followed by a calculation of the time to precipitation start using the process conditions for the austenite at that time interval. The relative values of the time for 5% recrystallization $t_{0.05X}$, time for 5% precipitation $t_{0.05p}$, and time for 95% recrystallization $t_{0.95X}$, control the behaviour of the material. It is assumed that for $t_{0.05X} < t_{0.05p} < t_{0.95X}$ recrystallization is retarded by precipitation, and for $t_{0.05X} > t_{0.05p}$ recrystallization does not take place and that there would be no further recrystallization once precipitation is predicted to occur. Figure 2.8 demonstrates the flow of calculations during modelling of microstructure evolution in the microalloyed niobium steels. RX in this figure stands for recrystallization, and t_p is the interpass time.

The relevant equations describing the kinetics of precipitation as well as recrystallization in niobium microalloyed steels, developed by various scientists, are given in Table 2.17. Notice that some of the equations describe the time for 5% recrystallization $t_{0.05X}$ and some the time for 50% recrystallization $t_{0.50X}$. One should remember that when $t_{0.05X}$ is used in the Avrami Eq. (2.69), the constant $k = 0.0513$. In Table 2.15, [Nb] represents the niobium content in the solution.

2.2.4 Phase Transformations

The problems associated with modelling phase transformations are best examined by studying the phase equilibrium diagrams. These diagrams show the phases and their compositions

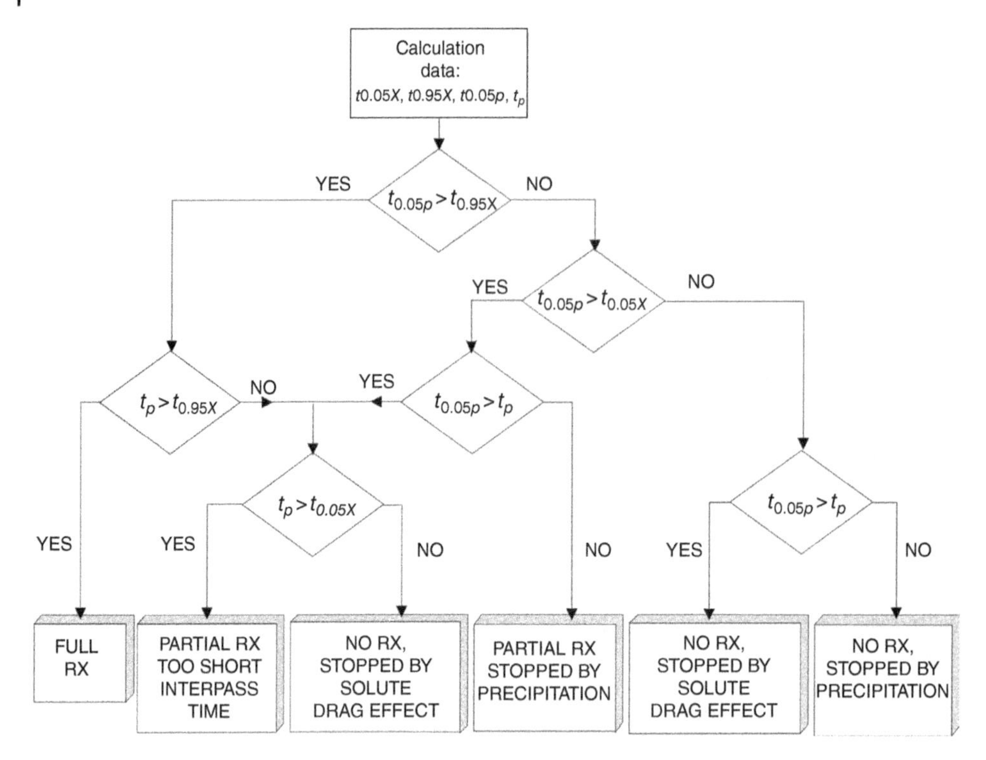

Figure 2.8 Flow chart of calculations of microstructure evolution in microalloyed niobium steels.

Table 2.17 Equations describing recrystallization and precipitation times for Nb microalloyed steels.

References	Recrystallization and precipitation times
[25]	$t_{0.05x} = 6.75 \times 10^{-20} D^2 \varepsilon^{-4} \exp \frac{300000}{RT} \exp \left\{ \left[\left(\frac{2.75 \cdot 10^5}{T} - 185 \right) \right] [\text{Nb}] \right\}$ (2.84)
	$t_{0.05p} = 3 \times 10^{-6} [\text{Nb}]^{-1} \varepsilon^{-1} Z^{-0.5} \exp \frac{270000}{RT} \exp \left(\frac{2.5 \cdot 10^{10}}{T^3 (\ln k_s)^2} \right)$ (2.85)
[121]	$t_{0.5x} = (-5.24 + 550[\text{Nb}]) \times 10^{-18} D^2 \varepsilon^{-4 + 77[Nb]} \exp \left(\frac{330000}{RT} \right)$ (2.86)
	$t_{0.05p} = 6 \times 10^{-6} [\text{Nb}]^{-1} \varepsilon^{-1} Z^{-0.5} \exp \frac{270000}{RT} \exp \left(\frac{2.5 \cdot 10^{10}}{T^3 (\ln k_s)^2} \right)$ (2.87)

at any temperature and alloy composition. The upper curve on the diagram represents the liquidus temperature above which the alloy is in the liquid phase. The liquid begins to solidify when the temperature cools below the liquidus temperature. During solidification, the amorphous liquid phase changes into a crystalline solid, and grains nucleate and grow. This transformation does not happen instantly. On reaching the liquidus, nuclei form at several locations in the melt. The temperature remains constant while the nuclei grow. In steels, the solid phase just formed is called austenite, designated by γ, and is a FCC. On further cooling, at the Ae3 temperature, the first ferrite (α) grains appear, and the steel is in the two-phase region. The structure of the ferrite grains is BCC, body centred cubic. As the temperature drops further, the transformation stops, and the steel becomes ferritic. This temperature is identified as the Ae1. Depending on the content of carbon and cooling rates, other phases like pearlite or bainite may appear, as well. The two temperatures Ae3 and Ae1 are affected by the chemical composition, pre-strain, cooling rate, and initial austenite grain size. While these effects are measurable, for most purposes, the phase diagram gives sufficient information regarding the equilibrium state of phases at various temperatures.

The solubility of carbon and other elements in various phases is different; therefore, phase transformations require transport of these elements on long distances by the mechanism of diffusion. The phase equilibrium diagrams determine concentrations of elements in phases at various temperatures. However, experimentally determined phase diagrams are usually available for binary systems only and to a limited extent for ternary systems. The thermodynamic properties controlling driving forces for phase transformations, boundary conditions, and kinetic parameters are complex temperature and chemical composition functions. The computational approach has emerged and been established over the last 30 years as the most efficient way of designing and studying complex systems.

The most advanced method of calculating phase transformations in complex systems is the CALPHAD (CALculation of PHAse Diagrams). The CALPHAD method is based on the fact that a phase diagram is a representation of the thermodynamic properties of a system. Thus, if the thermodynamic properties are known, it would be possible to calculate the multi-component phase diagrams. The computational thermodynamics enables modelling and numerically solving a number of different materials-dependent engineering problems. Some of them may be very complex. The application of tools such as ThermoCalc or DICTRA reduces the need for costly experiments. The thermodynamic models, which are the basis of ThermoCalc or DICTRA, are not discussed in this book. All equilibrium data used to model phase transformations in this section were obtained from the ThermoCalc software.

The phase transformation models adapted to varying temperature conditions used the equilibrium data predicted by the thermodynamic software described previously. The equilibrium state is a stationary state, which was reached by very slow changes in the temperature. The objective of phase transformation models, which are described in this section, is a prediction of the non-stationary state during fast changes of the temperature. The models predict the kinetics of transformations during temperature changes and also predict the non-stable states resulting from fast cooling. In the first part of the present section, a brief review of phase transformation models is presented. Then a more thorough analysis of conventional models based on JMAK type Eq. (2.69) [2–4, 45, 52] and on differential equation [121] follows.

Historically, various modifications of the JMAK type Eq. (2.69) were commonly used for simulations of phase transformations. In this approach, all attention is focused on the kinetics, and microstructural aspects are essentially ignored. However, even in the simplest JMAK model, nucleation and growth are recognized as being the two relevant and intrinsically different processes. The Avrami exponent is shown to be related to the nucleation conditions [125], even though only two modes of nucleation kinetics are considered: site saturation and continuous nucleation at a constant rate. In the case of site saturation, all nuclei are present and active at the start of the transformation, and their number and density remain constant during the entire transformation. In the case of continuous nucleation, the number of activated nucleation sites increases at a constant rate during the transformation, with a rate of nucleus formation depending on temperature and the fraction of parent phase still present. Suehiro et al. [121] developed a more advanced model based on differential equations describing the kinetics of transformation separately for the continuous nucleation and the site saturation.

More refined transformation models incorporate relevant features of the parent microstructure. The most straightforward approach considered the austenite grain as a sphere and the ferrite to nucleate uniformly along the outer surface. In this simplified model, the final ferrite grain size is necessarily identical to the prior austenite grain size, and assuming ferrite growth to be controlled by carbon diffusion in austenite provides a satisfactory description of austenite decomposition in most Fe-C alloys. However, more sophisticated growth parameters (i.e. interface mobility, solute-drag effect) have to be introduced as adjustable parameters to describe the growth rate of ferrite in more complex alloys, including Fe-C-Mn steels. Therefore, geometrically more refined models were proposed in which the austenite grain is assumed to be more complex geometrical figures [58]. These approaches allow incorporation of the ferrite nucleation site density per austenite grain as a modelling parameter to reproduce the grain size, depending on the cooling conditions.

In recent years, the phase-field approach has emerged as one of the most powerful methods for modelling many types of microstructure-evolution processes, including the austenite decomposition [58, 72, 73, 116, 117]. A detailed description of this method was presented by Mecozzi et al. [72]. Very briefly, the phase-field model treats a polycrystalline system, containing both bulk and boundary regions, in an integral manner. A set of continuous phase-field variables, each of them representing an individual grain of the system, are defined to have a constant value inside the grains and change continuously over a diffuse boundary. In this approach, each phase-field parameter of grain is equal to 1 inside the grain and 0 elsewhere. At the interface between the two grains, there is a gradual change of the two corresponding phase-field parameters from 0 to 1 such that the sum of all phase-field parameters holds 1 at each time step and at each point in the simulation domain. Following the formation of new (ferritic) nuclei in specific locations, depending on the cooling conditions, the microstructural evolution during the austenite to ferrite transformation is governed by the phase-field equations. The interface's mobility, interfacial energies, and the driving pressure for the transformation are parameters of these equations, and they determine the kinetics of the austenite decomposition. The driving pressure is calculated from the local carbon composition within the diffuse interface, controlled by the carbon diffusion within the austenite. Therefore, the phase-field model describes the ferrite growth via a mixed-mode approach, i.e. both the carbon diffusion and the apparent mobility of the austenite/ferrite interface are accounted for.

Since the early 1970s, the FEM has become the most popular simulation technique; see Chapter 3. In modelling phase transformations, this method was applied to simulations of carbon distribution in austenite and became an alternative for the phase-field models. An example of contribution to this research was presented by Pernach and Pietrzyk [95]. FE solution of the diffusion equation with a moving boundary (Stefan problem) was performed in that work for various shapes of austenite and ferrite grains. A similar approach was applied in [96] to simulation phase transformation during heating. In the late 1990s, such discrete methods as cellular automata, MD, or Monte Carlo began to be applied to modelling recrystallization and phase transformations during materials processing. These methods are discussed in the following chapters of this book.

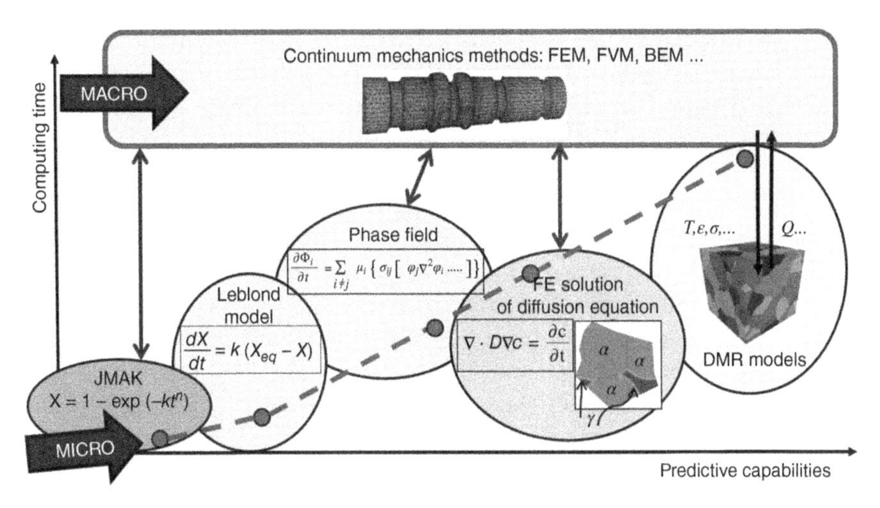

Figure 2.9 Classification of selected phase transformation models: computing costs versus predictive capabilities.

Classification of phase transformation models with respect to predictive capabilities and computing costs is presented in Figure 2.9. The first group (bottom left corner in Figure 2.9) contains models commonly used for fast simulations of industrial processes, and they are generally limited to the description of the kinetics of transformations and volume fractions of phases. The Scheil additivity rule [112] has to be applied in these models to account for the temperature changes during transformations. In the second group (the centre in Figure 2.9), differential equations or phase-field technique models are usually applied to technology design and optimization of processes. These models accurately describe transformations in varying temperatures. The next group (further right in Figure 2.9) includes models based on the FE solution of the diffusion equation with moving boundary [95, 96]. Beyond the mentioned earlier parameters, these models can predict the distribution of carbon concentration in austenite and the resulting hardness of bainite and martensite.

2.2.4.1 JMAK-Equation-Based Model

This basic time in Eq. (2.69) is not introduced in modelling phase transformations in steels, and one Avrami coefficient k is used:

$$X = 1 - \exp(-kt^n). \tag{2.88}$$

The model based on Eq. (2.31) will be further referred to as model A. Theoretical considerations show that according to a transformation type (nucleation and growth process, site saturation process) a constant value of coefficient n in Eq. (2.88) can be used. The values of n are introduced in the model as a_4, $a_{15,}$ and a_{24} for ferritic, pearlitic, and bainitic transformations, respectively. Coefficient k should map the form of a TTT (time–temperature–transformation) diagram. Following that observation, the k is defined as a temperature function $k = f(T)$. Various forms of k were tested by the authors. A simple function may cause low accuracy of the model, and a complex function may cause problems with the identification of the model and lack of the uniqueness of the solution. The function $k = f(T)$ has to be flexible enough to replicate complex phenomena of nucleation and growth controlled by diffusion, interface mobility, and solute-drag effect. These phenomena are reversibly dependent on the temperature.

Rate of nucleation increases with the temperature drop below A_{e3}. In contrast, diffusion becomes slower at lower temperatures. That inspired Donnay et al. [22] to propose a modified Gaussian function for parameter k, which is used for a ferritic transformation in this work:

$$k = k_{\max} \exp\left[-\left(\frac{T - T_{\max}}{a_7}\right)^{a_8}\right]. \tag{2.89}$$

The four coefficients in this function $k_{\max}$, $T_{\max}$, a_7, and a_8 allow to describe all shapes of the TTT curves in a quite intuitive way: $k_{\max}$ is the maximum value of k, $T_{\max}$ is a temperature position in °C of the nose of the modified Gaussian function and represents the temperature of the maximum rate of the transformation, a_8 is proportional to the nose width at mid-height, and a_7 is related to the sharpness of the curve.

Equation (2.89) is supposed to account for the influence of the austenite grain size D_γ at the beginning of the transformation. Thus, the following equations are used to calculate coefficients $k_{\max}$ and $T_{\max}$:

$$k_{\max} = \frac{a_5}{D_\gamma} \tag{2.90}$$

$$T_{\max} = A_{e3} + \frac{400}{D_\gamma} - a_6 \tag{2.91}$$

It was concluded from the primary model investigations that there is no need to introduce such a complex function $k(T)$ for pearlitic and bainitic transformations. Therefore, a slightly simpler function was selected for the bainitic transformation:

$$k = a_{23} \exp\left(a_{22} - \frac{a_{21}T}{100}\right) \tag{2.92}$$

Due to the low carbon content in the DP steels (Table 2.2), the effect of the pearlitic transformation is negligible in these steels. Therefore, the constant value of the coefficient k is introduced, and it is a coefficient a_{14} in the model. On the contrary, pearlitic transformation is the most important one, and the following function is used for the pearlitic transformation:

$$k = \frac{a_{14}}{D_\gamma^{a_{16}}} \exp\left(a_{13} - \frac{a_{12}T}{100}\right). \tag{2.93}$$

In the model, phases incubation times should be accounted for. Equation (2.88) combined with function (2.89) do not require the incubation time. It is assumed that ferritic transformation begins when the volume fraction of ferrite achieves 5%. Incubation times of the remaining pearlitic transformations (τ_p) and (τ_b) are calculated as:

$$\tau_P = \frac{a_9}{(A_{e1} - T_C)^{a_{11}}} \exp\left[\frac{a_{10} \times 10^3}{R(T + 273)}\right] \tag{2.94}$$

$$\tau_b = \frac{a_{17}}{(a_{20} - T)^{a_{19}}} \exp\left[\frac{a_{18} \times 10^3}{RT(T + 273)}\right] \tag{2.95}$$

Additional relationships in the model and equilibrium carbon concentrations are presented in Table 2.18. Notation in this table: c_γ – average carbon content in the austenite, c_α – carbon content in the ferrite, c_0 – carbon content in the steel, $c_{\gamma\alpha}$ – carbon concentration in the austenite at the γ-α boundary, $c_{\gamma\beta}$ – carbon concentration in austenite at the γ-*cementite* boundary, X_f – volume fraction of ferrite, X_{f0} – equilibrium (maximum) ferrite volume

Table 2.18 Additional equations in the model.

$c_\gamma = \dfrac{(c_0 - X_f c_\alpha)}{1 - X_f}$	$X_{f0} = \dfrac{c_{\gamma\alpha} - c_0}{c_{\gamma\alpha} - c_\alpha}$
$c_{\gamma\alpha} = c_{\gamma\alpha 0} + c_{\gamma\alpha 1}T,\quad c_{\gamma\beta} = c_{\gamma\beta 0} + c_{\gamma\beta 1}T,\ c_\alpha = f(T)$	

fraction in steel in a considered temperature. Equilibrium concentrations $c_{\gamma\alpha}$ and $c_{\gamma\beta}$ as well as carbon content in the ferrite c_α are introduced as temperature functions. These functions are polynomials, and they are determined using ThermoCalc software based on the information on the steel chemical composition.

2.2.4.2 Differential Equation Model

A more physically based model was proposed by Japanese scientists [121]. The model is based on the differential equation; therefore, it can be easily applied to varying temperature conditions. Beyond this, several physical parameters are introduced in this model. In this chapter, models based on differential equations are referred to as model B. The basic principles of this model are explained using ferritic transformation as an example. In the original version, the model was proposed without an incubation time, and the assumption was made that ferritic transformation begins when there is 5% of ferrite in a microstructure. Investigation performed by the authors of this book showed that modelling is more accurate when incubation time is introduced. The following equation was used:

$$\tau_f = \frac{a_1}{(A_{e3} - T)^{a_3}} \exp\left(\frac{1000a_2}{RT}\right). \tag{2.96}$$

Two differential equations are used to describe austenite-ferrite phase transformation kinetics:

- for the nucleation and growth

$$\frac{dX_f}{dt} = a_5\left(SIG^3\right)^{0.25}\left[\ln\left(\frac{1}{1-X_f}\right)\right]^{0.75}\left(1 - X_f\right) \tag{2.97}$$

- for the nucleation and growth

$$\frac{dX_f}{dt} = a_6 \times 10^{-12} \exp\left(\frac{a_8}{RT}\right)\frac{6}{D_\gamma}G\left(1 - X_f\right), \tag{2.98}$$

where: X_f – transformed volume fraction, I – rate of nucleation, G – rate of transformation, S – specific area of the grain boundary, D_γ – austenite grain size.

The equations used to calculate the parameters in the relationship (2.98) are given in Table 2.18 [42] with the following notation: $\Delta T = A_{e3} - T$, ΔG – Gibbs free energy calculated using ThermoCalc, r – radius of curvature of advancing phase, D – diffusion coefficient of carbon in austenite, ρ – density. The remaining equations used in model B are given in Table 2.19.

For both models, start temperatures for the bainitic (B_s) and martensitic (M_s) transformations are functions of chemical composition:

$$B_s[^\circ\mathrm{C}] = a_{20} - 425[\mathrm{C}] - 42.5[\mathrm{Mn}] - 31.5[\mathrm{Ni}]. \tag{2.99}$$

$$M_s[^\circ\mathrm{C}] = a_{25} - a_{26}C_\gamma. \tag{2.100}$$

Table 2.19 Equations describing parameters in model B.

$I = \frac{D}{\sqrt{T_K}} \exp\left(\frac{-a_7 \times 10^9}{RT_K \Delta G^2}\right)$	$G = \frac{1}{2r} D \frac{C_{\gamma\alpha} - C_\gamma}{C_\gamma - C_\alpha}$	$S = \frac{6}{D_\gamma^4}$
$r = \frac{1.14 D^{0.5}(C_\gamma - C_0)}{\sqrt{(C_\gamma - C_\alpha)(C_0 - C_\alpha)}} t^{0.5}$	$Q = \rho \Delta H \frac{\Delta X}{\Delta t}$	

Fraction of austenite, which transforms into martensite, calculated according to the model of Koistinen and Marburger [51], also described in [121], is:

$$X_m = 1 - \exp\left[-0.011(M_s - T)\right]. \tag{2.101}$$

Equation (2.101) represents volume fraction of martensite with respect to the whole volume of the austenite, which remains at the temperature Ms. The volume fraction of martensite with respect to the whole volume of the material is:

$$F_m = \left(1 - F_f - F_p - F_b\right)\{1 - \exp\left[-0.011(M_s - T)\right]\}, \tag{2.102}$$

where: F_f, F_p, F_b – fraction of ferrite, pearlite, and bainite with respect to the whole volume of material.

2.2.4.3 Numerical Solution

Simulation of the ferritic transformation in both models starts when the temperature drops below A_{e3}. In model A, calculations are performed with Eq. (2.88). In model B, calculations start with Eq. (2.97) and, when the value of derivative calculated from Eq. (2.98) becomes larger than that determined from Eq. (2.97), the simulation continues with Eq. (2.98). In both models, the transformed volume fraction X_f is calculated with respect to the maximum volume fraction of ferrite X_{f0} in the current temperature. Thus, this volume fraction of ferrite with respect to the whole volume of the body is $F_f = X_f X_{f0}$. During numerical simulation in the varying temperature, the current value of X_f calculated from Eqs. (2.88), (2.97), or (2.98) has to be corrected to account for the change of the equilibrium (maximum) volume fraction of ferrite X_{f0}, which is a function of temperature. Simulation continues until the transformed volume achieves 1. However, when carbon content in austenite achieves the limiting value $c_{\gamma\beta}$ (see Table 2.18), the austenite-pearlite transformation begins in the remaining volume of austenite.

There are several coefficients in the models. These coefficients are gathered in the vector $\mathbf{a} = \{a_1 \dots a_{27}\}$. The values of the coefficients are determined using inverse analysis of the dilatometric tests.

2.2.4.4 Additivity Rule

Modelling of phase transformations based on differential Eqs. (2.97) and (2.98) automatically accounts for changes of the temperature. On the other hand, as it has been mentioned previously, the classical modelling of phase transformations is often based on the Avrami theory [2–4] represented by Eq. (2.88). This equation was derived with the assumptions of uniform distribution of nucleation sites. It means further that a constant ratio between growth rate and nucleation is also assumed, and the progress of the transformation does not depend on the history of temperature changes. These conditions are called Avrami isokinetic conditions. For the transformation, which takes place in these conditions, the equation of kinetics

(2.88) derived for isothermal conditions can also be applied to the simulation of transformation, which takes place during continuous changes of the temperature. In other words, it means that the additivity condition is fulfilled.

The assumption that nucleation is homogeneous in the considered volume of the material constrains possibility of analysis of phase transformations to a certain case, which has no application to the majority of phase transformations in alloys. In general, the nucleation sites are expended at the early stage of the transformation, and further progress is due to growth. This observation led to the following condition, which defines when the additivity rule can be applied:

$$\frac{G(T)t_{0.5}}{D_\gamma} < 0.5, \tag{2.103}$$

where: T – temperature, D_γ – austenite grain size, $t_{0.5}$ – time for 50% transformation.

Contrary to model 3, basic equations of model 1 are based on the Avrami theory and are defined for the isothermal conditions. In the continuous cooling of, e.g., weld, the transformation occurs in nonisothermal conditions. In the experiments described next and in industrial welding processes, cooling rates between 4 and 300 °C/s are expected. Thus, the model of the transformation kinetics during welding has to account for changes in the temperature. The additivity rule, proposed by Scheil [112], is applied in this book for the prediction of the incubation time, as well as the kinetics of pearlitic and bainitic transformations in both models. Beyond this, the additivity rule is used in the simulation of the progress of the ferritic transformation using model A. According to this rule, the transformation begins when the sum of certain fractions reaches unity:

$$\int_0^\tau \frac{dt}{\tau_n(T)} = \int_0^\tau \frac{1}{\tau_n(T)} \frac{dt}{dT} dT = \int_0^\tau \frac{1}{C_r \tau_n(T)} dT = 1, \tag{2.104}$$

where: τ_n – incubation time at constant temperature T, C_r – cooling rate.

Additivity rule can be applied under certain conditions, which are discussed in [11, 112] and which are fulfilled in the experiment in this project. Briefly, if the progress of the phenomenon at any instant depends only on temperature, the process is additive. Thus, transformation during cooling after welding can also be modelled using the additivity rule. Since only k in Eq. (2.88) is temperature-dependent, Equation (2.88) can be written as:

$$X = 1 - \exp\left[-\int_{T_0}^{T} \frac{k(T)^{\frac{1}{n}}}{C_r} dT\right], \tag{2.105}$$

where: T_0 – temperature at the end of the incubation time.

Equation (2.48) is valid for the continuous cooling stage. Derivation of this equation, as well as its extension to the holding and heating stages of the process, is described in [11]. Dependence of the coefficient k on the temperature follows Eqs. (2.89), (2.92), or (2.93), depending on the transformation.

2.2.4.5 Phase Transformation During Heating

Information on modelling phase transformations during heating is scarce [88, 111], and it is in general limited to the ferritic-pearlitic microstructures; [111] contains a thorough analysis of this problem. Modelling of phase transformations in the heating part of the thermal cycle during continuous annealing or welding processes can be performed using various models. The conventional model based on the Avrami Eq. (2.88) is described briefly in this chapter. Simulation of the transformation begins at the equilibrium temperature A_{c1} and incubation time is calculated using the following equation:

$$\tau_a = \frac{b_1}{(A_{e3} - T)^{b_3}} \exp\left[\frac{1000b_2}{R(T + 273)}\right], \tag{2.106}$$

where: b_1, b_2, b_3 – coefficients.

Scheil additivity rule [112] was used to account for the temperature changes during both incubation time and the progress of the transformation. When the incubation time is finished, the kinetics of the transformation into austenite is calculated using Eq. (2.88). Coefficient k in this equation is introduced as the following function of temperature:

$$k = k_{BMP} b_4 \exp\left[\frac{-10^3 b_5}{R(T + 273)}\right]. \tag{2.107}$$

This model contains six coefficients, b_1, b_2, and b_3 in Eq. (2.106) describing incubation time and b_4 and b_5 in the equation describing the kinetics of transformation. Coefficient b_6 is the Avrami exponent n in Eq. (2.88). Coefficient k_{PBM} in Eq. (2.107) accounts for the influence of the pearlite, bainite, and martensite on the kinetics of transformation. This coefficient is equal to 1 during ferrite to austenite transformation.

2.2.4.6 Identification of the Model

Identification of the coefficients in the model is based on dilatometric tests and inverse analysis. Tests performed for identification of the phase transformation models in the present work were conducted with dilatometer DIL 805 at the Institute for Ferrous Metallurgy in Gliwice, Poland. This dilatometer is capable of deforming the sample prior to cooling. In this case, the sample is cylindrical, having dimensions $\phi 4 \times 7$ mm. The tubular samples $\phi 4/2 \times 10$ mm are used for experiments not involving deformation. Typically, the dilatometer is used for the development of TTT, CCT, and DCCT (with deformation) or TTT diagrams, which are referred to in the text as phase transformations diagrams. To develop CCT diagrams, the subsequent samples are austenitized at defined conditions (temperature and time) and cooled at different rates to ambient temperature. If the cooling stage is preceded by deformation, the phase transformation diagram is called the DCCT diagram. In contrast, for TTT diagram development, the sample is cooled at a high cooling rate to isothermal holding temperature and then held at this temperature until the phase transformation is completed. During phase transformation, the crystallographic structure of the parent phase changes, which results in a specific volume change. The dilatation curves are used for the identification of phase transformations [99]. The deviation of the dilatometric curve from the linearity is connected to the onset of phase transformation.

The phase transformation models contain several coefficients, which are grouped in vector **a**. The values of these coefficients are different for different steels. Beyond this, they depend on microstructure at the beginning of transformation and on the deformation of the austenite. Difficulties connected with the determination of these coefficients are the main factor, limiting the wide application of the model to simulation and control phase transformations in industrial processes. The method which allows fast and easy determination of components of the vector **a** is described in [99] and is used in the present work. Identification was performed for both models for various materials; see other authors' publications for results [69, 99]. The results of identification for model A only and for steels in Table 2.2 are described in this chapter.

Identification was performed using the inverse analysis. Basic principles of this method are described in a number of publications; see for example [123]. The most frequent applications of this method are connected with the determination of coefficients in rheological models of materials subjected to plastic deformation [29, 123, 124]. Torsion or compression plastometric tests are used as experiments in this analysis. Less information can be found on the application of the inverse approach to the phase transformation model. Some details are given in [53, 99], and the general idea of this algorithm is repeated briefly next. The mathematical model of an arbitrary phase transformation can be described by a set of equations:

$$\mathbf{d} = \mathrm{F}(\mathbf{a}, \mathbf{p}), \tag{2.108}$$

where: $\mathbf{d} = \{d_1, \dots d_r\}$ – vector of start and finish temperatures of transformations and volume fractions of structural components in the room temperature, which are measured in the dilatometric tests performed with constant cooling rates, $\mathbf{a} = \{a_1, \dots a_r\}$ – vector of coefficients of the model, $\mathbf{p} = \{p_1, \dots p_r\}$ – vector of such process parameters as cooling rates, austenite grain size, and deformation of austenite.

When vectors $\mathbf{p}$ and $\mathbf{a}$ are known, the solution of the problem (2.108) is called a direct solution. Inverse solution of the problem (2.108) is defined as the determination of the components of vector $\mathbf{a}$ for known vectors $\mathbf{d}$ and $\mathbf{p}$. For some simple liner problems, the inverse function F^{-1} can be found analytically. For phase transformations, these relations are strongly nonlinear, and optimization techniques are used to solve the inverse task.

The objective of the inverse analysis is the determination of the optimum components of the vector $\mathbf{a}$. It is achieved by searching for the minimum, with respect to the vector $\mathbf{a}$, of the objective function defined as a square root error between measured and calculated components of the vector $\mathbf{d}$:

$$\Phi(\mathbf{a}, \mathbf{p}) = \sum_{i=1}^{n} \beta_i \left[\mathbf{d}_i^c(\mathbf{a}, \mathbf{p}_i) - \mathbf{d}_i^m\right]^2, \tag{2.109}$$

where: $\mathbf{d}_i^m$ – vector containing measured values of output parameters, $\mathbf{d}_i^c$ – vector containing calculated values of output parameters, β – weights of the points, $(i = 1 \dots n)$, n – number of measurements.

Measurements are obtained from the dilatometric tests carried out with constant cooling rates. Components are calculated using one of the models of the direct problem, which are described previously.

Identification of parameters of the phase transformation model is composed of two parts. The first is a solution of the direct problem based on the model. The second part is the solution of the inverse problem, in which optimization techniques are used.

Results of dilatometric tests (see Section 3.2), including measurements of the start and end temperatures for transformation and volume fractions of phases after cooling to room temperature, are used as an input to the inverse analysis. Thus, in the particular case of phase transformations, the objective function (2.52) is defined as:

$$\Phi(\mathbf{a}, \mathbf{p}) = \sqrt{\frac{1}{n}\sum_{i=1}^{n}\left(\frac{T_{im} - T_{ic}}{T_{im}}\right)^2 + \frac{1}{k}\sum_{i=1}^{k}\left(\frac{X_{im} - X_{ic}}{X_{im}}\right)^2}, \tag{2.110}$$

Table 2.20 Coefficients in the phase transformation model determined using inverse analysis for the three steels in Table 2.1.

DP 1	b_1	b_2	b_3	b_4	b_5	b_6	a_4	a_5	a_6	a_7
	977.6	0.15	2.78	8×10^9	255.5	1.23	1.69	0.858	187.9	39.06
	a_8	a_9	a_{10}	a_{11}	a_{12}	a_{13}	a_{14}	a_{15}	a_{16}	a_{17}
	1.78	64.76	1.106	0.618	0.153	–	–	0.128	–	1600
	a_{18}	a_{19}	a_{20}	a_{21}	a_{22}	a_{23}	a_{24}	a_{25}	a_{26}	
	64.64	3.5	669.2	0.118	0.074	0.344	1.04	421.7	1.83	
DP 2	b_1	b_2	b_3	b_4	b_5	b_6	a_4	a_5	a_6	a_7
	977.6	0.15	2.78	8×10^9	255.5	1.23	3.15	0.626	219.9	77.4
	a_8	a_9	a_{10}	a_{11}	a_{12}	a_{13}	a_{14}	a_{15}	a_{16}	a_{17}
	4.56	64.76	1.106	0.618	0.153	–	–	0.128	–	1842
	a_{18}	a_{19}	a_{20}	a_{21}	a_{22}	a_{23}	a_{24}	a_{25}	a_{26}	
	66.59	3.49	692.7	0.18	0.074	0.406	1.05	434.8	1.73	
900A	a_8	a_9	a_{10}	a_{11}	a_{12}	a_{13}	a_{14}	a_{15}	a_{16}	a_{17}
	1.98	0.078	122.8	2.634	0.134	8.913	0.013	0.063	0.862	1399
	a_{18}	a_{19}	a_{20}	a_{21}	a_{22}	a_{23}	a_{24}	a_{25}	a_{26}	
	61.98	2.537	887.6	0.942	0.399	0.386	1.08	218.1	1.65	

where: T_{im}, T_{ic} – measured and calculated start and end temperatures of phase transformations, n – number of temperature measurements, X_{im}, X_{ic} – measured and calculated volume fractions of phases at room temperatures, k – number of measurements of volume fractions of phases.

Any phase transformation model can be identified using the technique described previously; see for example [53, 69, 99]. Identification of model A described in this section is presented next. Due to a large number of coefficients in this model, problems of effectiveness and efficiency of the optimization techniques need to be considered. The uniqueness of the solution should be discussed as well. Some light on the solution of these problems can be put by performing the sensitivity analysis of the output of the model with respect to the coefficients of this model. This method deals with the question – Which factors of the physical model or computer simulation are really important? Details of the sensitivity analysis for the phase transformation model were presented by Szeliga [122] and are not repeated in this book.

Dilatometric tests were performed for the DP steels and eutectoid steel with the chemical composition given in Table 2.1. Identification of the phase transformation model for these steels was performed using the inverse analysis. Values of coefficients **a** obtained from this analysis for the objective function (2.110) are given in Table 2.20. Coefficients for austenitic transformation during heating were the same for the two DP steels. Since it is not a part of the technological process of manufacturing of rods or rails, the heating process was not considered for the eutectoid steel 900A.

The general model for all transformations during cooling contains altogether 27 coefficients. However, only 23 of the coefficients are active in model A for the considered DP steels. After the sensitivity analysis, the number of coefficients was reduced to 20. Comparison of predicted and calculated CCT diagrams is shown in Figure 2.10, where: A – austenite, F – ferrite, P – pearlite, B – bainite, and M – martensite.

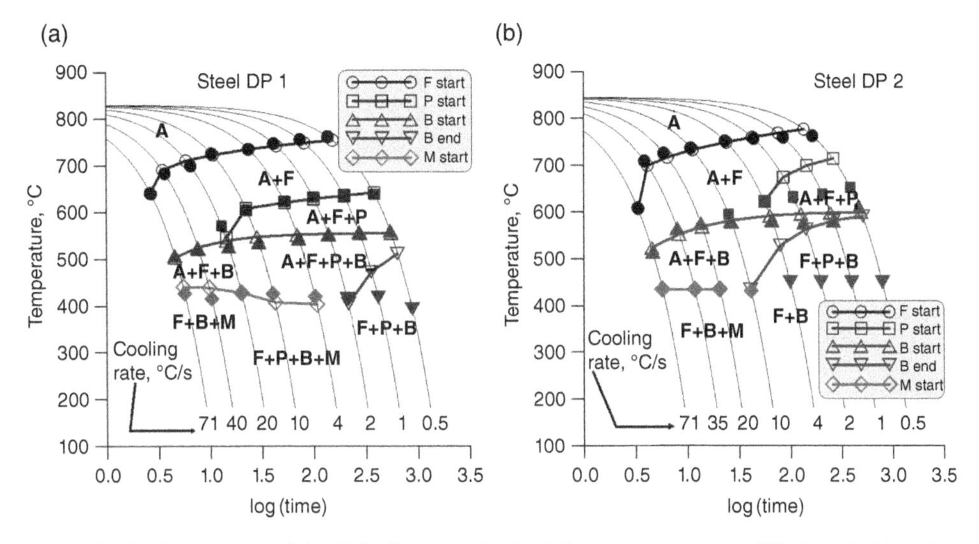

Figure 2.10 Comparison of the CCT diagrams obtained from measurements (filled symbols) and calculated by model A with the coefficients in Table 2.18 (open symbols) for steels DP 1 (a) and DP 2 (b).

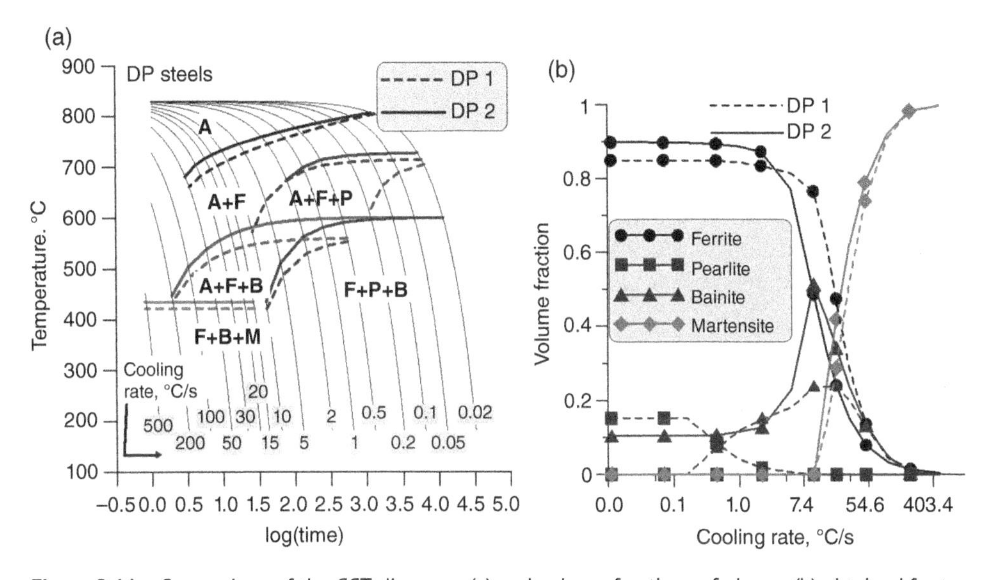

Figure 2.11 Comparison of the CCT diagrams (a) and volume fractions of phases (b) obtained for two DP steels.

A comparison of the CCT diagrams obtained for two DP steels is shown in Figure 2.11a. A similar comparison for volume fractions of phases is shown in Figure 2.11b. It is seen that there are some differences between the two steels. Higher content of carbon (DP 1) results in lower start temperatures for ferritic bainitic and martensitic transformation. This steel is also characterized by a higher volume fraction of pearlite at low cooling rates and a lower volume fraction of bainite at medium cooling rates. The dependence of the volume fraction of martensite on the steel composition is negligible in the investigated range of carbon content.

In model A, for the eutectoid steel, 19 of the coefficients are active. Coefficients dedicated to the ferritic transformation are removed from the model, but on the other hand, a more

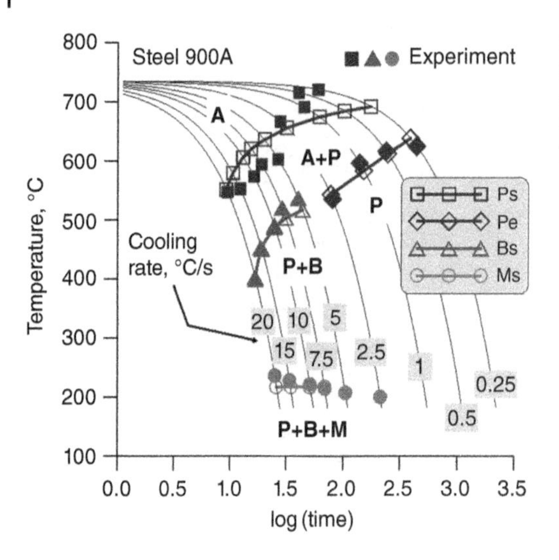

Figure 2.12 Comparison of the CCT diagrams obtained from measurements (filled symbols) and calculated by model A with the coefficients in Table 2.18 (open symbols) for the eutectoid steel 900A.

complex Eq. (2.93) with four coefficients is used to describe the coefficient k dependence on the temperature for pearlitic transformation. A comparison of predicted and calculated CCT diagrams for the eutectoid steel 900A is shown in Figure 2.12.

2.2.4.7 Case Studies

The main physical aspects of phase transformations are discussed in this section, and simple models based on closed-form equations describing the kinetics of transformation are presented. These models are an efficient tool for the simulation of phase transformations when information concerning the kinetics of transformation and volume fractions of phases is satisfactory. Typical results of the simulation of the continuous annealing process are shown in Figure 2.13. A phase transformation model for heating with coefficients in Table 2.18 was used in this simulation. It is a typical intercritical annealing process in which the transformation of ferrite into austenite during heating was not completed. After cooling, 78% of ferrite was obtained in the microstructure.

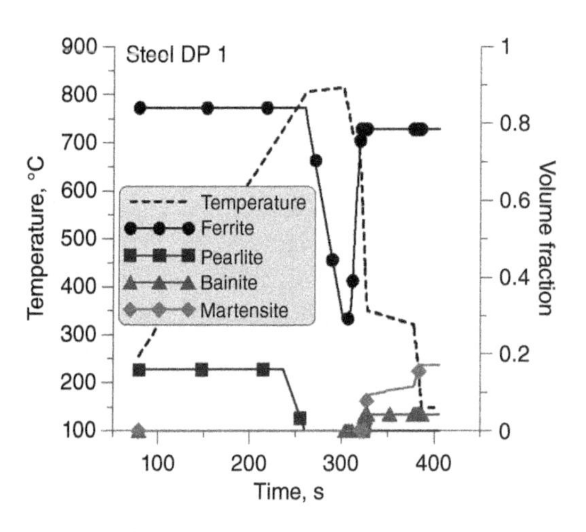

Figure 2.13 Thermal profile (dotted line) and changes of volume fractions of phases for a typical continuous annealing cycle for DP steel.

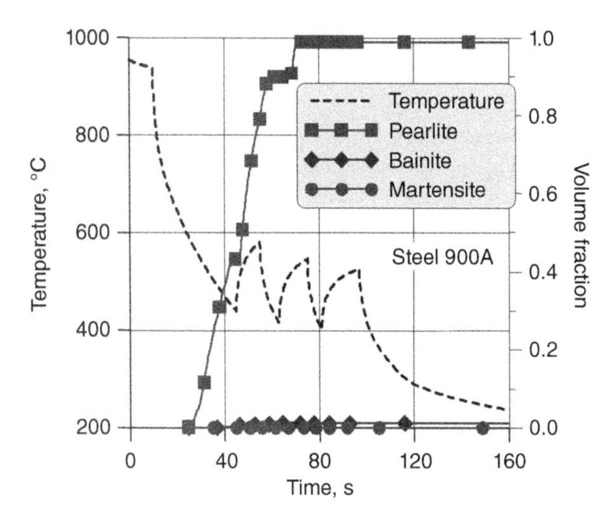

Figure 2.14 Thermal profile (dotted line) and changes of volume fractions of phases for a typical controlled cooling cycle for rails.

Controlled cooling of rails is the second case study made to demonstrate the capabilities of the conventional phase transformation models. In this process, the head of the rail is subsequently immersed in the cooling liquid and pulled out into the air. The resulting thermal profile in the area of the working surface of the rail is shown by the dotted line in Figure 2.14. The objective of this cooling schedule is to maintain the pearlitic transformation at low temperatures and to avoid dropping the temperature below the start temperature of the bainitic transformation. The lower the temperature of the pearlitic transformation, the better is wear resistance of the railhead. It is seen in Figure 2.14 that after cooling, the purely pearlitic microstructure was obtained, and the amount of bainite in the microstructure was negligible. The temperature of pearlitic transformation was very low, an average about 550 °C.

Presented results of simulations confirm good predictive capabilities of conventional models when the kinetics of transformation and volume fractions of phases are of interest. Examples of advanced phase transformation models based on the cellular automata method are presented in Chapter 5.

2.2.5 Fracture

Fracture (crack) is a partial or total splitting of material into parts. Statement fracture is usually used in the situation when real damage appears in materials, while crack is used to describe damage defined by a mathematical model. In other words, the name of crack is used for describing this process in mathematical models, where some simplifications can be introduced, e.g. cracks observed in the material have a zero rounding radius. In reality, this radius is always bigger than zero. The second difference is associated with the crack tip and fracture zone. The shape of a crack area has a regular form. In comparison, the term 'fracture area' is used when damage has a more complicated form closer to reality.

Fracture is one among various structural failure models. Seldom does a fracture occur due to an unforeseen overload of an undamaged structure. Usually, it is caused by a structural flaw or a crack. Due to repeated or sustained normal service loads, a crack may develop starting from a flaw or a stress concentration and grow slowly in size. Cracks and defects impair the strength. Thus, during the continuing development of a crack, structural strength decreases until it becomes so low that service loads cannot be carried anymore, and fracture occurs.

Fracture mechanics is a broad area covering several disciplines, and it is a challenge to simulate fracture initiation and propagation. There may be different reasons for failure, although some of the phenomena involved in various failure mechanisms are similar. Thus, a consistent description of such distinct aspects of this problem in this book is not possible. Therefore, the emphasis will be put on the description of the continuum damage mechanics (CDM) often used in numerical modelling to describe material at the microscale in a homogenized manner. The application part in this chapter is constrained to elastic-plastic fracture, in which the cracking of materials is preceded by plastic deformation.

The basic concepts of fracture modelling are based on the energy approach. Evaluation of energy release rate from strain energy, compliance method, and potential energy are presented.

2.2.5.1 Fundamentals of Fracture Mechanics and Classical Fracture and Failure Hypotheses

The fracture usually develops in three stages. The occurrence of a structural flaw or a microcrack is the first stage. Development and connections of these faults is the second stage and the occurrence of a dominant crack leading to damage of a material is the last stage. The fracture may occur during typical exploitation of a part or during plastic deformation in forming processes. Fracture caused by cyclic dynamic loading is called fatigue. Loads leading to fatigue damage are usually much smaller than the strength of the material.

There are several classifications of types of fracture. The contribution of the plastic deformation is the main factor used in classification, which distinguishes two types:

- Brittle fracture – which occurs without noticeable contribution of plastic deformation.
- Ductile fracture – in which propagation of cracks is combined with intensive plastic deformation.

Structural criterion is the second classification of fractures (Figure 2.15):

- Transcrystalline fracture, which develops through grains.
- Intercrystalline fracture, which propagates along the grain boundaries.

Up to a certain limit, the response of many materials is essentially elastic. As shown in Figure 2.16, ductile behaviour is characterized through plastic deformation, which occurs

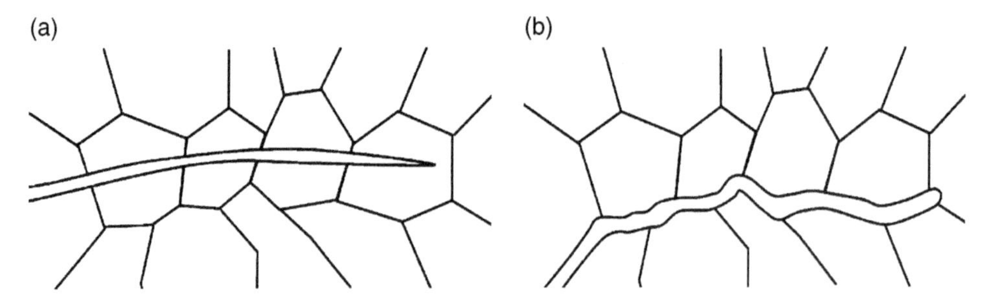

Figure 2.15 Transcrystalline fracture (a) and intercrystalline fracture (b).

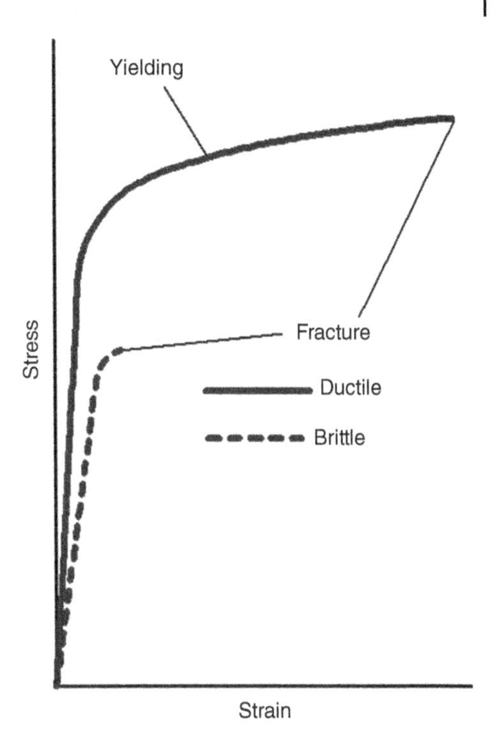

Figure 2.16 Ductile and brittle material behaviour.

when the stress exceeds the yield stress σ_y. In this case, the ultimate stress at fracture will be attained only after sufficiently large inelastic deformations. In contrast, brittle material behaviour is characterized by negligible inelastic deformation prior to fracture. The ductile or brittle behaviour is not a pure material property; it also depends on the stress state. In general, a hydrostatic stress state does not lead to inelastic deformations of most materials, and under certain loading conditions, such material can behave as absolutely brittle.

As presented, the fracture phenomenon is a complex process; thus for a better understanding of fracture mechanics, a new field of science named fracture mechanics (FM) was introduced in the twentieth century. This branch of science takes into account investigations related to fracture toughness in materials. Fracture mechanics is based on quantitative relationships between fracture toughness, critical fracture size, size of the element, or type of applied loading conditions. Three important points of fracture mechanics are usually used in material science:

- Empirical fracture criteria: Major drawbacks of these models for simulation fracture is that they require well-defined calibration procedures based on the experimental data.
- Fracture mechanics: Linear fracture mechanics describes quantitative fracture propagation in size of elastic and linear material deformation. Sometimes it can be used to describe fracture during plastic deformation in a situation when a crack tip is much smaller than the size of the investigated sample. It can be used only for calculating brittle fracture. Nonlinear fracture mechanics describes fracture in materials where plastic deformation plays an important role, and material behaviour cannot be calculated by means of linear fracture mechanics.
- CDM describes fracture initiation and propagation in the framework of continuum numerical-based models.

The most basic difference between FM and CDM is that the latter falls within the standard continuum mechanics framework with a continuous displacement field (hence, FEM implementation is rather easy). In contrast, in the FM the displacement is discontinuous through the crack face, so special techniques such as remeshing, meshless, or XFEM are needed to model this discontinuity in the displacement field. In other words, according to CDM, you smear out the crack over a failure zone where the displacement is still continuous but involves a high strain. In this book, the emphasis is put on modelling fracture development during forming processes on the basis of CDM [35].

2.2.5.2 Empirical Fracture Criteria

Freudenthal [30] proposed first crack criteria based on the critical value state variable per unit volume. Generally, crack initiation criteria can be classified into four different types:

- stress and strain crack criteria,
- energy-type crack criteria,
- damage crack criteria, and
- empirical crack criteria.

Modelling the fracture with these criteria is possible in most commercial numerical applications. Information needed for these criteria, e.g. stress or strain distributions, are calculated through the FEM or an alternative approach. Defining critical crack values usually has to be proceeded by a series of simple experiments.

Most of these models have empirical character and are based on integrals, which are functions of the stress and strain distributions. The following equation shows the general form of the empirical crack criterion:

$$\int_0^{\varepsilon_{eq}(t)} f(\text{process parameters})d\varepsilon_i \geq C, \tag{2.111}$$

where: C – critical value of the function resulting in crack initiation.

The most commonly used approach that can be classified into this group is the Latham-Cockcroft criterion [16]:

$$\int_0^{\varepsilon_{eq}(t)} \sigma_{\max} d\varepsilon_{eq} \geq C, \tag{2.112}$$

where: $\sigma_{\max}$ – maximal principal stress.

In this criterion, crack initiates when integral is equal or larger than damage parameter C. The simplest versions of these criteria are based on the effective stress ($\sigma_i \geq C$), mean stress ($\sigma_m \geq C$), maximal shear stress ($\tau_{\max} \geq C$), effective stress ($\varepsilon_i = \geq C$), maximal principal stress ($\sigma_{11,\ 22,\ 33} \geq C$), and maximal principal strain ($\varepsilon_{11,\ 22,\ 33} \geq C$). Beyond this, a more general crack initiation criterion used for the ductile fracture can be specified as:

$$\int_0^{\varepsilon_{eq}^p} G d\varepsilon_{eq}^p \geq C, \tag{2.113}$$

where: G – parameter, which can take a different form depending on the available process parameters, e.g.:

- for Mohr and Coulomb criterion [76]:

$$G = \phi \frac{\sigma_{11}}{\sigma_i} + (1 - \phi) \frac{\sigma_{11} - \sigma_{33}}{\sigma_i}, \tag{2.114}$$

where:

$$\phi = \frac{2c_1}{\sqrt{1 + c_2} + c_1}, \tag{2.68}$$

c_1 – a material parameter often referred to as the friction, σ_{11}, σ_{33} – principal stresses, σ_i – effective stress.

- for the Rice and Tracey criterion [105]:

$$G = \alpha \exp\left(\frac{3}{2}\frac{\sigma_m}{\sigma_i}\right), \tag{2.115}$$

where: α – material constant.

- for Bao and Wierzbicki criterion [5]:

$$G = \frac{1}{\eta}, \tag{2.116}$$

where: η – stress triaxiality.

- for Lou criterion [68]:

$$G = \left(\frac{2\tau_{\max}}{\sigma_i}\right)^a \left(\frac{\langle 1 + 3\eta \rangle}{2}\right)^b, \tag{2.117}$$

where: $\tau_{\max}$ – maximum shear stress, a, b – material-dependent parameters characterizing the overall mechanisms of the ductile damage.

These criteria have been used in many applications when material damage plays an important role during structural fracture. A majority of these criteria have been implemented in commercial FE codes that are used to model material behaviour during deformation.

2.2.5.3 Fracture Mechanics

Fracture mechanics problems are well discussed in [35], and only basic information is repeated in this book. As mentioned, linear fracture mechanics describes quantitatively fracture propagation in the elastic region of deformation. Several mathematical theories were proposed to describe this phenomenon, among which Griffith energy crack criterion [34] should be mentioned. This theory, which is appropriate for the perfectly brittle materials only, is based on the assumption that the growth of a crack requires the development of two new surfaces and then an increase in the surface energy. Thus, the Griffith crack propagation criterion is formulated as follows. The critical crack propagation occurs when released elastic energy $W_E = \pi x^2 h \sigma^2 / E$ is higher than surface energy growth $W_S = 4xh\gamma_S$:

$$\frac{\partial W}{\partial x} = 0 \quad \rightarrow \quad W = W_E + W_S, \tag{2.118}$$

where: x – crack length, h – thickness of the sample, E – Young modulus, σ – external stress, γ_S – specific surface energy.

Accounting for these relations, the Griffith crack criterion for plain stress and plain strain conditions, respectively, is described as:

$$\sigma_{cr} = \sqrt{\frac{2E\gamma_S}{\pi x}} \qquad \sigma_{cr} = \sqrt{\frac{2E\gamma_S}{\pi x(1-\nu^2)}}, \tag{2.119}$$

where: ν – Poisson coefficient.

Griffith criterion was further extended by Irwin and Orowan, who included plastic strain energy. However, in all these criteria, when stress reaches some critical value, the stage of the uncontrolled crack propagation initiates, and then crack propagates in a catastrophic way. Therefore, a series of force crack criteria were developed. The stress field near the crack tip was modified, and stress distribution in the point situated at some distance from the crack was calculated. Such a concept allowed improving criteria (2.119).

When a ductile fracture is taken into account, criteria (2.119) are not sufficient. The crack tip opening criterion (CTOD) was proposed to deal with ductile failure. The main assumption of this method is to consider microcrack in the area where material deformation occurs in plastic. The method defines the size of the crack in the investigated part of the material and parameters controlling the maximum crack dimension that can occur due to external stresses. In the CTOD, the crack starts to propagate when crack opening reaches the critical value of:

$$\delta_{cr} = \frac{\pi \sigma_{cr}^2 x}{E\sigma_y} \quad , \tag{2.120}$$

where: x – crack length, h – thickness of the sample, E – Young modulus, σ_y – yield stress, σ_{cr} – stress calculated from Eq. (2.119) with specific surface energy γ_S substituted by product of the yield stress and initial gap opening.

Further research to crack modelling led to the development of the integral energy method, which is devoted to elastic-plastic materials. Changes in the potential energy as a result of the crack propagation give an opportunity to define the parameter, which characterizes material resistance to fracture:

$$J = \int_{\Gamma} \left[\left(\int_{0}^{\varepsilon} \sigma : d\varepsilon \right) dx_2 - \left(\mathbf{n} \cdot \boldsymbol{\sigma} \frac{\partial \mathbf{u}}{\partial x_1} \right) dS \right], \tag{2.121}$$

where: Γ – domain of the integration, $\mathbf{n}$ – unit vector normal to Γ, x_1, x_2 – coordinates, dS – increment of Γ, $\boldsymbol{\sigma}$ – stress tensor.

The calculation of the integral J is shown in Figure 2.17. The integral J is a method of description of the stress field on the basis of the energy at the crack tip. The theoretical basis for this approach was created by Rice in [105], who proved that integral J does not depend on the integration time.

2.2.5.4 Continuum Damage Mechanics (CDM)

Crack initiation in materials starts in the smaller length scales, namely nanoscale or microscale. Taking into account the initial stages of the crack initiation and their influence on the

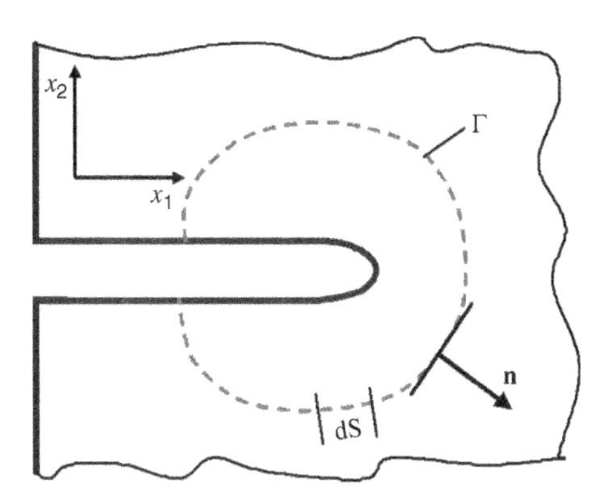

Figure 2.17 Interpretation of the J integral.

macroscale failure problems is crucial in damage modelling. In real materials, multiple heterogeneities occurring at the nanoscale or microscale directly influence crack initiation, while at the macroscale, the material behaves in a homogenous manner. Such microstructure features like voids, fibres, or inclusions are locations where loss of continuity under deformation begins. That is why modelling of the damage at the macroscale requires information provided by the lower length scales. Such micro–macro transition is usually proceeded by application of the appropriate averaging technique at the microscale level, e.g. homogenization approach [42]. As a result of this approach, mentioned microstructural features like microcracks or microvoids have a direct influence on the macroscopic material stiffness or strength, which can be referred to as the damage.

One of the commonly used methods that is based on the homogenization approach in multimodelling of damage considers representative volume elements (RVE) [92]. The main assumption of this approach is that a specific section of the material can be considered representative for the entire investigated domain. The size of this section is described by volume with dimension. The volume must contain a specific number of defects to be representative. Volume dimension cannot also exceed the characteristic length scale of the microstructure. On the other hand, the volume has to be small enough to be regarded as a material point at the macroscopic level. A schematic illustration of the RVE approach is presented in Figure 2.18.

In practice, the continuum damage mechanism is based on the concept of the quasi-continuum. Real variables like σ and ε are replaced by the effective state variables $\langle\sigma\rangle$ and $\langle\varepsilon\rangle$. Those variables are needed for modelling of the evolution of the average damage parameters inside the RVE without its explicit representation. Several approaches to model damage initiation and propagation were proposed.

The CDM is often based on the geometrical measure of damage proposed by Kachanov [46]. He suggested the new damage tensor D and a new definition of stresses in the material. Three different definitions of stresses for material, not deformed, deformed, and weakened by damage, were proposed. In material where deformation is not applied, the apparent stress is defined as $\sigma^{\mathbf{a}}$. When deformation begins, the stress accounts for the change of area A. Finally, when damage is initiated in the material, then stress is defined by the true stress σ^{*}, which is a function of the damage variable ω. A scalar measure of damage ω expresses the current reduction of cutting flank $A^{*} = A{-}A_{\omega}$, relative to the nominal surface (Figure 2.19).

Chaboche [13] proposed another definition of the geometrical measure of damage, based on the measurement of macroscopic damage features. He introduced new stress and strain

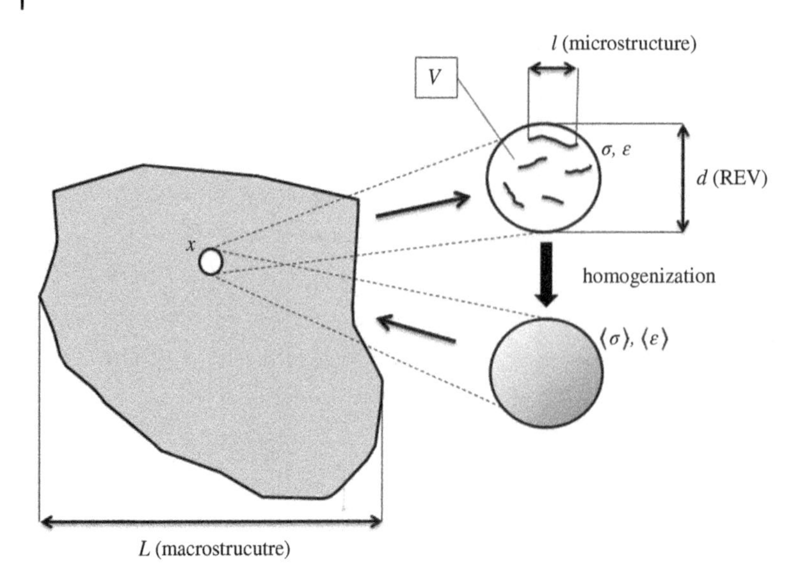

Figure 2.18 Schematic illustration of the RVE approach based on homogenization scheme (courtesy of K. Perzyński).

definitions called effective stress/strain parameters. Such effective stresses $\widetilde{\sigma}$ and effective strains $\widetilde{\varepsilon}$ are calculated in the pseudo not damaged material. Stress $\widetilde{\sigma}$, when applied to the pseudo-no damaged material, leads to the same deformation as the real damaged sample subject to σ, see Figure 2.20. Proposed damage parameter D in this model is measured as a change of the macroscopic properties for a damaged material expressed by the reduction of the elastic modulus $\widetilde{E} < E$. Effective elastic module decreases during the growth of D parameter. When $D \rightarrow 1$ material loses stability and undergoes complete damage $\widetilde{E} \rightarrow 0$.

The assumptions discussed previously are based on the empirical fracture criteria, fracture mechanics, and continuum damage models, which are commonly used in numerical models.

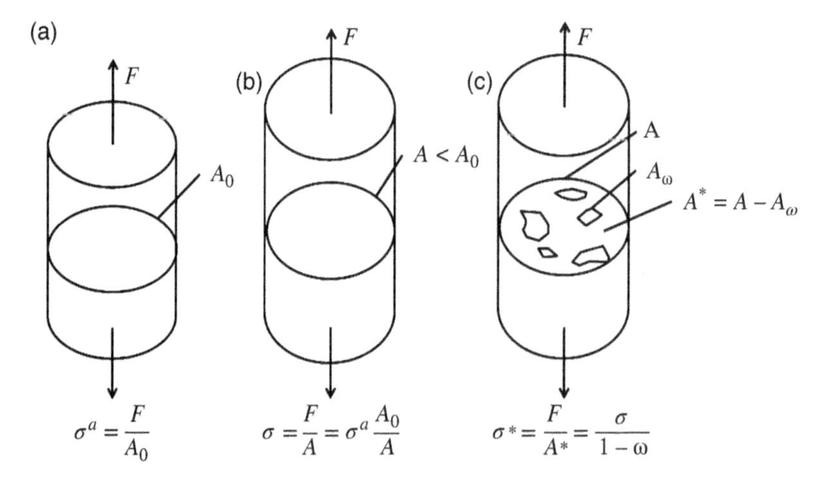

Figure 2.19 Geometric measure of damages proposed by Kachanov [46] for material not deformed (a), material deformed (b), and material deformed with damages (c).

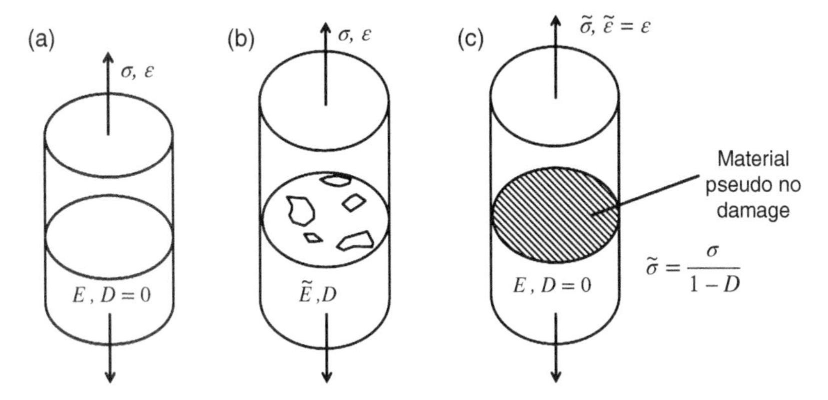

Figure 2.20 Macroscopic features of damages proposed by Chaboche for material not damaged (a), material deformed with damages (b), and for pseudo no damaged material in terms of CDM (c). Source: Based on Chaboche and Lesne [13].

Another set of damage models considers cracks occurring due to the presence of voids in ductile materials. The contribution of voids is usually represented by the void volume fraction or porosity:

$$f_v = \frac{V_p}{V}, \tag{2.122}$$

where: V_p – is the total volume of voids within the volume V in the RVE.

In order to investigate damage behaviour for the ductile materials, the porosity f_v is used for defining void growth. Gurson model [36] and its improvement by Tvergaard-Needleman [86] define the evolution of circular voids in porous ductile materials. The plastic surface is defined in this model as a function of the average stress σ_m, equivalent stress, and actual voids volume fraction f_v:

$$F\left(\sigma_{ij}, \sigma_s, f_v\right) = \frac{\sigma_i^2}{\sigma_S^2} + 2q_1 f_v^{eff} \cos\left(\mathbf{h}\right)\left(\frac{3q_2\sigma_m}{\sigma_s}\right) - \left(1 + q_2 f_v^{eff}\right)^2 = 0, \tag{2.123}$$

where: q_1, q_2, q_3, –the parameters added by Tvergaard and Needleman [62], which were omitted by Gurson in his model, σ_s – actual yield strength for rigid-plastic material, f_v^{eff} – effective voids volume fraction define as:

$$f_v^{eff} = \begin{cases} f_v & \text{for} \quad f_v \leq f_{cr} \\ f_{cr} + \kappa(f_v - f_{cr}) & \text{for} \quad f_v > f_{cr} \end{cases}, \tag{2.124}$$

where: κ – material state, f_{cr} – critical void volume fraction, after which voids begin to form cracks and material loses stability.

Other parameters from Eq. (2.123) are defined as:

$$\sigma_i = \left(\frac{3}{2}\sigma_{ij}\sigma_{ij}\right)^{0.5}, \tag{2.125}$$

where: σ_{ij} – stress deviator, σ_m – average stress defined as:

$$\sigma_m = \frac{1}{3}\sigma_{kk}, \tag{2.126}$$

where: σ_{kk} – principal stresses.

The Gurson-Tvergaard-Needleman model for the ductile materials with geometric damages is phenomenological. It is widely used in practical applications because of its simplicity and straightforward implementation in numerical applications.

In the previously mentioned models, the crucial part is always related to the crack initiation criterion. When GTN or CDM models are used for fracture simulation in selected steels, a well-defined calibration procedure based on the sophisticated experimental data is of importance to determine crack initiation and propagation criteria. This is a limiting factor in the wide application of these approaches. That is why for industrial applications, simple uncoupled damage models are preferred. These models are discussed in the next section.

2.2.6 Creep

It is concluded from previous sections that material damage can be classified according to the dominant macroscopic phenomenon such as brittle damage, ductile damage, creep damage, and fatigue damage [35]. The prevailing mechanism of brittle damage is the formation and growth of microcracks as it takes place, for instance in ceramics, geomaterials, or concrete. In contrast, ductile damage and creep damage in metals is essentially due to the nucleation, growth, and coalescence of microvoids. The source of fatigue damage are microcracks which are formed at stress concentrators in the course of microplastic cyclic loading and which more and more grow and coalesce. The basic information on creep and fatigue damage is given in this and the next section.

Various materials display a time-dependent behaviour, which is the source of phenomena such as creep or relaxation. These processes typically take place quasi-statically, i.e. so slow that inertia forces do not play any role. If a component made of such material contains a crack and is loaded, time-dependent deformations occur especially in the vicinity of the crack tip due to the locally high stresses. This may cause a delay of crack initiation until some critical crack-tip deformation is attained. Creep of the material at the crack tip, however, may also lead directly to creep crack growth.

Typical examples of material response to loading can be shown on the basis of the simplest mechanical models: ideal elastic and ideal viscous. These models are represented by a spring and by cylinders with a piston filled with a viscous fluid (Figure 2.21). The constitutive laws for these materials are:

$$\sigma = E\varepsilon \qquad \sigma = \eta \frac{d\varepsilon}{dt}, \tag{2.127}$$

where: σ – stress, ε – strain, E – Young modulus, η – viscosity, t – time.

Various rheological models were built using these basic material models, and the simplest are Kelvin material model and Maxwell material model. The former is a parallel connection

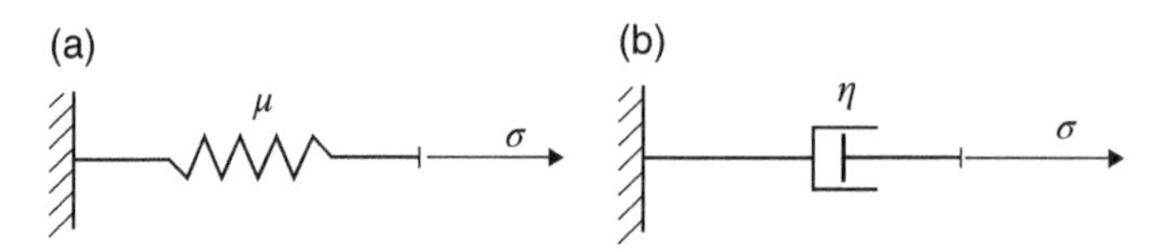

Figure 2.21 Mechanical models of ideal elastic (a) and ideal viscous (b) materials.

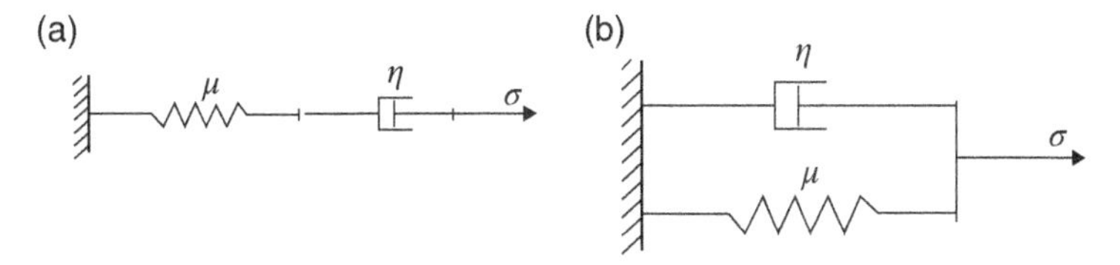

Figure 2.22 Kelvin (a) and Maxwell (b) material rheological models.

of spring with a cylinder, and the latter is a connection of spring with a cylinder in series (Figure 2.22).

Kelvin and Maxwell material models expose different physical properties. Maxwell model represents delayed elasticity, and it is not discussed here. Kelvin model is used to explain creep and stress relaxation phenomena.

If the sample is loaded with a strain ε_0 and fixed, the stress increases to $\sigma_0 = E\varepsilon_0$, and after that, it decreases gradually as far as the elastic strain is substituted by a viscous strain. The total strain is always a sum of an elastic strain and viscous strain: $\varepsilon_0 = \varepsilon_e + \varepsilon_v$. Since the total strain remains constant, the total strain rate has to be equal to zero and:

$$\dot{\varepsilon} = \frac{d\varepsilon}{dt} = \frac{1}{\mu}\sigma + \frac{1}{E}\frac{d\sigma}{dt} = 0. \tag{2.128}$$

Thus:

$$\frac{d\sigma}{\sigma} = -\frac{E}{\mu}dt. \tag{2.129}$$

In this process, the total strain remains constant, and the total stress decreases. This phenomenon is called stress relaxation.

If the sample is loaded with stress σ_0, the strain increases to $\varepsilon_0 = \sigma_0/E$, and after that, it further increases due to a viscous strain. The total strain is always a sum of an elastic strain and viscous strain:

$$\varepsilon = \varepsilon_0 + \varepsilon_v = \frac{\sigma_0}{E} + \frac{\sigma_0}{\mu}t. \tag{2.130}$$

In this process, the total stress remains constant, and the total strain increases with time. This phenomenon is called creep.

Both these phenomena, stress relaxation and creep, can be explained on the basis of micromechanics. They can also be well described by discrete microscale models (cellular automata etc.). Thermally activated creep of metals is connected with void growth at grain boundaries. In the vicinity of a macroscopic crack tip, this leads to the formation of microcracks, and ultimate fracture takes place by their coalescence. Thus, creep is especially pronounced at a crack tip. In certain cases, the time-dependent inelastic behaviour is restricted even to the immediate vicinity of the crack tip, whereas the material can otherwise be regarded as linear elastic. Small-scale creep conditions prevail if the creep zone is sufficiently small. The crack-tip state then can be characterized by parameters of linear elastic fracture mechanics, which, however, may now be time dependent. Parameters of linear elastic fracture mechanics can also be employed in the case of creep in larger regions (e.g. creep of a whole component), provided that the material can be described as linear viscoelastic. In the case of nonlinear material

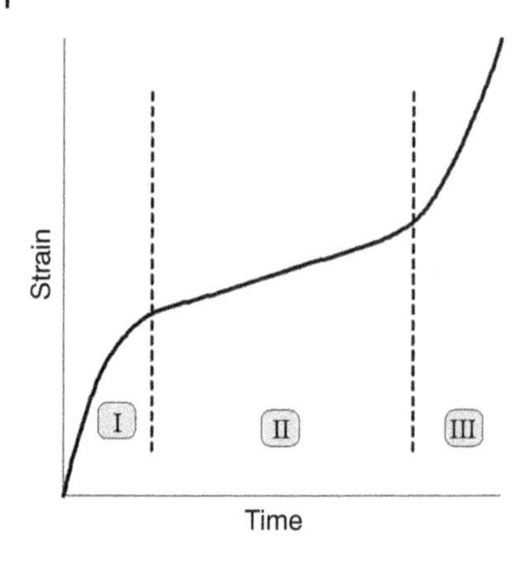

Figure 2.23 Typical creep curve with marked first (I), second (II), and third (III) stage of the creep.

behaviour, parameters based on integral quantities are useful and frequently applied, for instance, the C– or C*–integral, which are closely related to the J integral of elastic-plastic fracture mechanics.

As it was shown, creep is a process of strain increase under constant stress. For metals, this process occurs mainly at temperatures in the range between 0.3 and 0.7 melting temperature, but it is also possible in lower temperatures (e.g. creep of the high voltage electrical tractions). A typical plot of strain changes under constant stress is shown in Figure 2.23.

The first stage is called transient creep and is characterized by a nonlinear change of strain with time. Strain increases and the rate of the strain change decreases in this stage. The total strain in the stage I is low, and it is often neglected. When the strain rate stabilizes, the second stage, called steady-state creep, begins. This stage is usually the longest and may lead to large strains. The rate of strain in this stage is a characteristic parameter of the creep. The third stage, called accelerated creep, begins when the strain rate increases rapidly. This stage is short, but large strains can be achieved.

The creep curve shape depends on the level of stress. For large stress, the time before the damage is short and a large strain is achieved. Damage is ductile in nature and is usually preceded by necking. For lower stress, the time before the damage is long, and the strain rate is small. Damage is brittle in nature, and low strain is achieved. Damage may occur suddenly without observable changes in the sample.

Relation between time to damage (t_f) and stress is usually presented in a logarithmic scale, and the results are constrained by two parallel lines (Figure 2.24). The first line, for large stress, corresponds to ductile damage, and the second for low stress corresponds to brittle damage. The former has a lower inclination than the latter.

Modelling of creep and validation of micromodels are often done using the fundamental work of Monkman and Grant [77], who proposed a relation between strain rate in the stage II of creep and the time to damage:

$$\dot{\varepsilon}_{II}^{\beta} t_f = C, \tag{2.131}$$

where: β, C – material parameters, which depend on temperature and strain rate.

Parameter β is often taken as 1, and then parameter C is called Monkman-Grant constant (C_{MG}). This constant has a dimension of strain and represents the strain to the damage in the second stage of creep. C_{MG} is usually between 1 and 20. Low values refer to easy initiation of failure in the strain concentration points.

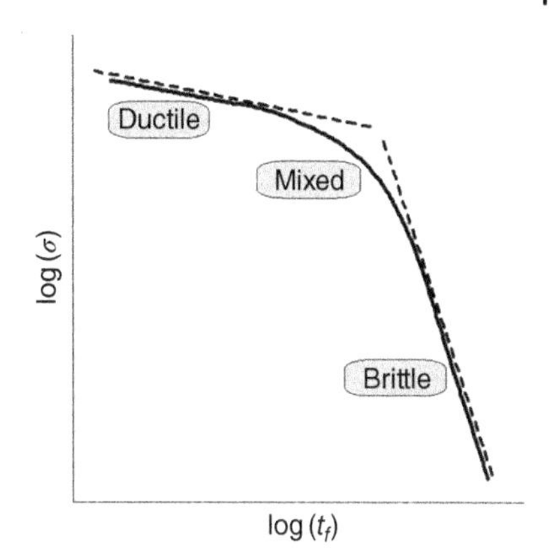

Figure 2.24 Schematic relation between loading stress and the time to damage.

Description of the creep curve (Figure 2.19) is based on three independent variables: time, stress, and temperature. Generally, an influence of these variables is separated, and:

$$\varepsilon = f_1(t) f_2(\sigma) f_3(T) \tag{2.132}$$

Function $f_1(t)$, which fits to three stages of creep, is:

$$f_1(t) = \left(1 + b_1 t^{1/3}\right) \exp\left(b_2 t\right) - 1, \tag{2.133}$$

where: b_1, b_2 – material parameters, which depend on temperature and strain rate.

Since the second stage is the most important part of creep, the following relation of this part on the stress was proposed:

$$\dot{\varepsilon}_{II} = b_3 \sigma^n, \tag{2.134}$$

where: b_3, n – material parameters. Parameter n is called creep index and is usually between 3 and 7.

As far as the influence of the temperature is considered, the well-known Arrhenius relation is used (see e.g. [57]):

$$\dot{\varepsilon} = b_4 \exp\left(\frac{-Q}{RT}\right), \tag{2.135}$$

where: b_4 – material parameter, R – gas constant, Q – activation energy.

As it has been mentioned, ductile or brittle damage is possible in creep. The basic models for these creep mechanisms were proposed by Hoff [41] and Kachanov [46] for ductile and brittle damage, respectively. The former is characterized by large strain (10–20%) and an occurrence of necking. According to Hoff [41], his power law (2.80) leads to a decrease of the sample cross-section and, consequently, to an increase in the strain rate. Even small changes of the cross-section may cause instability of flow and damage and make the constant stress test difficult. The time necessary for the cross-section area to decrease to zero is calculated from the constant volume condition:

$$t_f = \frac{1}{n\dot{\varepsilon}_0} = \frac{1}{n b_3 \sigma^n}, \tag{2.136}$$

where: σ – initial stress, $\dot{\varepsilon}_{II}$ – strain rate during the steady-state creep.

In a logarithmic scale, Equation (2.136) is represented by a straight line inclined $1/n$ with respect to the axis. Thus, Monkman-Grant constant for the ductile damage model proposed by Hoff is:

$$C_{MG} = \frac{1}{n}, \tag{2.83}$$

where: σ_0 – initial stress, $\dot{\varepsilon}_{II}$ – strain rate during the steady-state creep.

The brittle damage in creep occurs at low levels of stress, low strain rates, and in a long time and is characterized by small strain (few percent). The development of internal defects is mainly responsible for this damage. The fundamental model of his phenomenon was proposed by Kachanov [46], who introduced the parameter of material continuity ψ. For the material, which is free of defects, $\psi = 1$, and for the damaged material $\psi = 0$. The equation describing the development of defects is:

$$\dot{\omega} = C\left(\frac{\sigma}{1-\omega}\right)^m, \tag{2.137}$$

where: $\omega = 1{-}\psi$, C, m – material parameters.

By integration of Eq. (2.137) for ω changing from 0 to 1, time to brittle damage is calculated as:

$$t_f = \frac{1}{C\sigma^m(1+m)}. \tag{2.138}$$

In a logarithmic scale, Equation (2.138) is represented by a straight line inclined $1/m$ with respect to the axis. This line crosses the line for the brittle fracture at:

$$\sigma_{D-B} = \left[\frac{C(m+1)}{nb_3}\right]^{\frac{1}{n-m}}. \tag{2.86}$$

By calculation of the time to damage from the Kachanov Eq. (2.138), but accounting for the decrease of the cross-section as it is done in the Hoff model, time to the damage for the mixed mechanism is calculated; see [87] for details.

Damage parameter ω is a measure of weakening of the cross-section due to development of defects; see for example Kachanov model [46] in Figure 2.16. The stress acting on the non-defected part of the cross-section is:

$$\sigma_\omega = \frac{\sigma}{1-\omega}. \tag{2.139}$$

Substituting σ by σ_ω in the continuum mechanics equations is the basis of the CDM. Damage parameter ω becomes a new state variable with its own evolution equation. Thus, a constitutive law for damage mechanics can be written as:

$$\dot{\varepsilon} = f(\sigma, \varepsilon, \omega) \qquad \dot{\omega} = g(\sigma, \varepsilon, \omega). \tag{2.140}$$

Structural parameters for creep were proposed in [102]:

$$\dot{\varepsilon} = f(\sigma, T, s_i) \qquad \dot{s} = g(\sigma, T, s_i), \tag{2.141}$$

where: T – temperature, S_i – structural parameters.

Equation (2.141) which describes the development of strains in the second stage becomes:

$$\dot{\varepsilon}_{II} = b_3\left(\frac{\sigma}{1-\omega}\right)^n. \tag{2.142}$$

Equation (2.142) written for three stages of creep is:

$$\dot{\varepsilon}_c = b_5\left(\frac{\sigma}{1-\omega}\right)^n \varepsilon_c^{-\nu}, \tag{2.91}$$

where: b_5 – material parameter, ν – hardening exponent in the first stage of creep.

Presented constitutive law can be adapted to the state of stresses in triaxiality. In this state, scalar state variables become tensors. However, it is assumed in the present approach that the damage parameter is isotropic, and it can be described by the scalar function. Consequently, the creep strain rate depends on the effective stress (σ_i) only:

$$\dot{\varepsilon}_{ij}^c = \frac{3}{2} b_3 \frac{\sigma_i^{n-1}}{(1-\omega)^n} s_{ij}, \tag{2.143}$$

where: b_5 – material parameter, ν – hardening exponent in the first stage of creep.

More advanced descriptions of conventional models can be found in [9, 35, 87]. Continuum mechanics defines all variables in a material point and requires continuity of all variables. The objective of this section was to supply basic knowledge for building multiscale models, which account for the discontinuous structure of materials.

2.2.7 Fatigue

Fatigue is a set of phenomena occurring in materials subjected to repeated cycles of mechanical or thermal stress. The fatigue may lead to damage of the part under stresses much lower than the strength of the considered material. The insidious feature of fatigue failure is that there is no obvious warning, and a crack initiates and develops without appreciable deformation of the structure. It makes detection of the presence of growing cracks difficult. Mechanics of fracture due to fatigue is similar to that observed in creep. Failure can be influenced by a number of factors, including size, shape, and design of the component, condition of the surface, or operating environment. The number of load cycles before failure is calculated from fatigue criteria.

Fatigue is usually divided into two groups:

- Low cycle fatigue (LCF): Material damage usually occurs below 1000 cycles. This type of fatigue is described by parameters based on strains. Machines with load frequencies around 0.01–5 Hz are used to investigate this fatigue.
- High cycle fatigue (HCF): Material damage usually occurs around 100 cycles. This type of fatigue is described by parameters based on stresses. Machines with load frequencies around 20–250 Hz are used to investigate this fatigue.

Three mechanisms are distinguished during thermomechanical fatigue:

- Creep – deformation of the material under constant load discussed in the previous section.
- Fatigue – initiation of microcracks and their propagation and coalescence caused by repeating loads.
- Oxidation – changes of the chemical composition caused by environment. Oxidized material is more brittle and susceptible to cracking. This mechanism is not discussed in the present book.

Each of the previously listed mechanisms contributes to material damage during cycling loading. Creep prevails when thermal and mechanical loads are applied simultaneously.

The combination of high temperature and mechanical loads fosters the development of creep. Fatigue and oxidation are dominant mechanisms when small mechanical strains are combined with high temperatures. The latter mechanism weakens the material surface and creates microcracks and flaws. New crack surfaces occur and are oxidized, leading to further weakening of the material. Fatigue is the only one active mechanism when high-stress amplitudes are applied at low temperatures. Creep and oxidation are slow, and they do not develop before failure due to fatigue occurs.

The development of microcracks depends, to a large extent, on the type of loads acting on the material. Three types of loading are distinguished: I (tension), II (transverse shear), III (longitudinal shear). Each type can act separately, or a combination of various types can be applied.

The fracture criteria are divided as follows:

- empirical criteria,
- criteria based on the fracture mechanics,
- criteria based on the continuum mechanics, and
- criteria accounting for the microstructure.

Empirical criteria describe a number of load cycles before damage as a function of such parameters as stress, strain, temperature, etc. Lack of physical basis is the main disadvantage of this approach. Criteria based on the fracture mechanics usually use J integral to describe the initiation and propagation of cracks. The most important criteria in this group were proposed by Chaboche [13] and Lemaitre [59]. Crack in these criteria develops depending on the actual value of the damage parameter. Criteria accounting for the microstructure use the physical basis of microcracks initiation and propagation.

The most commonly used fatigue criteria are:

- Mansona-Coffina-Basquina empirical criterion, described in [74]:

$$\Delta\varepsilon_m = c_1 N_f^{c_2} + \frac{\Delta\sigma_m}{E} N_f^{c_3}, \tag{2.144}$$

where: $\Delta\varepsilon_m$ – total amplitude of strains (elastic and plastic), c_1 – empirical coefficient of plastic fatigue, N_f – number of cycles before damage, $\Delta\sigma_m$ – amplitude of mean stress, E – Young modulus, c_2 – empirical plastic exponent, c_2 – empirical elastic fatigue exponent. Difficulties with the determination of empirical coefficients are the main limitation of this criterion.

- Ostergren criterion:

$$N_f = c_4 \left(\sigma_{\max} \Delta\varepsilon_p\right)^n, \tag{2.145}$$

where: c_4 – empirical coefficient, $\sigma_{\max}$ – maximum tensile stress, $\Delta\varepsilon_p$ – total amplitude of plastic strain.

This criterion belongs to energetic criteria with low number of empirical coefficients. It can be applied to various types of fatigue mechanisms.

- Paris criterion, which describes a number of cycles giving an increase of the length of the crack by Δd:

$$N_d = \frac{\Delta d}{c_5 \left(\sqrt{\Delta K_I^2 + 2\Delta K_I^2}\right)^m}, \tag{2.146}$$

where:

$$\begin{aligned} \Delta K_I &= \Delta K_I^{\max} - \Delta K_I^{\min} \quad \Delta K_{II} = \Delta K_{II}^{\max} - \Delta K_{II}^{\min} \\ K_I &= \sqrt{2\pi l}[p_\tau \cos(\varphi) - p_n \sin(\varphi)] \quad , \\ K_{II} &= \sqrt{2\pi l}[p_\tau \sin(\varphi) + p_n \cos(\varphi)] \end{aligned}$$

K_I, K_{II} – effective stress coefficients for loads type I and II, ΔK – difference between the maximum and minimum effective stress coefficient in one cycle, l – length of the boundary element, p_τ, p_n – tangent and normal stress at the tip of the crack, φ – angle of the crack development:

$$\varphi = 2\text{arc}\tan\left[\frac{1}{4}\frac{K_I}{K_{II}}\right] \pm \frac{1}{4}\sqrt{\left(\frac{K_I}{K_{II}}\right)^2 + 8}. \tag{2.147}$$

Following the Paris criterion, numerical calculations of the number of cycles before damage due to fatigue follow the sequence:

- Calculation of distributions of temperature and strain, accounting for thermal and mechanical loads.
- Calculations of the effective stress coefficients.
- Calculation of the number of cycles, after which crack length increases by Δd.
- Summation of the Δd increments and calculation of the number of cycles before damage.

Presented conventional fatigue criteria are selected and can be used for approximate evaluation of the maximum number of cycles depending on the stress and temperature amplitudes and on the process parameters. More advanced simulation of fatigue phenomenon is also possible using multiscale models.

References

1 Abajingin, D.D. (2012). Solution of Morse potential for face center cube using Embedded Atom Method. *Advances in Physics Theories and Applications* 8: 36–44.

2 Avrami, M. (1939). Kinetics of phase change. I. General theory. *The Journal of Chemical Physics* 7: 1103–1112.

3 Avrami, M. (1940). Kinetics of phase change. II. Transformation-time relations for random distribution of nuclei. *The Journal of Chemical Physics* 8: 212–224.

4 Avrami, M. (1941). Kinetics of phase change. III. Granulation, phase change, and microstructure. *The Journal of Chemical Physics* 9: 177–184.

5 Bao, Y. and Wierzbicki, T. (2004). On fracture locus in the equivalent strain and stress triaxiality space. *International Journal of Mechanical Sciences* 46: 81–98.

6 Baskes, M.I. (1992). Modified Embedded-Atom potentials for cubic materials and impurities. *Physical Review B* 46: 2727–2742.

7 Baskes, M.I. and Johnson, R.A. (1994). Modified Embedded atom potentials for HCP metals. *Modelling and Simulation in Material Science and Engineering* 2: 147–163.

8 Brenner, D.W., Shenderova, O.A., Harrison, J.A. et al. (2002). A second-generation reactive empirical bond order (REBO) potential energy expression for hydrocarbons. *Journal of Physics: Condensed Matter* 14: 783–802.

9 Broberg, K.B. (1999). *Cracks and Fracture.* London: Academic Press.

10 Burczyński, T., Mrozek, A., Górski, R., and Kuś, W. (2010). The Molecular Statics coupled with the subregion Boundary Element Method in multiscale analysis. *International Journal for Multiscale Computational Engineering* 8 (3): 319–330.

11 Cahn, J.W. (1956). Transformation kinetics during continuous cooling. *Acta Metallurgica* 4: 572–575.

12 Car, R. and Parinello, M. (1985). Unified approach for molecular dynamics and density functional theory. *Physical Review Letters* 55: 2471–2474.

13 Chaboche, J.L. and Lesne, P.M. (1988). A non-linear continuous fatigue damage model. *Fatigue and Fracture of Engineering Materials and Structures* 11: 1–17.

14 Chenoweth, K., van Duin, A.C.T., and Goddard, W.A. (2008). ReaxFF reactive force field for molecular dynamics simulations of hydrocarbon oxidation. *The Journal of Physical Chemistry A* 112: 1040–1053.

15 Choquet, P., Fabregue P., and Giusti J. et al. (1990). Modelling of forces, structure and final properties during the hot rolling process on the hot strip mill. Proc. Mathematical Modelling of Hot Rolling of Steel, ed., Yue S., Hamilton. 34–43.

16 Cockcroft, M.G. and Latham, D.J. (1968). Ductility and the workability of metals. *Journal of the Institute of Metals* 96: 33–39.

17 Cox, H., Johnston, R.L., and Murrell, J.N. (1997). Modelling of surface relaxation and melting of aluminium. *Surface Science* 373: 67–84.

18 Davenport, S.B., Silk, N.J., Sparks, C.N., and Sellars, C.M. (1999). Development of constitutive equations for the modelling of hot rolling. *Materials Science and Technology* 16: 1–8.

19 Daw, M.S. and Baskes, M.I. (1984). Embedded-atom method: derivation and application to impurities, surfaces, and other defects in metals. *Physical Review B* 29: 6443–6453.

20 Daw, M.S., Foiles, S.M., and Baskes, M.I. (1993). The embedded-atom method: a review of theory and applications. *Materials Sience Reports* 9 (7–8): 251–310.

21 Dirac, P.A.M. (1930). Note on exchange phenomena in the Thomas-Fermi atom. *Cambridge Philosophical Society* 26 (3): 376–385.

22 Donnay, B., Herman, J.C., and Leroy, V. et al. (1996). Microstructure Evolution of C-Mn Steels in the Hot Deformation Process: The STRIPCAM Model, Proceedings of the Second Conference Modelling of Metal Rolling Processes, (eds), Beynon J.H., Ingham P., Teichert H., Waterson K., London, 23–35.

23 Dreizler, R.M. and Gross, E.K.U. (1990). *Density Functional Theory.* Berlin: Springer-Verlag.

24 van Duin, A.C.T., Dasgupta, S., Lorant, F., and Goddard, W.A. (2001). ReaxFF: a reactive force field for hydrocarbons. *The Journal of Physical Chemistry A* 105 (41): 9396–9409.

25 Dutta, B. and Sellars, C.M. (1987). Effect of composition and process variables on Nb(C,N) precipitation in niobium microalloyed austenite. *Materials Science and Technology* 3: 197–206.

26 Dutta, B., Valdes, E., and Selars, C.M. (1992). Mechanism and kinetics of strain induced precipitation of Nb(C,N) in austenite. *Acta Metallurgica and Materialia* 40: 653–662.

27 Elizondo, A. (2007). Horizontal coupling in continuum atomistics. PhD thesis.

28 Foiles, S.M., Baskes, M.I., and Daw, M.S. (1986). Embedded atom method functions for fcc metals Cu, Ag, Au, Ni, Pd, Pt and their alloys. *Physical Review B* 33: 7983–7991.

29 Forestier, R., Massoni, E., and Chastel, Y. (2002). Estimation of constitutive parameters using an inverse method coupled to a 3D finite element software. *Journal of Materials Processing Technology* 125: 594–601.

30 Freudenthal, A.M. (1950). *The Inelastic Behavior in Solids*. New York: Wiley.

31 Gavrus, A., Massoni, E., and Chenot, J.L. (1996). An inverse analysis using a finite element model for identification of rheological parameters. *Journal of Materials Processing Technology* 60: 447–454.

32 Girifalco, L.A. and Weizer, V.G. (1959). Application of the Morse potential function to cubic metals. *Physical Review* 114: 687–690.

33 Griebel, M., Knapek, S., and Zumbusch, G. (2007). *Numerical Simulation in Molecular Dynamics: Numerics, Algorithms, Parallelization, Applications*, Texts in Computational Science and Engineering, vol. 5. Springer.

34 Griffith, A.A. (1921). The phenomena of rupture and flow in solids. *Philosophical Transaction of the Royal Society of London, Series A* 221: 163–198.

35 Gross, D. and Seelig, T. (2011). *Fracture Mechanics: With an Introduction to Micromechanics*. Berlin, Heidelberg: Springer-Verlag.

36 Gurson, A.L. (1977). Continuum theory of ductile rupture by void nucleation and growth: Part I - Yield criteria and flow rules for porous ductile media. *Journal of Engineering Materials Technology* 99: 2–15.

37 Hadasik, E. (2004). Determination of plasticity characteristics in the hot torsion tests. In: *Plasticity of Metallic Materials* (ed. E. Hadasik and I. Schindler), 39–66. Gliwice: Publ. Politechnika Śląska.

38 Haile, J.M. (1992). *Molecular Dynamics Simulation*, Elementary Methods. USA: Wiley.

39 Hansel, A. and Spittel, T. (1979). *Kraft- und Arbeitsbedarf Bildsamer Formgebungs-verfahren*. Leipzig: VEB Deutscher Verlag fur Grundstoffindustrie.

40 Hodgson, P.D. and Gibbs, R.K. (1992). A Mathematical model to predict the mechanical properties of hot rolled C-Mn and microalloyed steels. *ISIJ International* 32: 1329–1338.

41 Hoff, N.J. (1953). The necking and the rupture of rods subjected to constant tensile loads. *Journal of Applied Mechanics* 20: 105–108.

42 Hornung, U. (1997). *Homogenization and Porous Media*. New York: Springer.

43 Interatomic Potentials Repository Project (2021). www.ctcms.nist.gov/potentials ().

44 Jiang, J.-W., Park, H.S., and Rabczuk, T. (2013). Molecular dynamics simulations of single-layer molybdenum disulphide (MoS2): Stillinger-Weber parametrization, mechanical properties, and thermal conductivity. *Journal of Applied Physics* 114: 064307.

45 Johnson, W.A. and Mehl, R.F. (1939). Reaction kinetics in processes of nucleation and growth. *Transactions AIME* 135: 416–442.

46 Kachanow, L.M. (1958). O wremienii razrusheniya w usloviach polzychesti, Izv. Akad. Nauk. SSR OTN, no. 8, 26–31 (in Russian).

47 Kandemir, A., Yapicioglu, H., Kinaci, A. et al. (2016). Thermal transport properties of MoS2 and MoSe2 monolayers. *Nanotechnology* 27: 055703.

48 Karjalainen, L.P. and Perttula, J. (1996). Characteristics of static and metadynamic recrystallization and strain accumulation in hot-deformed austenite as revealed by the stress relaxation method. *ISIJ International* 36: 729–736.

49 Kirsch, A. (1996). *An Introduction to the Mathematical Theory of Inverse Problems*. Springer.

50 Kocks, U.F. (1976). Laws for work-hardening and low-temperature creep. *Transactions ASME, Journal of Engineering for Industry* 98: 76–85.

51 Koistinen, D.P. and Marburger, R.E. (1959). A general equation prescribing the extent of the austenite-martensite transformation in pure iron-carbon alloys and plain carbon steels. *Acta Metallurgica* 7: 59–69.

52 Kolmogorov, A. (1937). A statistical theory for the recrystallisation of metals. *Izvestiya Akademii Nauk USSR, Seriya Matematicheskaya* 1: 355–359.
53 Kondek, T., Kuziak, R., and Pietrzyk, M. (2003). Identification of parameters of phase transformation models for steels. *Steel GRIPS* 1: 59–66.
54 Kuziak, R., Cheng, Y.-W., Glowacki, M., and Pietrzyk, M. (1997). Modelling of the Microstructure and Mechanical Properties of Steels during Thermomechanical Processing, NIST Technical Note 1393, Boulder.
55 Laasraoui, A. and Jonas, J.J. (1991). Recrystallization of austenite after deformation at high temperature and strain rates - analysis and modelling. *Metallurgical Transactions A* 22A: 151–160.
56 Large-scale Atomic/Molecular Massively Parallel Simulator (2021): www.lammps.org ().
57 Larson, F.R. and Miller, J. (1952). A time-temperature relationship for rupture and creep stresses. *Transactions ASME* 74: 765–775.
58 van Leeuwen, Y., Kop, T.A., Sietsma, J., and van der Zwaag, S. (1999). Phase transformations in low-carbon steels; modelling the kinetics in terms of the interface mobility. *Journal of Physics IV* 9: 401–409.
59 Lemaitre, J. (1984). How to use damage mechanics. *Nuclear Engineering and Design* 80: 233–245.
60 Lenard, J.G., Pietrzyk, M., and Cser, L. (1999). *Mathematical and Physical Simulation of the Properties of Hot Rolled Products.* Amsterdam: Elsevier.
61 Lennard-Jones, L.E. (1924). On the determination of molecular fields. *Proceedings of the Royal Society of London A* 106 (738): 463–477.
62 Lennard-Jones, L.E. (1931). Cohesion. *Proceedings of the Physical Society* 43 (5): 461–482.
63 Liang, T., Phillpot, S.R., and Sinnot, S.R. (2009). Parametrization of a reactive many-body potential for Mo-S systems. *Physical Review B* 79: 245110.
64 Liang, T., Phillpot, S.R., and Sinnot, S.R. (2012). Erratum: Parametrization of a reactive many-body potential for Mo-S systems. *Physical Review B* 85: 199903(E).
65 Liu, X. (1995). New model of potential energy functions for atomic solids and application to silicon crystal. *Chinese Physical Letters* 12 (9): 527–529.
66 Liu, X., Zhen, Z., Cox, H., and Murrell, N. (1998). New potential-energy functions for Cu, Ag and Au solids and their applications to computer simulations on metallic surfaces. *Science in China B* 41 (6): 566–574.
67 Lloyd, L.D. and Johnston, R.L. (1998). Modelling aluminium clusters with an empirical many-body potential. *Chemical Physics* 236: 107–121.
68 Lou, Y.S., Huh, H., Lim, S., and Pack, K. (2012). New ductile fracture criterion for prediction of fracture forming limit diagrams of sheet metals. *International Journal of Solids and Structures* 49 (25): 3605–3615.
69 Macioł, P., Gawąd, J., Kuziak, R., and Pietrzyk, M. (2010). Internal variable and cellular-automata finite element models of heat treatment. *International Journal for Multiscale Computational Engineering* 8: 267–285.
70 Martin, R.M. (2004). *Electronic Structure.* UK: Cambridge University Press.
71 Marx, D. and Hutter, J. (2000). Ab initio molecular dynamics: theory and implementation. In: *Modern Methods and Algorithms of Quantum Chemistry*, proceedings, NIC Series, 2e, vol. 3 (ed. J. Grotendorst), 329–477. John von Neumann Institute for Computing in cooperation with Arbeitsgemeinschaft f¨ur Theoretische Chemie.
72 Mecozzi, M.G., Militzer, M., Sietsma, J., and van Der Zwaag, S. (2008). The role of nucleation behavior in phase-field simulations of the austenite to ferrite transformation. *Metallurgical and Materials Transactions A* 39A: 1237–1247.

73 Militzer, M. (2011). Phase field modeling of microstructure evolution in steels. *Current Opinion in Solid State and Materials Science* 15: 106–115.

74 Minichmayr, R., Riedler, M., Winter, G. et al. (2008). Thermo-mechanical fatigue life assessment of aluminium components using the damage rate model of Sehitoglu. *International Journal of Fatigue* 30: 298–304.

75 Mishin, Y., Farkas, D., Mehl, M.J., and Papaconstantoopoulos, D.A. (1999). Interatomic potentials for monoatomic metals form experimental data and ab-initio calculations. *Physical Review B* 59 (5): 3393–3407.

76 Mohr, D. and Henn, S. (2007). Calibration of stress-triaxiality dependent crack formation criteria: a new hybrid experimental-numerical method. *Experimental Mechanics* 47: 805–820.

77 Monkman, F.C. and Grant, N.J. (1956). An empirical relationship between rupture life and minimum creep rate in creep-rupture test. *Proceedings-American Society for Testing and Materials* 56: 593–620.

78 Morse, P.M. (1929). Diatomic molecules according to the wave mechanics. II. Vibrational levels. *Physical Review* 34: 57–64.

79 Mortazavi, B., Ostadhossein, A., Rabczuk, T., and van Duin, A.C.T. (2016). Mechanical response of all-MoS2 single-layer heterostructures: a ReaxFF investigation. *Physical Chemistry Chemical Physics* 18 (34): 23695–23701.

80 Mrozek, A. (2019). Basic mechanical properties of 2H and 1T single-layer molybdenum disulfide polymorphs: a short comparison of various atomic potentials. *International Journal for Multiscale Computational Engineering* 17 (3): 339–359.

81 Mrozek, A., Kuś, W., and Burczyński, T. (2015). Nano level optimization of graphene allotropes by means of a hybrid parallel evolutionary algorithm. *Computational Materials Science* 106: 161–169.

82 Mrozek, A., Kuś, W., and Burczyński, T. (2017). Method for determining structures of new carbon-based 2D materials with predefined mechanical properties. *International Journal for Multiscale Computational Engineering* 15 (5): 379–394.

83 Murrell, J.N. and Mottram, R.E. (1990). Potential energy functions for atomic solids. *Molecular Physics* 69 (3): 571–585.

84 Murrell, J.N. and Rodriguez-Ruiz, J.A. (1990). Potential energy functions for atomic solids. II. Potential functions for diamond-like structures. *Molecular Physics* 71 (4): 823–834.

85 Nanba, S., Kitamura, M., Shimada, M. et al. (1992). Prediction of microstructure distribution in the through-thickness direction during and after hot rolling in carbon steels. *ISIJ International* 32: 377–386.

86 Needleman, A. and Tvergaard, V. (1991). An analysis of dynamic ductile crack growth in a double edge cracked specimen. *International Journal of Fracture* 49: 41–67.

87 Nowak, K. (2009). zastosowanie automatów komórkowych w opisie rozwoju uszkodzeń w warunkach pełzania. PhD thesis. Cracow University of Technology, Kraków (in Polish).

88 Oliveira, F.L.G., Andrade, M.S., and Cota, A.B. (2007). Kinetics of austenite formation during continuous heating in a low carbon steel. *Materials Characterization* 58: 256–261.

89 Ordon, J., Kuziak, R., and Pietrzyk, M. (2000). History dependent constitutive law for austenitic steels. Proceedings of the Metal Forming 2000, eds, Pietrzyk M., Kusiak J., Majta J., Hartley P., Pillinger I., Publ. A. Balkema, Krakow, 747–753.

90 Ordon, J., Pietrzyk, M., Kędzierski, Z., and Kuziak, R. (2002). Constitutive model based on two internal variables for constant and changing deformation conditions. Proceedings of the Thermomechanical Processing: Mechanics, Microstructure and Control, eds, Palmiere E.J., Mahfouf M., Pinna C., Sheffield, 33–39.

91 Ostadhossein, A., Rahnamoun, A., Wang, Y. et al. (2017). ReaxFF reactive force-field study of molybdenum disulfide (MoS2). *The Journal of Physical Chemistry Letters* 8 (3): 631–640.

92 Ostoja-Starzewski, M. (2005). Scale effect in plasticity of random media: status and challenges. *International Journal of Plasticity* 21: 1119–1160.

93 Parinello, M. (2000). Simulating complex systems without adjustable parameters. *Computer Science and Engineering* 2: 22–27.

94 Parr, R.D. and Yang, W. (1989). *Density-Functional Theory of Atoms and Molecules*. New York: Oxford University Press.

95 Pernach, M. and Pietrzyk, M. (2008). Numerical solution of the diffusion equation with moving boundary applied to modeling of the austenite-ferrite phase transformation. *Computational Materials Science* 44: 783–791.

96 Pernach, M., Bzowski, K., and Pietrzyk, M. (2012). Application of numerical solution of the diffusion equation to modelling phase transformation during heating of DP steels in the continuous annealing process. *Computer Methods in Materials Science* 12: 183–196.

97 Pietrzyk, M. (1990). Finite element based model of structure development in the hot rolling process. *Steel Research* 61: 603–607.

98 Pietrzyk, M. (1994). Numerical aspects of the simulation of hot metal forming using internal variable method. *Metallurgy and Foundry Engineering* 20: 429–439.

99 Pietrzyk, M. and Kuziak, R. (2012). Modelling phase transformations in steel. In: *Microstructure Evolution in Metal Forming Processes* (ed. J. Lin, D. Balint and M. Pietrzyk), 145–179. Oxford: Woodhead Publishing.

100 Pietrzyk, M., Kuziak, R., Pidvysots'kyy, V. et al. (2013). Computer-aided design of manufacturing chain based on closed die forging for hardly deformable Cu-based alloys. *Metallurgical and Materials Transactions A* 44A: 3281–3302.

101 Plimpton, S. (1995). Fast parallel algorithms for short-range molecular dynamics. *Journal of Computational Physics* 117: 1, 1–19.

102 Rabotnow, J.N. (1966). *Polzuchest elementom konstrukcji*. Moscow: Publ. Nauka.

103 Rapaport, D. (2004). *The Art of Molecular Dynamics Simulation*. UK: Cambridge University Press.

104 Rappe, A.K. and Goddard, W.A. (1991). Charge equilibration for molecular dynamics simulations. *Journal of Chemical Physics* 95 (8): 3358–3363.

105 Rice, J.R. and Tracey, D.M. (1969). On the ductile enlargement of voids in triaxial stress fields. *Journal of Mechanics and Physics of Solids* 17: 201–217.

106 Roberts, W., Sandberg, A., Siwecki, T., and Welefors T. (1983). Prediction of microstructure development during recrystallization hot rolling of Ti-V steels. Proc. Conf. HSLA Steels, Technology and Applications, Philadelphia, ASM, 67–84.

107 Roucoules, C. and Hodgson, P.D. (1995). Post-dynamic recrystallization after multiple peak dynamic recrystallization in C-Mn steels. *Materials Science and Technology* 11: 548–556.

108 Roucoules, C., Hodgson, P.D., Yue, S., and Jonas, J.J. (1994). Softening and microstructural change following the dynamic recrystallization of austenite. *Metallurgical and Materials Transactions A* 25A: 389–400.

109 Roucoules, C., Pietrzyk, M., and Hodgson, P.D. (2003). Analysis of work hardening and recrystallization during the hot working of steel using a statistically based internal variable method. *Materials Science and Engineering* A339: 1–9.

110 Sakai, T. (1995). Dynamic recrystallization microstructures under hot working conditions. *Journal of Materials Processing Technology* 53: 349–361.

111 Savran, V.I. (2009). Austenite formation in C-Mn steel. PhD thesis. University of Delft.

112 Scheil, E. (1935). Anlaufzeit der Austenitumwandlung. *Archiv für Eissenhüttenwesen* 12: 565–567.

113 Sellars, C.M. (1979). Physical metallurgy of hot working. In: *Hot Working and Forming Processes* (ed. C.M. Sellars and G.J. Davies), 3–15. London: The Metals Society.

114 Sellars, C.M. (1990). Modelling microstructural development during hot rolling. *Materials Science and Technology* 6: 1072–1081.

115 Sharma, K.S. and Kachhava, C.M. (1979). Application of Morse Potential to metals in the molecular-metallic-framework. *Physical Review A* 45 (5): 423–431.

116 Simmons, J.P., Shen, C., and Wang, Y. (2000). Phase field modeling of simultaneous nucleation and growth by explicitly incorporating nucleation events. *Scripta Materialia* 43: 935–942.

117 Singer-Loginova, I. and Singer, H.M. (2008). The phase field technique for modeling multiphase materials. *Reports on Progress in Physics* 71: 1–32.

118 Stillinger, F. and Weber, T.A. (1985). Computer simulation of local order in condensed phases of silicon. *Physical Review B* 31 (8): 5262–5271.

119 Stuart, S.J., Tutein, A.B., and Harrison, J.A. (2000). A reactive potential for hydrocarbons with intermolecular interactions. *Journal of Chemical Physics* 112 (14): 6472–6486.

120 Suehiro, M., Sato, K., Yada, H. et al. (1988). Mathematical model for predicting microstructural changes and strength of low carbon steels in hot strip rolling. Proceedings of the THERMEC'88, ed., Tamura I., Tokyo, 791–798.

121 Suehiro, M., Senuma, T., Yada, H., and Sato, K. (1992). Application of mathematical model for predicting microstructural evolution to high carbon steels. *ISIJ International* 32: 433–439.

122 Szeliga, D. (2012). Selection of the efficient phase transformation model for steels. *Computer Methods in Materials Science* 12: 70–84.

123 Szeliga, D., Gawąd, J., and Pietrzyk, M. (2006). Inverse analysis for identification of rheological and friction models in metal forming. *Computer Methods in Applied Mechanics and Engineering* 195: 6778–6798.

124 Szeliga, D., Kuziak, R., and Pietrzyk, M. (2011). Rheological model of Cu based alloys accounting for the preheating prior to deformation. *Archives of Civil and Mechanical Engineering* 11: 451–467.

125 Tamura, I., Ouchi, C., Tanaka, T., and Sekine, H. (1988). *Thermomechanical Processing of High Strength Low Alloy Steels*. London: Butterworth & Co. Press.

126 Tersoff, J. (1988). New empirical approach for the structure and energy of covalent systems. *Physical Review B* 37: 6991–7000.

127 Tersoff, J. (1989). Modeling solid-state chemistry: Interatomic potentials for multicomponent systems. *Physical Review B* 39: 5566–5568.

128 Tully, J. (1998). Nonadiabatic dynamics. In: *Modern Methods for Multidimensional Dynamics Computations in Chemistry* (ed. D. Thompson), 34–79. Singapore: World Scientific.

129 Urcola, J.J. and Sellars, C.M. (1987). Effect of changing strain rate on stress-strain behaviour during high temperature deformation. *Acta Metallurgica* 35: 2637–2647.

130 Vervynckt, S., Verbeken, K., Thibaux, P., and Houbaert, Y. (2010). Characterization of the austenite recrystallization by comparing double deformation and stress relaxation tests. *Steel Research International* 81: 234–244.

131 Vink, R.L.C., Barkema, G.T., and van der Weg, W.F. (2001). Fitting the Stillinger-Weber potential to amorphous silicon. *Journal of Non-Crystalline Solids* 282 (2–3): 248–255.

132 Yada, H. (1987). Prediction of microstructural changes and mechanical properties in hot strip rolling. Proceedings of the Symposium Accelerated Cooling of Rolled Steel, (eds), Ruddle, G.E. and Crawley, A.F., Pergamon Press, Winnipeg, 105–119.

3

Computational Methods

The useful computational methods for multiscale modelling and optimization are described in this chapter. Firstly, the commonly used macroscale and microscale analysis methods are described in detail. Secondly, the incorporation of these methods into the optimization procedures is presented.

3.1 Computational Methods for Continuum

The classic computational methods like finite element method (FEM) and boundary element method (BEM) can be successfully used in the multiscale computations of various physical phenomena. This chapter describes the two most commonly used methods: FEM and BEM in the multiscale context. Additionally, the new method extended finite element method (XFEM) is also described to highlight the variety of available numerical methods. The computational homogenization as a tool for coupling different analyses scales is discussed in this chapter.

3.1.1 FEM and XFEM

The FEM and its generalizations are still the most powerful computer-oriented methods of analysis of practical scientific and engineering problems. Finite element analysis is an integral and major component in the fields of engineering, design, and manufacturing. The main principles of this method are discussed in the following section. The assumption of material continuity is the basis for the finite element model. On the other hand, accounting for evolving discontinuities in the materials, resulting from static or dynamic instabilities, grain boundaries, etc. is a new challenge for numerical modelling. Extended finite element method (XFEM) is capable of capturing these discontinuities, and this method is also discussed below.

3.1.1.1 Principles of Computational Modelling Using FEM

The FEM is a numerical method of solving partial differential equations. This method is more general and powerful in applications to real-world problems compared to other methods. A number of publications and books have been written on the FEM, among which those published by Olgierd C. Zienkiewicz are noteworthy [125, 126]. Only basic principles of the FEM are summarized in this book, with particular emphasis on solving thermomechanical metal-forming problems.

Multiscale Modelling and Optimisation of Materials and Structures, First Edition. Tadeusz Burczyński, Maciej Pietrzyk, Wacław Kuś, Łukasz Madej, Adam Mrozek, and Łukasz Rauch.

In the most general problem of simulation, the behaviour of the investigated system depends on:

- the geometry of the domain of this system,
- properties of the material, and
- boundary and initial conditions.

Since real systems in engineering are usually very complex, modelling of the geometry of the domain is essential for the accuracy and robustness of the solution. In the FEM, the solution domain is represented by a collection of subdomains [100]. Over each subdomain, the solution is approximated by traditional variational methods. Individual segments of the solution have to fit with their neighbour's, which means that function and possibly its derivatives up to a selected order have to be continuous at connections between subdomains. There are few features of the FEM which confirm its superiority over other computing methods. One of these features is that the geometrically complex solution domain Ω is represented by a collection of simple subdomains, as is shown in Figure 3.1. This action is called meshing.

Meshing is performed to discretize the geometry into small pieces called elements. The rationale behind this is that the problem, which is very complex and varies in an unpredictable way using functions across the whole domain, can be easily approximated when divided into small elements. Mesh generation is a very important task of the preprocessor in finite element codes. It can be time-consuming and should be completely automated. A number of publications of mesh generation methods can be found in the scientific literature; see for example [71], where a concise and comprehensive guide to the application of finite element mesh generation over 2D domains, curved surfaces, and 3D domains was presented.

Each finite element Ω_e (Figure 3.1) is viewed as an independent domain itself. Over each element, algebraic equations among the quantities of interest are developed using the governing equations of the problem. Following this, the equations from all elements are assembled together using common degrees of freedom in these elements.

Defining the properties of the materials is the next important task that has to precede simulations. Many engineering systems consist of more than one material. Property of the material can be defined for a group of elements or for each individual element if needed. Different

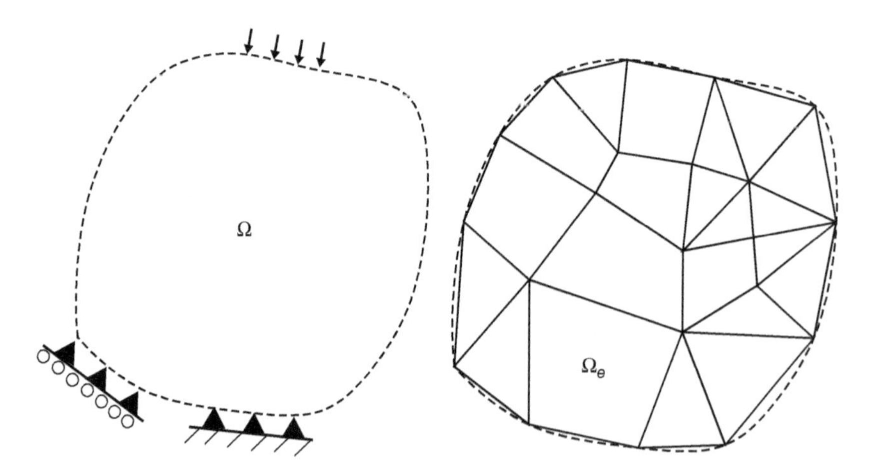

Figure 3.1 Representation of a 2D domain by a collection of triangles and quadrilaterals.

properties are needed depending on the simulated problem, ranging from thermo-physical properties through elastic modules to advanced nonlinear rheological and flow stress models. These properties should be determined in experiments for the investigated material. As an alternative, there are commercially available material databases where properties of popular materials can be found.

Boundary and initial conditions play a decisive role in solving physical problems. Inputting these conditions is usually done through a preprocessor and is often interfaced with graphics. The boundary and initial conditions vary from problem to problem. Therefore, to accurately simulate these conditions for actual engineering problems requires experience, knowledge, and proper engineering judgement.

The element equations are derived using the basic idea that any continuous function within a certain domain can be represented by its interpolation:

$$u(\mathbf{x}) = \sum_{i=1}^{n} n_i(\mathbf{x}) u_i, \tag{3.1}$$

where: n_i – interpolation functions, which are called shape functions, u_i – values of the solution in nodes of the element, n – number of nodes in the element.

Since shape functions are interpolation functions, they have to satisfy the condition $n_i(\mathbf{x}_j) = \delta_{ij}$, where δ_{ij} is Kronecker delta.

3.1.1.2 Principles of Computational Modelling Using FEM

The governing problems considered in this book concern mass transport (diffusion) or heat transport and strains and stresses. In the former problem, the unknown u is a scalar function, which represents concentration or temperature:

$$\nabla \cdot (k \nabla u) + Q = A\left(\frac{\partial u}{\partial t} + \mathbf{v} \cdot \nabla u\right), \tag{3.2}$$

where: k – conductivity or diffusion coefficient, Q – rate of heat generation or mass generation in the element, t – time, $A = 1$ for diffusion and $A = \rho c_p$ for heat transport, ρ – density, c_p – specific heat, $\mathbf{v}$ – velocity field.

The left-hand side of Equation (3.2) represents the classical transport mechanism (diffusion or conduction) and the right-hand side terms, with the velocity vector $\mathbf{v}$, represent transport by convection.

In the problem of strains and stresses, the governing equation is known as the equation of stress equilibrium:

$$\nabla \cdot \boldsymbol{\sigma} + \rho \mathbf{g} = 0, \tag{3.3}$$

where: $\boldsymbol{\sigma}$ – Cauchy stress tensor, $\mathbf{g}$ – gravity acceleration vector.

Both Eqs. (3.2) and (3.3) are solved using either variational method or weighted residual method, which transfers partial differential equation into integral form. Let's begin with the transport Eq. (3.2). The solution of this equation has to satisfy the specified boundary conditions. There are three commonly used types of boundary conditions that are applicable to the simulation of thermomechanical processes; see for example [67]:

- The function u is prescribed along the boundary surface and, generally, it is a function of both time and position (Dirichlet boundary condition):

$$u(\mathbf{x}, t) = u_D(t) \quad \mathbf{x} \in \Gamma_D, \tag{3.4}$$

where: $\mathbf{x}$ – vector of coordinates, u_D – known value of the solution at the boundary, Γ_D – part of the boundary on which Dirichlet boundary condition is imposed.

- The normal derivative of the function u is prescribed at the boundary surface, and it may be a function of both time and position (Neumann boundary condition):

$$k\frac{\partial u(\mathbf{x},t)}{\partial \mathbf{n}} = q \quad \mathbf{x} \in \Gamma_N, \tag{3.5}$$

where: $\mathbf{n}$ – unit vector normal to the surface, q – heat or mass flux through the surface, Γ_N – part of the boundary on which Neumann boundary condition is imposed.

- The third boundary condition is applied to thermal problems. The normal derivative of the function u (temperature) is proportional to the difference between value of this function at the surface and in the surroundings (Fourier boundary condition):

$$k\frac{\partial u(\mathbf{x},t)}{\partial \mathbf{n}} = \alpha(u_a - u) \quad \mathbf{x} \in \Gamma_F, \tag{3.6}$$

where: α – heat transfer coefficient, u_a – ambient temperature, Γ_F – part of the boundary on which Fourier boundary condition is imposed.

The typical solution of Eq. (3.2) is based on the variational principle [67, 125], which states that a minimum of the general functional:

$$J = \int_\Omega F\left(x, y, z, u, \frac{\partial u}{\partial x}, \frac{\partial u}{\partial y}, \frac{\partial u}{\partial z}\right) d\Omega + \int_\Gamma G\left(x, y, z, u, \frac{\partial u}{\partial x}, \frac{\partial u}{\partial y}, \frac{\partial u}{\partial z}\right) d\Gamma, \tag{3.7}$$

where: Ω – domain of the solution, Γ – boundary of the domain, which is obtained when the field $u(x, y, z)$ satisfies the Euler equations:

$$\frac{\partial F}{\partial u} - \nabla\left[\frac{\partial F}{\partial(\nabla u)}\right] = 0 \qquad \mathbf{x} \in \Omega. \tag{3.8}$$

The solution of Eq. (3.2) with relevant boundary conditions, Equations (3.5) and (3.6), then reduces to searching for a functional, for which the Euler equation is identical to (3.7). It is quite straightforward to show that, when convection in the equation is neglected ($\mathbf{v} = 0$), the functional

$$J = \int_\Omega \left\{\frac{k}{2}\left[\left(\frac{\partial u}{\partial x}\right)^2 + \left(\frac{\partial u}{\partial y}\right)^2 + \left(\frac{\partial u}{\partial z}\right)^2\right] - \hat{Q}T\right\} d\Omega - \int_\Gamma \left[\alpha\left(u_a - \frac{u}{2}\right)u + qu\right] d\Gamma, \tag{3.9}$$

gives, on minimization, the satisfaction of the problem set in Eq. (3.2) as well as the boundary conditions given by Eqs. (3.5) and (3.6). The algebraic manipulations verifying the previously mentioned equations are presented in [67, 125]. In Eq. (3.9), $\hat{Q}$ is given by:

$$\hat{Q} = Q - A\frac{\partial u}{\partial t}. \tag{3.10}$$

Discretization is performed in the usual finite element manner. The function u inside an element is presented as a function of nodal values according to the interpolation formula. Substitution of the relationship (3.10) into Eq. (3.9) yields:

$$J = \int_{\Omega} \left\{ \frac{k}{2} \left[\left(\sum \frac{\partial n_i}{\partial x} u_i \right)^2 + \left(\sum \frac{\partial n_i}{\partial y} u_i \right)^2 + \left(\sum \frac{\partial n_i}{\partial z} u_i \right)^2 \right] - \hat{Q} \sum n_i u_i \right\} d\Omega$$
$$- \int_{\Gamma} \left[\alpha \left(u_a - \frac{1}{2} \sum n_i u_i \right) \sum n_i u_i + q \sum n_i u_i \right] d\Gamma \tag{3.11}$$

Minimization of the functional (3.11) requires the calculation of the partial derivatives with respect to nodal values of u, and it results in the following set of linear equations:

$$\frac{\partial J}{\partial u_j} = \int_{\Omega} k \left[\sum \left(\frac{\partial n_i}{\partial x} u_i \right) \frac{\partial n_j}{\partial x} + \sum \left(\frac{\partial n_i}{\partial y} u_i \right) \frac{\partial n_j}{\partial y} \right] d\Omega - \int_{\Omega} \hat{Q} n_j d\Omega - \int_{\Gamma} (\alpha u_a + q) n_j d\Gamma$$
$$+ \int_{S} \alpha \sum (n_i u_i) n_j d\Gamma = 0. \tag{3.12}$$

Equation (3.12) written in a matrix form is:

$$\mathbf{Hu} = \mathbf{p}, \tag{3.13}$$

where:

$$H_{ij} = \int_{\Omega} (\nabla n_i)^T k (\nabla n_j) d\Omega + \int_{\Gamma} \alpha n_i n_j d\Gamma$$

$$p_i = \int_{\Gamma} \left[\alpha u_a n_i + \sum_j \left(n_j q_j \right) n_i \right] d\Gamma + \int_{\Omega} \sum_j \left(n_j \hat{Q}_j \right) n_i \, d\Omega.$$

Equation (3.13) is a set of linear equations, a solution of which gives the values of the nodal values u_i during a stationary thermal state when $\partial u / \partial t = 0$. However, non-stationary heat or mass transfer problems are usually involved in the modelling of thermomechanical processing. A time-varying solution is usually obtained by the modal method or by direct temporal integration using the finite difference method. Due to the nonlinearity of the problem and possible sharp transients, the latter approach is preferable in coupled thermal-mechanical solutions [67], and it is discussed briefly below. During finite but reasonably short time intervals, the partial derivatives of u in the non-steady state can be considered as functions of the $\mathbf{x}$ coordinates only. Then the solution of Eq. (3.2) for $\mathbf{v} = 0$ is obtained as:

$$\mathbf{Hu} + \mathbf{C} \frac{\partial}{\partial t} \mathbf{u} - \mathbf{p} = 0, \tag{3.14}$$

where:

$$C_{ij} = \int_{\Omega} n_i A n_j d\Omega.$$

C_{ij} is called the heat capacity matrix in the heat transfer problems ($A = \rho c_p$), and in the mass transfer problems, when $A = 1$, it is a geometrical matrix without physical meaning.

Different integration schemes can be applied to solve Eq. (3.14) with the initial condition $\mathbf{u}(t = 0) = \mathbf{u}_0$. In general, the simplest integration schemes are one-step schemes in which a two-level difference approximation is chosen for the time derivative of u and a linear variation of **H**, **C**, and **p** is assumed during Δt. Thus, from Eq. (3.14) we have:

$$\frac{\partial \mathbf{u}}{\partial t} = \frac{\mathbf{u}_{i+1} - \mathbf{u}_i}{\Delta t} = \left(-\mathbf{C}_i^{-1}\mathbf{H}_i\mathbf{u}_i + \mathbf{C}_i^{-1}\mathbf{p}_i\right)(1-\theta) + \left(-\mathbf{C}_{i+1}^{-1}\mathbf{H}_{i+1}\mathbf{u}_{i+1} + \mathbf{C}_{i+1}^{-1}\mathbf{p}_{i+1}\right)\theta. \tag{3.15}$$

Equation (3.15) contains coefficient θ, which can be selected by an analyst. The value of this coefficient determines the type of integration scheme. Rearranging of Eq. (3.14) gives the vector of nodal temperatures at the end of time interval Δt as:

$$\mathbf{u}_{i+1} = \left(\mathbf{I} + \Delta t\theta \mathbf{C}_{i+1}^{-1}\mathbf{H}_{i+1}\right)^{-1}\left[\mathbf{u}_i + \Delta t\theta\left(\mathbf{C}_{i+1}^{-1}\mathbf{p}_{i+1}\right) + \Delta t(1-\theta)\left(-\mathbf{C}_i^{-1}\mathbf{H}_i\mathbf{u}_i + \mathbf{C}_i^{-1}\mathbf{p}_i\right)\right]. \tag{3.16}$$

Assuming further that:
$\mathbf{H}_{i+1} = \mathbf{H}_i = \mathbf{H}$; $\mathbf{C}_{i+1} = \mathbf{C}_i = \mathbf{C}$; and $\mathbf{p}_{i+1} = \mathbf{p}_i = \mathbf{p}$,

Equation (3.16) is written as:

$$\mathbf{C}\frac{\mathbf{u}_{i+1} - \mathbf{u}_i}{\Delta t} = (-\mathbf{H}\mathbf{u}_i + \mathbf{p})(1-\theta) + (-\mathbf{H}\mathbf{u}_{i+1} + \mathbf{p})\theta. \tag{3.17}$$

Rearranging Eq. (3.17) yields:

$$\hat{\mathbf{H}}\mathbf{u} = \hat{\mathbf{p}}, \tag{3.18}$$

where:

$$\hat{\mathbf{H}} = \left[2\mathbf{H} + \frac{3}{\Delta t}\mathbf{C}\right] \qquad \hat{\mathbf{p}} = \left[-\mathbf{H} + \frac{3}{\Delta t}\mathbf{C}\right]\mathbf{u}_i - 3\mathbf{p}.$$

Values of θ and relevant equations obtained for various integration schemes are given in Table 3.1. These equations describe all matrices for the set of Eq. (3.18) for various schemes. More details on the integration schemes and analysis of their stability is given in [109].

Let's now return to the problem of stress–strain field modelling. According to the weighted residual method, the weak form of the equilibrium Eq. (3.3) is expressed as:

$$\int_\Omega (\nabla \cdot \boldsymbol{\sigma} + \rho \mathbf{g}) \cdot \mathbf{w} d\Omega = 0, \tag{3.19}$$

where: **w** – vector of weight functions.

Table 3.1 Coefficient θ and matrices $\hat{\mathbf{H}}$ and $\hat{\mathbf{p}}$ for various integration schemes.

θ	Matrix $\hat{\mathbf{H}}$	Vector $\hat{\mathbf{p}}$	Scheme
0	$\hat{\mathbf{H}} = \frac{1}{\Delta t}\mathbf{C}$	$\hat{\mathbf{p}} = \left[\mathbf{H} - \frac{1}{\Delta t}\mathbf{C}\right]\mathbf{u}_i - \mathbf{p}$	Euler explicit (forward) Conditionally stable
$\frac{1}{2}$	$\hat{\mathbf{H}} = \left[\mathbf{H} + \frac{2}{\Delta t}\mathbf{C}\right]$	$\hat{\mathbf{p}} = \left[\mathbf{H} - \frac{2}{\Delta t}\mathbf{C}\right]\mathbf{u}_i - 2\mathbf{p}$	Trapezoidal or Crank-Nicholson (Midpoint rule) unconditionally stable
$\frac{2}{3}$	$\hat{\mathbf{H}} = \left[2\mathbf{H} + \frac{3}{\Delta t}\mathbf{C}\right]$	$\hat{\mathbf{p}} = \left[\mathbf{H} - \frac{3}{\Delta t}\mathbf{C}\right]\mathbf{u}_i - 3\mathbf{p}$	Galerkin Unconditionally stable
1	$\hat{\mathbf{H}} = \left[\mathbf{H} + \frac{1}{\Delta t}\mathbf{C}\right]$	$\hat{\mathbf{p}} = \left[\frac{1}{\Delta t}\mathbf{C}\right]\mathbf{u}_i - \mathbf{p}$	Fully implicit or Euler backward Unconditionally stable

Application of the integration by parts (Green theorem) to the first term in the Eq. (3.19) gives [117]:

$$\int_{\Omega} \nabla \cdot \boldsymbol{\sigma} \cdot \mathbf{w} d\Omega = \int_{\Gamma} \boldsymbol{\sigma} \cdot \mathbf{n} \cdot \mathbf{w} d\Gamma - \int_{\Omega} \boldsymbol{\sigma} \cdot \nabla \mathbf{w}^T d\Omega. \tag{3.20}$$

Each weight function in vector $\mathbf{w}$ is selected in a way that it is zero on the boundary Γ_D, where the Dirichlet condition is applied. Thus, for the surface integral term in Eq. (3.20), the boundary conditions that $\boldsymbol{\sigma} \bullet \mathbf{n} = \mathbf{f}$ on Γ_N (suppressible boundary condition) and $\delta \mathbf{u} = 0$ on the remaining part of the surface Γ_D (essential boundary condition), are imposed:

$$\int_{\Omega} \nabla \boldsymbol{\sigma} \cdot \mathbf{w} d\Omega = \int_{\Gamma_N} \mathbf{f} \cdot \mathbf{w} d\Gamma_N - \int_{\Omega} \boldsymbol{\sigma} \cdot \nabla \mathbf{w}^T d\Omega, \tag{3.21}$$

where: $\mathbf{u}$ – vector of displacements (solid formulation) or velocities (flow formulation), $\delta \mathbf{u}$ – an arbitrary variation in $\mathbf{u}$.

Substituting Eq. (3.21) into Eq. (3.19) and neglecting the gravity forces in the further analysis ($\mathbf{g} = 0$) gives:

$$\int_{\Omega} \boldsymbol{\sigma} \cdot \nabla \mathbf{w}^T d\Omega = \int_{\Gamma_N} \mathbf{f} \cdot \mathbf{w} d\Gamma_N. \tag{3.22}$$

If $\mathbf{w}$ is considered to be the virtual motion ($\mathbf{w} = \delta\mathbf{u}$; $\nabla \mathbf{w}^T = \delta\dot{\boldsymbol{\varepsilon}}$), Equation (3.22) is the expression of the virtual work-rate principle. Using the symmetry of the stress tensor and the divergence theorem, we obtain:

$$\int_{\Omega} \boldsymbol{\sigma} \delta \dot{\boldsymbol{\varepsilon}} d\Omega - \int_{\Gamma_N} \mathbf{f} \cdot \delta \mathbf{u} d\Gamma_N = 0. \tag{3.23}$$

Finite elements based on the standard velocity formulation are vulnerable to volumetric locking in the analysis of problems with incompressible deformations. This creates difficulties in the simulation of elastoplastic and rigid-plastic cases [102]. Introduction of the mixed finite element formulation in which we split the stresses into the deviatoric part $\boldsymbol{\sigma}'$ and pressure $\boldsymbol{\sigma}_m$: $\boldsymbol{\sigma} = \boldsymbol{\sigma}' + \mathbf{I}p$ can be a solution for this problem. After substituting this sum into Eq. (3.23) and assuming further that $\boldsymbol{\sigma}' = \sigma_i = \sigma_p$, $\sigma_m = \text{trace}(\sigma)$ and $\dot{\boldsymbol{\varepsilon}} = \dot{\varepsilon}_i$ we obtain:

$$\int_{\Omega} \sigma_p \delta \dot{\varepsilon}_i d\Omega + \int_{\Omega} p \delta \dot{\varepsilon}_V d\Omega - \int_{\Gamma_N} \mathbf{f} \delta \mathbf{u} d\Gamma_N = 0, \tag{3.24}$$

where: σ_i – effective stress σ_p – flow stress, $\dot{\varepsilon}_i$ – effective strain rate, $\dot{\varepsilon}_V$ – volumetric strain rate.

In Eq. (3.24) the second term represents the incompressibility condition and in this term p is a variable. Independent interpolations for velocities and pressure are used, and mixed equations are obtained [102]. This approach is known as flow formulation, and full description of its application to metal-forming processes can be found in [59]. A minimum of the power functional (3.24) is searched assuming that the material obeys Huber-Mises yield criterion and associated Levy-Mises flow rule:

$$\boldsymbol{\sigma} = D\dot{\boldsymbol{\varepsilon}} \quad D = \frac{2}{3} \frac{\sigma_p}{\dot{\varepsilon}_i}, \tag{3.25}$$

where: $\boldsymbol{\sigma}$ – vector with components of the stress tensor, $\dot{\boldsymbol{\varepsilon}}$ – vector with components of the strain rate tensor, σ_p – flow stress, which according to the Huber-Mises yield criterion is equal to the effective stress σ_i.

After discretization, the nodal velocities and pressures are calculated by searching for a minimum of the following functional:

$$J = \int_{\Omega} \sigma_p \sqrt{\frac{2}{3} \mathbf{v}^T \mathbf{K} \mathbf{v}} d\Omega + \int_{\Omega} Q d\Omega - \int_{\Gamma_N} \mathbf{f} \mathbf{N}^T \mathbf{u} d\Gamma_N, \tag{3.26}$$

where:

$$\mathbf{K} = \mathbf{B}^T D \mathbf{B}. \tag{3.27}$$

Newton-Raphson method is used to solve nonlinear equations, and the following set of linear equations is obtained [89, 90]:

$$\overline{\mathbf{K}} \begin{Bmatrix} \Delta \mathbf{v} \\ p \end{Bmatrix} = \overline{\mathbf{f}}, \tag{3.28}$$

where:

$$\overline{\mathbf{K}} = \begin{bmatrix} \dfrac{\partial^2 J}{\partial \mathbf{v}^T \partial \mathbf{v}} & \mathbf{b} \\ \mathbf{b} & 0 \end{bmatrix} \quad \mathbf{f} = \begin{Bmatrix} \dfrac{\partial J}{\partial \mathbf{v}} \\ b^T \mathbf{v} \end{Bmatrix} \quad b = \int_{\Omega} B^T. \tag{3.29}$$

In Equation (3.28), $\Delta \mathbf{v}$ represents a vector of increments of the nodal velocities. In the flow formulation, nodal velocities are calculated in each time step. The new locations of nodes are calculated using explicit/implicit time integration scheme, depending on θ in Table 3.1 using the following equation:

$$x_{t+\Delta t} = x_i + [\theta v_t (1-\theta) v_{t+\Delta t}]. \tag{3.30}$$

Rigid-plastic flow formulation is an efficient simulation method for metal-forming processes.

3.1.1.3 Extended Finite Element Method

The approach for solving a problem with strong discontinuities, such as crack or shear bands development, is more frequently used [4, 86]. A method based on special finite elements with the special points emerging from the strong discontinuity analysis is presented in [86]. Using those special points makes it possible to combine a continuum approach with the discrete approach, which is based on discrete constitutive equations. The author of [86] considered the existence of a body Ω subjected to strong discontinuity conditions along with the discontinuity path S (Figure 3.2).

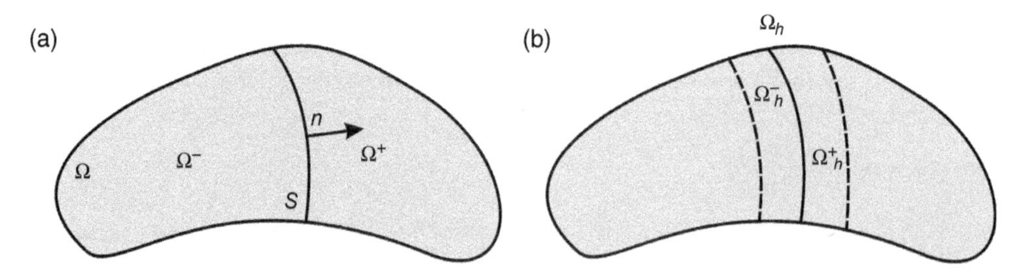

Figure 3.2 Discontinuity path (a) and domains (b) created in [86]. *Source:* Based on Olivier [86].

S is a discontinuity path, which is fixed at the reference configuration and characterized by the normal vector **n**. The S path is introduced in a way that divides the Ω domain into Ω^+ and Ω^- parts, and a Heaviside function $H_s(x)$ is defined on Ω:

$$H_S(x) = \begin{cases} 1 & \forall x \in \Omega^- \\ 0 & \forall x \in \Omega^+ \end{cases} \tag{3.31}$$

Illustration of the body with the discontinuity path S is presented in Figure 3.2, where Γ_u and Γ_σ are the Ω boundaries subjected to the usual essential and natural boundary conditions, respectively, and obey three conditions:

$$\Gamma_u \cup \Gamma_\sigma = \partial\Omega, \quad \Gamma_u \cap \Gamma_\sigma = 0, \quad \Gamma_u \cap \Omega^h = 0. \tag{3.32}$$

S_h^+ and S_h^- are the two boundaries of the Ω_h^+ and Ω_h^- subdomains, respectively, that surrounds the S path. It is also assumed that a function $\phi^h(x)$ is defined as:

$$\phi^h(x) = \begin{cases} 0 & \forall x \in \Omega^- \setminus \Omega_h^- \\ 1 & \forall x \in \Omega^+ \setminus \Omega_h^+ \end{cases}. \tag{3.33}$$

In [86] a unit jump function, which takes the zero value everywhere in Ω excluding Ω_h, was introduced:

$$M_S^h(x) = H_S(x) - \phi^h(x). \tag{3.34}$$

Such a function exhibits a jump across the discontinuity path S. An expression for the displacement field $u(x, t)$ under a strong discontinuity condition on the S is:

$$u(x,t) = \hat{u}(x,t) + M_S^h(x)[[u]](x,t), \tag{3.35}$$

where: $\hat{u}$ – the conventional part of the displacement field, $[[u]](x, t)$ – displacement jump function.

A strain field is calculated based on the symmetric part of the gradient of the displacement field:

$$\varepsilon = (\nabla u)^S = \overline{\varepsilon} + \delta_S([[u]] \times n)^S, \tag{3.36}$$

where: $\overline{\varepsilon}$ – conventional part of the strain field, S – superscript denoting the symmetric part of the vector, δ_S – the Dirac delta function along S.

This decomposition of the displacement field is illustrated in Figure 3.3.

Based on this knowledge [86], a finite element approximation of the strong discontinuity problem is illustrated in Figure 3.4, where: l_e – length of a straight line in the element, n_e – the normal vector to S_e.

By extending Eq. (3.35) to the whole boundary S, an approximation of the displacement field in the FE method is expressed as (Figure 3.5):

$$u^h(x,t) = \hat{u}^h(x,t) + \sum_{e=1}^{n_{el}} M_{Se}^h(x)[[u]]^h(t) \tag{3.37}$$

$$M_{Se}^h(x) = H_{Se}(x) - N_{ke}(x). \tag{3.38}$$

The strain field is then finally calculated as:

$$\varepsilon^h(x,t) = \left(\nabla\hat{u}^h\right)^S + \sum_{e=1}^{n_{el}} \left(\nabla M_{Se}^h \times [[u]]^h(t)\right)^S \Rightarrow \textit{regular} + \textit{enhanced}. \tag{3.39}$$

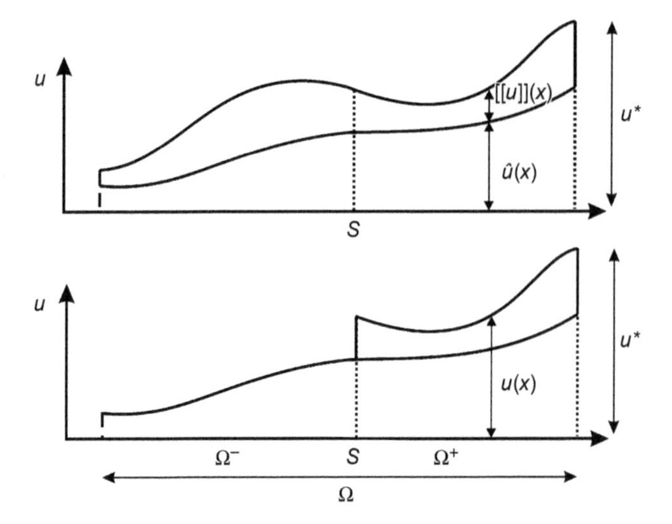

Figure 3.3 Illustration of the displacement field decomposition [86]. *Source:* Based on Olivier [86].

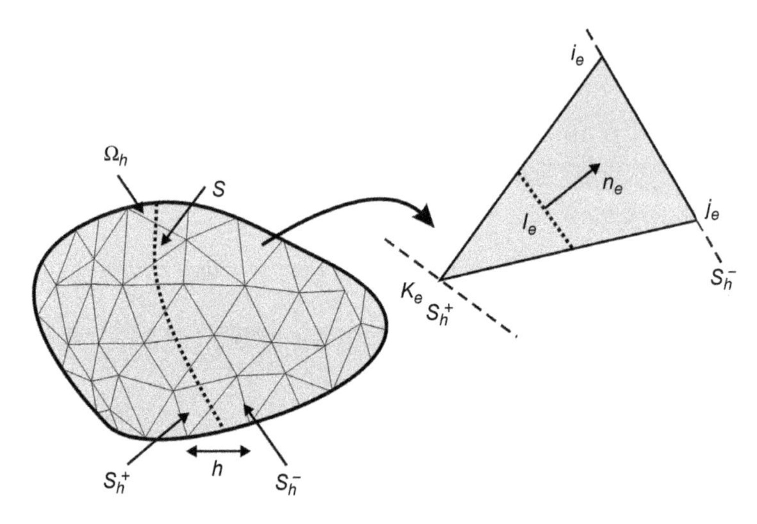

Figure 3.4 Illustration of the finite element approximation [86]. *Source:* Based on Olivier [86].

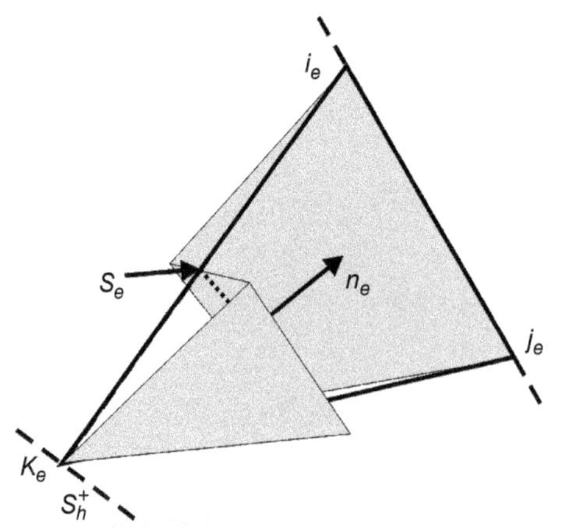

Figure 3.5 Illustration of the unit jump function finite element approximation [86]. *Source:* Based on Olivier [86].

Based on this approach, interesting simulation results, including uniaxial tension or shear tests, are shown in [4, 86]. However, an artificial method has been used to trigger a discontinuity, such as shear band formation. Artificially, peak stress value of the upper element in the band has been reduced in [86]. The same procedure was used in [4]. So the initiation of the strain localization is not a natural outcome of the processes taking place during deformation, which is the main disadvantage of the methods which involve enhanced finite elements.

3.1.2 BEM and FEM/BEM Coupling

The BEM, just as the FEM, enables solving numerically boundary-value and boundary-initial-value problems. It has some advantages, especially when dealing with infinite or semi-infinite structures or bodies with complex geometries. For most problems, only the boundary of the body is discretized, which leads to fewer elements than FEM. For some specific problems, like elastoplasticity, internal cells must be introduced. The details about BEM can be found in [7, 9, 15, 16, 18–20, 35, 45, 108]. The BEM is still under development, and new approaches like fast multipole BEM [92] have been recently introduced and allow overcoming some BEM disadvantages.

3.1.2.1 BEM

Let us consider a body Ω bounded by the boundary Γ shown in Figure 3.6. The body is loaded using internal forces $\mathbf{b}$ in the area Ω, tractions $\mathbf{p}_0$ act on the boundary segment Γ_p, and the displacements $\mathbf{u}_0$ are prescribed on the boundary segment Γ_u, whereby $\Gamma_p \cup \Gamma_u = \Gamma$ and $\Gamma_p \cap \Gamma_u = \emptyset$.

The displacement equation of static elastic problem can be formulated as:

$$\mathbf{L}_s\mathbf{u} = \mathbf{b}; \quad \mathbf{x} \in \Omega. \tag{3.40}$$

In the case of isotropic material, L_s operator is defined as:

$$\mathbf{L}_s = -\begin{bmatrix} \mu\Delta + (\mu+\lambda)\dfrac{\partial^2}{\partial x_1^2} & (\mu+\lambda)\dfrac{\partial^2}{\partial x_1 \partial x_2} & (\mu+\lambda)\dfrac{\partial^2}{\partial x_1 \partial x_3} \\ (\mu+\lambda)\dfrac{\partial^2}{\partial x_1 \partial x_2} & \mu\Delta + (\mu+\lambda)\dfrac{\partial^2}{\partial x_2^2} & (\mu+\lambda)\dfrac{\partial^2}{\partial x_2 \partial x_3} \\ (\mu+\lambda)\dfrac{\partial^2}{\partial x_1 \partial x_3} & (\mu+\lambda)\dfrac{\partial^2}{\partial x_2 \partial x_3} & \mu\Delta + (\mu+\lambda)\dfrac{\partial^2}{\partial x_3^2} \end{bmatrix}. \tag{3.41}$$

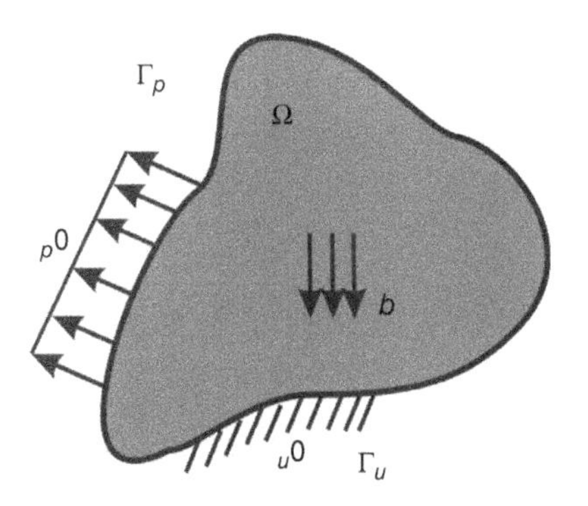

Figure 3.6 The elastic body bounded by the boundary.

Using the Maxwell-Betti reciprocal work, the theorem Somigliana identity can be obtained:

$$\mathbf{u}(\mathbf{x}) = \int_{\Gamma} \mathbf{U}^*(\mathbf{x},\mathbf{y})\mathbf{p}(\mathbf{y})d\Gamma(\mathbf{y}) - \int_{\Gamma} \mathbf{P}^*(\mathbf{x},\mathbf{y})\mathbf{u}(\mathbf{y})d\Gamma(\mathbf{y}) + \int_{\Omega} \mathbf{U}^*(\mathbf{x},\mathbf{y})\mathbf{b}(\mathbf{y})d\Omega(\mathbf{y}), \quad (3.42)$$

where: $\mathbf{U}^*(\mathbf{x}, \mathbf{y})$, $\mathbf{P}^*(\mathbf{x}, \mathbf{y})$ – fundamental solutions which have a form of symmetric tensors.

Components of the tensor are displacements of the point **y** in direction *j* in the infinite elastic medium due to unit point load in the point **x** in the direction *i*. Elements of tensor describe the elements of stress vector in the point **y** in direction *j* in the infinite elastic medium due to the point load in point **x** in the direction *i*. Equation (3.42) allows computing the displacements inside the body when the boundary displacements and tractions are known. Stress tensor coefficients can be obtained using the equation:

$$\sigma_{ij}(\mathbf{x}) = \int_{\Gamma} D^*_{ijk}(\mathbf{x},\mathbf{y})p_k(\mathbf{y})d\Gamma(\mathbf{y}) - \int_{\Gamma} S^*_{ijk}(\mathbf{x},\mathbf{y})u_k(\mathbf{y})d\Gamma(\mathbf{y}) + \int_{\Omega} D^*_{ijk}(\mathbf{x},\mathbf{y})b_k(\mathbf{y})d\Omega(\mathbf{y}), \quad (3.43)$$

where:

$$D^*_{ijk}(x,y) = C_{ijlm}\frac{\partial}{\partial x_m}U^*_{lk}(x,y) \quad (3.44)$$

$$S^*_{ijk}(x,y) = C_{ijlm}\frac{\partial}{\partial x_m}P^*_{lk}(x,y). \quad (3.45)$$

The fundamental solutions $\mathbf{U}*(\mathbf{x}, \mathbf{y})$ and $\mathbf{P}*(\mathbf{x}, \mathbf{y})$ for the isotropic 2D problems are given by:

$$U^*_{ij}(x,y) = -\frac{1}{8\pi(1-\nu)\mu}\left[(3-4\nu)\ln(r)\delta_{ij} - r_{,i}r_{,j}\right] \quad (3.46)$$

$$P^*_{ij}(x,y) = -\frac{1}{4\pi(1-\nu)r}\left\{\left[(1-2\nu)\delta_{ij} + 2r_{,i}r_{,j}\right]\frac{\partial r}{\partial \mathbf{n}} - (1-2\nu)\left(r_{,i}n_j - r_{,j}n_i\right)\right\}, \quad (3.47)$$

where: ν – a Poisson's ratio, r – a distance defined as:

$$r(x,y) = \left(r_i r_j\right)^{1/2}; r_i = x_i(y) - x_i(x) \quad (3.48)$$

and:

$$r_{,i} = \frac{\partial r}{\partial x_i(y)} = \frac{r_i}{r}. \quad (3.49)$$

The Somigliana identity (3.42) can be used when all displacements and tractions are known on the boundary of the body. In real boundary-value problems, only parts of the displacements and tractions are known. If the point x tends to the boundary, the Somigliana identity (3.42) takes the form of a boundary integral equation. The equation can be modified by assuming the area near boundary as the point **x** surrounded by boundary Γ_ε with radius ε as shown in Figure 3.7. The Γ boundary can be expressed as a sum:

$$\Gamma = (\Gamma - \Gamma_\varepsilon) + \Gamma^*_\varepsilon \quad (3.50)$$

In this case, the Somigliana identity takes the form:

$$u(x) = \lim_{\varepsilon \to 0}\left\{\int_{\Gamma-\Gamma_\varepsilon+\Gamma^*_\varepsilon} U^*(x,y)p(y)\mathrm{d}\Gamma(y) - \int_{\Gamma-\Gamma_\varepsilon+\Gamma^*_\varepsilon} P^*(x,y)u(y)\mathrm{d}\Gamma(y) + \int_{\Omega'} U^*(x,y)b(y)\mathrm{d}\Omega(y)\right\}, \quad (3.51)$$

where: Ω' – a spherical area near the source point **x** with radius ε.

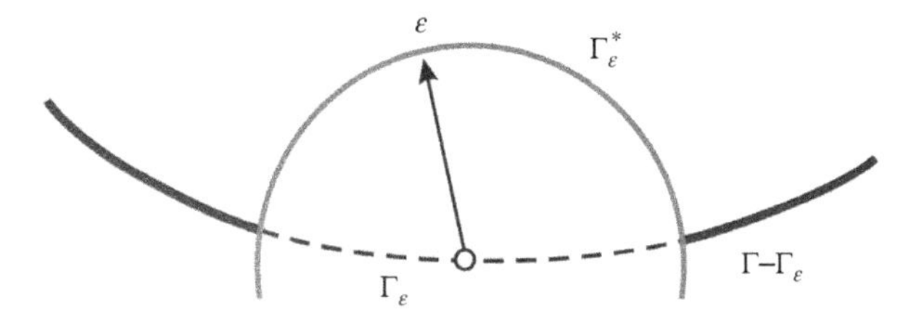

Figure 3.7 An area near the boundary of the body.

The first and third integral are improper due to non-uniqueness of U^*. The second integral can be expressed as a sum of two integrals:

$$u(x) = \int_\Gamma U^*(x,y)p(y)\mathrm{d}\Gamma(y) + \int_\Omega U^*(x,y)b(y)\mathrm{d}\Omega(y) + \\ - \lim_{\varepsilon\to 0}\left\{ \int_{\Gamma-\Gamma_\varepsilon} P^*(x,y)u(y)\mathrm{d}\Gamma(y) + \int_{\Gamma_\varepsilon^*} P^*(x,y)u(y)\mathrm{d}\Gamma(y) \right\} \tag{3.52}$$

and the last integral in the Eq. (3.52) can be formulated as:

$$\lim_{\varepsilon\to 0}\int_{\Gamma_\varepsilon^*} P^*(x,y)u(y)\mathrm{d}\Gamma(y) = \lim_{\varepsilon\to 0}\int_{\Gamma_\varepsilon^*} P^*(x,y)[u(y)-u(x)]\mathrm{d}\Gamma(y) \\ + \lim_{\varepsilon\to 0} u(x)\int_{\Gamma_\varepsilon^*} P^*(x,y)\mathrm{d}\Gamma(y), \tag{3.53}$$

where the first integral is equal to zero due to the continuity of displacements. Equation (3.53) after the rearrangement can be formulated as:

$$\mathbf{c}(\mathbf{x}) = \mathbf{I} + \lim_{\varepsilon\to 0}\int_{\Gamma_\varepsilon} \mathbf{P}^*(\mathbf{x},\mathbf{y})\mathrm{d}\Gamma(\mathbf{y}), \tag{3.54}$$

where: $\mathbf{I}$ – a unit matrix.

Taking into account Eq. (3.54), the Somigliana identity finally takes the form of the boundary integral equation, which can be written in the vector form:

$$\mathbf{c}(\mathbf{x})\mathbf{u}(\mathbf{x}) + \int_\Gamma \mathbf{P}^*(\mathbf{x},\mathbf{y})\mathbf{u}(\mathbf{y})\mathrm{d}\Gamma(\mathbf{y}) = \int_\Gamma \mathbf{U}^*(\mathbf{x},\mathbf{y})\mathbf{p}(\mathbf{y})\mathrm{d}\Gamma(\mathbf{y}) + \int_\Omega \mathbf{U}^*(\mathbf{x},\mathbf{y})\mathbf{b}(\mathbf{y})\mathrm{d}\Omega(\mathbf{y}) \tag{3.55}$$

or using index notation:

$$c_{ij}(\mathbf{x})u_i(\mathbf{x}) + \int_\Gamma P_{ij}{}^*(\mathbf{x},\mathbf{y})u_j(\mathbf{y})\mathrm{d}\Gamma(\mathbf{y}) = \int_\Gamma U_{ij}{}^*(\mathbf{x},\mathbf{y})p_j(\mathbf{y})\mathrm{d}\Gamma(\mathbf{y}) + \int_\Omega U_{ij}{}^*(\mathbf{x},\mathbf{y})b_j(\mathbf{y})\mathrm{d}\Omega(\mathbf{y}). \tag{3.56}$$

The boundary of a 2D structure is discretized by means of linear or curvilinear elements Γ_e, as shown in Figure 3.8.

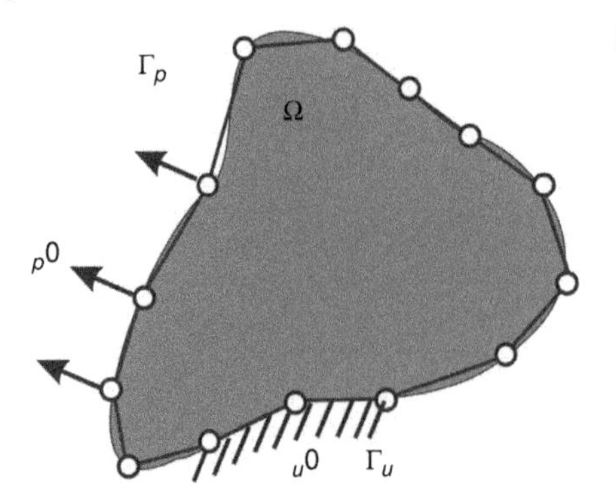

Figure 3.8 Discretization with boundary elements.

Each boundary element Γ_e has W_e nodes. In the case of 2D problems and linear shape functions $W_e = 2$ while $W_e = 3$ for quadratic shape functions. The points in the element are described using nodes locations and shape functions:

$$x_i(\xi) = M_w(\xi)(x_i)^w; i = 1, 2; w = 1, ..., W_e. \tag{3.57}$$

The linear shape functions for the 2D problem are expressed as:

$$M_1(\xi) = \frac{1}{2}(1-\xi), M_2(\xi) = \frac{1}{2}(1+\xi). \tag{3.58}$$

Boundary displacements and tractions are expressed using nodal values and shape functions as:

$$\begin{aligned} &\mathbf{u}(\mathbf{x}(\xi)) \approx M_w(\mathbf{u})^w; \mathbf{x} \in \Gamma_e \\ &\mathbf{p}(\mathbf{x}(\xi)) \approx M_w(\mathbf{p})^w; \mathbf{x} \in \Gamma_e \end{aligned}. \tag{3.59}$$

Due to local formulation in direction, the Jacobian of coordinate system transfer should be used:

$$\mathrm{d}\Gamma(\mathbf{y}) = \mathbf{J}(\xi)\mathrm{d}\xi, \tag{3.60}$$

where:

$$\mathbf{J}(\xi) = \left[\left(\frac{\partial x_1}{\partial \xi}\right)^2 + \left(\frac{\partial x_2}{\partial \xi}\right)^2\right]^{1/2}. \tag{3.61}$$

Equation (3.55) is reformulated taking into account the discretization and shape functions:

$$\begin{aligned} \mathbf{c}(\mathbf{x})\mathbf{u}(\mathbf{x}) = &\sum_{e=1}^{ne}\sum_{w=1}^{W_e}(\mathbf{u})_\mathbf{e}^\mathbf{w}\int_{\Gamma_e}\mathbf{P}^*[\mathbf{x},\mathbf{y}(\xi)]\mathbf{M_w}(\xi)\mathbf{J}(\xi)d\xi + \\ &-\sum_{e=1}^{ne}\sum_{w=1}^{W_e}(\mathbf{p})_\mathbf{e}^\mathbf{w}\int_{\Gamma_e}\mathbf{U}^*[\mathbf{x},\mathbf{y}(\xi)]\mathbf{M_w}(\xi)\mathbf{J}(\xi)d\xi + \mathbf{B}(\mathbf{x}) \end{aligned}. \tag{3.62}$$

$\mathbf{B}(\mathbf{x})$ is present in the case of body loads and it is the only integral over the area of body. In many problems, $\mathbf{B}(\mathbf{x})$ vanishes or can also be expressed as a boundary integral:

$$\mathbf{B}(\mathbf{x}) = \int_\Omega \mathbf{U}^*(\mathbf{x},\mathbf{y})\mathbf{b}(\mathbf{y})\mathrm{d}\Omega(\mathbf{y}). \tag{3.63}$$

The integral over boundary element is understood as:

$$\int_{\Gamma_e} [\bullet] d\xi = \int_{node(1)}^{node(2)} [\bullet] d\xi. \tag{3.64}$$

Equation (3.62) can be expressed in the matrix form:

$$\mathbf{Hu} = \mathbf{Gp} + \mathbf{B}, \tag{3.65}$$

where matrices contain the values from integrals. Taking into account boundary conditions, the equation (3.65) is converted to the form of a system of the algebraic equation:

$$\mathbf{AX} = \mathbf{F}, \tag{3.66}$$

where matrix **A** contains a part of the values from matrices **H** and **G**, **X** contains unknown displacements and tractions, and vector **F** contains known values of displacements and tractions multiplied by part of the **H** and **G** matrices.

3.1.2.2 Coupling FEM and BEM

It is possible to use the advantages of BEM and FEM by coupling both methods. FEM is an efficient approach for linear and nonlinear problems, but it needs mesh generation for the interior of the structure. In the case of structures with infinite or semi-infinite volume, FEM is very inconvenient.

The application of FEM in such a case is complicated. It is possible to add an artificial boundary in the structure and treat it as a structure with finite boundary, or specialized finite elements can be proposed. BEM treats infinite structures by simply defining only interior boundaries. The boundary of the structure is meshed with the use of boundary elements if the problem is linear. The coupled BEM and FEM is a useful tool for infinite structures with local nonlinearities.

In such a case, the infinite domain is modelled using boundary elements. The structures near the areas with nonlinearities are modelled by using finite elements [9].

The body is discretized into finite elements and boundary element regions (Figure 3.9). Ω_1 denotes region discretized with finite elements, and Ω_2 is the region discretized by boundary elements. The regions discretized using finite elements can contain nonlinearities (e.g. plastic strains). The common nodes are placed on the common boundary between finite and boundary elements regions.

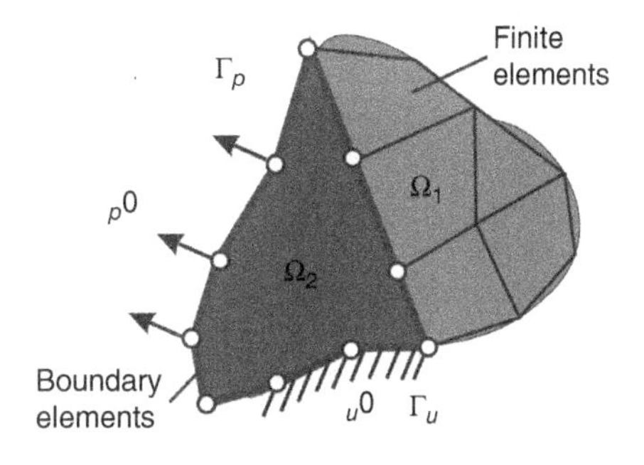

Figure 3.9 A body discretized using finite and boundary elements.

The boundary integral equation for BEM region takes the form:

$$\mathbf{cu} = \int_{\Gamma} \mathbf{U}^{*}\mathbf{p}d\Gamma - \int_{\Gamma} \mathbf{P}^{*}\mathbf{u}d\Gamma, \tag{3.67}$$

where: $\mathbf{u}$, $\mathbf{p}$ – displacement and traction vectors, respectively, $\mathbf{U}^{*}$, $\mathbf{P}^{*}$ – fundamental solutions, $\mathbf{c}$ – a coefficient which depends on the boundary smoothness.

The boundary integral Eq. (3.67) after discretization is expressed as follows:

$$\mathbf{Hu} = \mathbf{Gp}. \tag{3.68}$$

Tractions $\mathbf{p}$ are converted into nodal forces by multiplying the Eq. (3.68) by shape function matrix $\mathbf{M}$:

$$\mathbf{MG}^{-1}\mathbf{Hu} = \mathbf{Mp}. \tag{3.69}$$

Equation (3.69) is transformed to the form:

$$\mathbf{K}'\mathbf{u} = \mathbf{f}', \tag{3.70}$$

where $\mathbf{f}' = \mathbf{Mp}$ is the vector representing external forces and $\mathbf{K}' = \mathbf{MG}^{-1}\mathbf{H}$ is treated as FEM stiffness matrix.

Because of the presence of nonlinearities in the finite element region, the iterative method should be used to solve the problem.

The structure is divided into finite element region near the interior hole, and the boundary element models the infinite structure (Figure 3.10). The boundary elements are located on the outer boundary of the finite elements region.

3.1.3 Computational Homogenization

Solids have generally non-homogeneous structure. Multiscale modelling as a science concept allowing consideration of the material structure at different levels enables to analyze systems in various scales taking into account coupling between scales (see Figure 1.1).

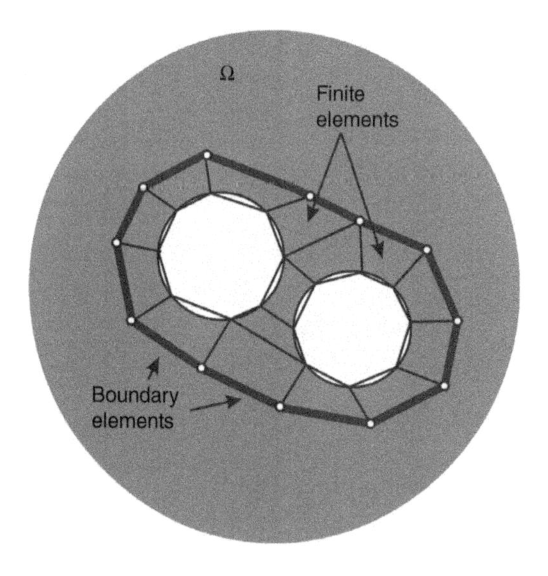

Figure 3.10 The infinite body discretized using finite and boundary elements.

The primary purpose of the homogenization is to find a homogeneous model for the non-homogeneous medium and thus to determine the relationship between the macroscopic quantities, in practice the mean values, and the efficient material constants. Figure 3.11 shows an exemplary body under thermal and mechanical loads before and after homogenization.

The relationship between model behaviour in different scales is determined by physical laws corresponding to those scales. That relationship is described by the material constitutive law. Homogenization constitutes a part of multiscale calculations. It allows obtaining macroscopically equivalent homogeneous material properties (effective properties) based on the microstructure analysis. Constitutive relations are typically evaluated by relating averaged field variables. Homogenization makes it possible to determine the macroscopic response of linear and nonlinear materials. To solve a homogenization problem, analytical or numerical methods can be used.

Analytical methods are the most effective in terms of computational effort, but usually can be applied only for simplified microstructure geometry and simple materials models. There are several approaches, e.g. semi-analytical mean-field homogenization, asymptotic homogenization, and fast Fourier transform homogenization.

One of the numerical techniques which enables multiscale analysis of structures is a computational homogenization. The detailed description of the computational homogenization can be found in [38, 47, 60, 84]. Computational homogenization has the widest range of application among homogenization methods. The main idea of computational homogenization is to represent material microstructure by a statistically representative fragment of its geometry, called representative volume element (RVE) (Figure 3.12). A particular type of RVE containing one inhomogeneity is called a unit cell. The stress analysis of RVE allows obtaining

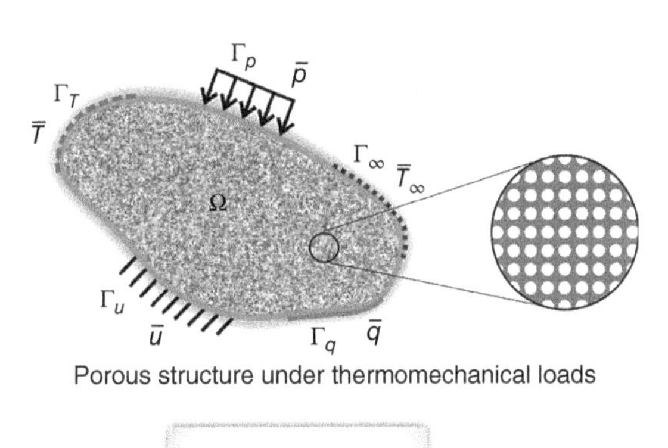

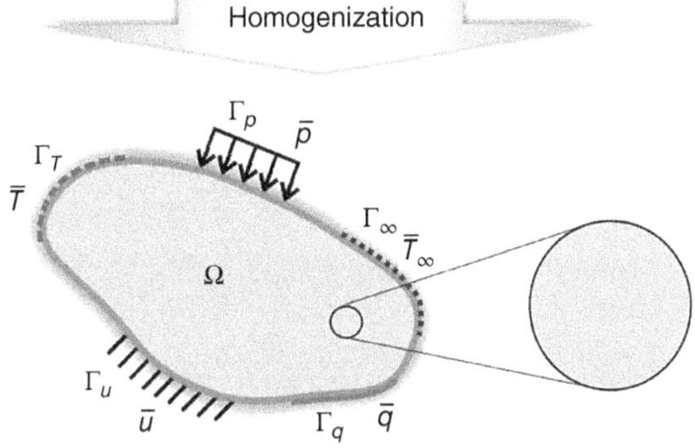

Figure 3.11 Homogenization of the non-homogeneous structure.

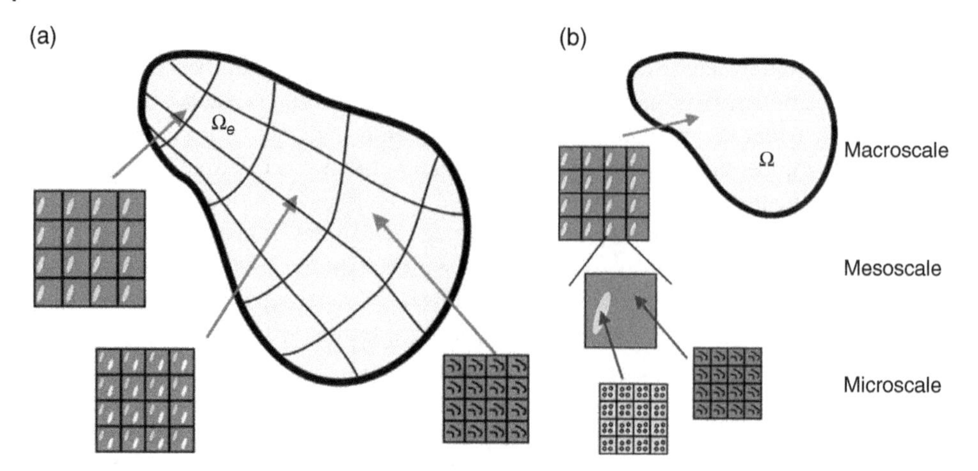

Figure 3.12 Models of a structure with locally periodical microstructures: (a) two-scale model and (b) three-scale model.

specific information about material behaviour on the macroscale. The analysis can be done by means of various numerical methods, like the finite element method [125, 126] or boundary element method [16, 19].

It is typically assumed that inhomogeneous material consists of two or more homogeneous phases, which satisfy the laws of continuum mechanics. Structures with a local periodicity are considered. The local periodicity means there are areas of structure with the same microstructure. The example of such a structure is presented in Figure 3.12a. The microstructures can also be built from lower scale, locally periodic microstructures like in Figure 3.12b. The goal of the computational homogenization is analysis of the structure taking into account the local periodicity of microstructures. The main advantage of the computational homogenization is analysis in few scales which allow to use models with at least a few orders of degrees of freedom lower than the model created in one scale.

Generally, for the thermoelasticity problems, the material constants may include elastic constants, thermal conductivity constants, and thermal expansion coefficients. The computational homogenization idea for thermoelasticity is presented below. The linear thermoelasticity problem is described by differential equation of heat conduction and elasticity taking into account thermal strains:

$$kT_{,ii} = 0 \tag{3.71}$$

$$\mu u_{i,jj} + (\mu + \lambda)u_{j,ji} - (3\lambda + 2\mu)\alpha_T T_{,i} = 0, \tag{3.72}$$

where: k – thermal conductivity, T – temperature, u – displacement, α_T – linear expansion coefficient, μ, λ – Lame's constants.

These equations have to be supplemented by mechanical and thermal boundary conditions:

$$\Gamma_t : t_i = \overline{t_i} \quad \Gamma_u = u_i = \overline{u_i} \tag{3.73}$$

$$\Gamma_T : T_i = \overline{T_i}, \quad \Gamma_q : q_i = \overline{q_i}, \quad \Gamma_c : q_i = \alpha(T_i - T^{\infty}), \tag{3.74}$$

where: $\overline{u}_i, \overline{t}_i, \overline{T}_i, \overline{q}_i, \alpha, T^{\infty}$ – known: displacements, tractions, temperatures, heat fluxes, heat conduction coefficient, and ambient temperature, respectively, $\Gamma_t, \Gamma_u, \Gamma_T, \Gamma_q, \Gamma_c$, – parts of the boundary, where mechanical and thermal boundary conditions are specified (Figure 3.11).

An application of numerical homogenization using RVE takes into consideration the following principles:

- The separation of scales

$$\frac{l}{L} << 1, \tag{3.75}$$

where: l, L are characteristic lengths of the structure for RVE and macroscale, respectively.

- Volume averaging (avg.) is done accordingly as follows:

$$\langle\cdot\rangle = \frac{1}{|\Omega_{RVE}|}\iiint(\cdot)d\Omega_{RVE}, \tag{3.76}$$

where $\langle\cdot\rangle$ denotes the average of a given field over the volume Ω_RVE of the RVE

- The condition for equivalence of energetically and mechanically defined effective properties of heterogeneous materials (Hill's condition):

$$\langle\sigma_{ij}\varepsilon_{ij}\rangle = \langle\sigma_{ij}\rangle\langle\varepsilon_{ij}\rangle, \tag{3.77}$$

where: σ_{ij}, ε_{ij} – stress and strain tensors, respectively.

For the heat conduction problem, the Hill condition takes the form:

$$\langle T_{,i}q_i\rangle = \langle T_{,i}\rangle\langle q_i\rangle, \tag{3.78}$$

where: $T_{,i}$, q_i – temperature gradient and heat flux, respectively.

For the elasticity and heat conduction problems in microscale and macroscale, the numerical homogenization by means of FEM is used in the book.

Periodic boundary conditions are applied for the RVE. After FEM analysis for the RVE, average stresses and heat fluxes are used to calculate effective constants' values, according to relation (6). Constitutive equation and Fourier's law in the microscale take the form:

$$\langle\sigma_{ij}\rangle = c'_{ijkl}\langle\varepsilon_{ij}\rangle \tag{3.79}$$

$$\langle q_i\rangle = -k'_{ij}\langle T_{,i}\rangle. \tag{3.80}$$

The tensor of elastic constants c'_{ijkl} is symmetric and has nine independent material constants, for the 3D RVE:

$$c'_{ij} = \begin{bmatrix} c_{11} & c_{12} & c_{13} & 0 & 0 & 0 \\ c_{21} & c_{22} & c_{23} & 0 & 0 & 0 \\ c_{31} & c_{32} & c_{33} & 0 & 0 & 0 \\ 0 & 0 & 0 & c_{44} & 0 & 0 \\ 0 & 0 & 0 & 0 & c_{55} & 0 \\ 0 & 0 & 0 & 0 & 0 & c_{66} \end{bmatrix}. \tag{3.81}$$

For anisotropic non-crystalline materials, the tensor of heat conduction coefficients takes the form:

$$k'_{ij} = \begin{bmatrix} k_{11} & 0 & 0 \\ 0 & k_{22} & 0 \\ 0 & 0 & k_{33} \end{bmatrix}. \tag{3.82}$$

In order to calculate nine independent elastic constants, six analyses of linear elasticity are needed. Each column (or row) of the tensor c'_{ij} is obtained by applying a consecutive initial strain to the RVE model. Respectively, three analyses of the heat conduction problem have to be performed in order to calculate all coefficients of k'_{ij}. For the 2D analysis, numerical homogenization consists of three analyses of the elasticity and two analyses of the heat conduction problem. The microscale and macroscale are typically connected by stress/strain state at each point of the structure.

The material parameters for each integration point in finite elements depend on the solution of a RVE in the lower scale. The RVE is a model of the microstructure, voids, and inclusions, and other properties of microstructure can be included in the model. The RVE is in most cases modelled as a cube or a square. A numerical method like FEM is used to solve the boundary value problem for RVE. The periodic displacement boundary conditions are taken into account. The strains from the higher level are prescribed as additional boundary conditions. The RVE for each integration point of the higher level model must be created and stored for the next iteration steps if the nonlinear problem with plasticity is considered. The transfer of information both from lower to higher and higher to lower scales is needed in most cases. The one-way transfer of results (from lower to higher scales) is possible if the linear problem is considered. The transfer of average strains and stresses between scales is shown in Figure 3.13.

The material parameters for the higher scale are obtained on the basis of solving a few direct problems for RVE in the lower scale. The homogenized material parameters depend on average stress values in RVE obtained after applying average strains to RVE. The stress–strain relation obtained using RVE is used in the higher level model. The average strains are strains in the integration point from the higher level.

A scheme of the computational homogenization is presented in Figure 3.13.

Validation of the developed procedures of numerical homogenization which are used in this book is presented in [33].

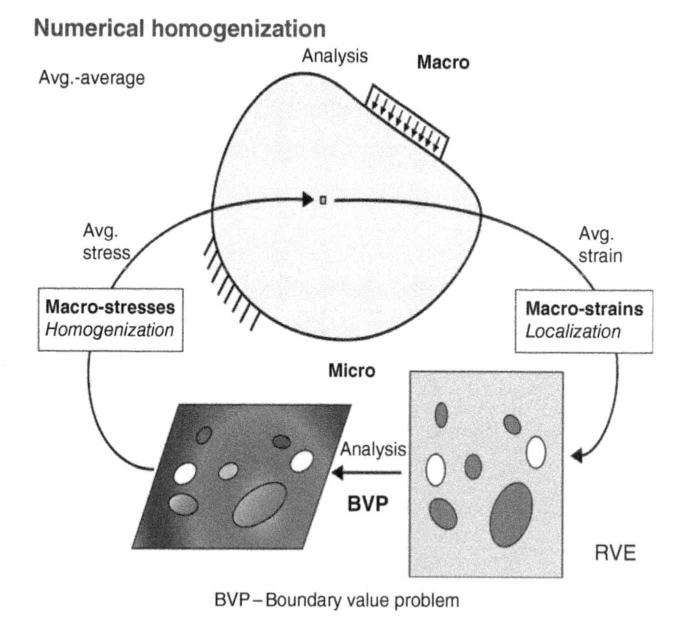

Figure 3.13 Computational homogenization scheme: the average strain and stress transfer between scales.

3.2 Computational Methods for Nano and Micro

3.2.1 Classical Molecular Dynamics

The development of molecular simulation methods is related to the development and increasing power of computers. First documented computer simulations using the molecular dynamics (MD) method were performed in the late 1950s [1, 2]. The Lennard-Jones potential was used in MD for the first time by Rahman [98] during the investigation of the motion of atoms in the liquid argon. In 1967, Verlet introduced neighbour lists and time integration algorithm, which became essential tools even in modern applications of MD methods [115]. Further development of MD results with simulations at constant temperature or pressure [3, 10] and introduction of the many-body interaction models such as embedded atom method [29]. Nowadays, the computational power of modern computers allows performing MD simulations with 10^{11} atoms which provide information not only about fundamental microscopic structural properties but also macroscopic quantities such as stresses, energies, and heat capacity.

As an alternative approach to the atomistic simulations, the molecular statics (MS) method is presented in Section 3.2.2. The proposed algorithm [81] is based on the equations of equilibrium of the interatomic forces [65] and describes the behaviour of the atomic system at zero temperature. Similar methods of equilibration-minimization of the potential energy, usually based on the conjugate gradient or 'damped dynamics' [12, 106] techniques, are built into many commercial or scientific MD software packages. Another MS method adapts the FEM framework: the finite element truss model serves as an atomistic lattice, and the rod elements have nonlinear characteristics given by atomic potential [34].

This work does not cover all aspects of the molecular simulations, only the basics necessary to understand the methods that are presented and explained. More detailed descriptions, e.g. in terms of statistical mechanics, software implementation, and parallelization of the computations, can be found in books dedicated to MD such as [42, 51, 54, 55, 99].

3.2.1.1 Equations of Motion

A molecular system with N atoms placed in the domain Ω is taken into consideration. All atoms are characterized by the masses $\{m_1,...m_N\}$ and coordinates $\{\mathbf{r}_1,...\mathbf{r}_N\}$. The corresponding velocities $\{\mathbf{v}_1,...\mathbf{v}_N\}$ and momenta $\{\mathbf{p}_1,...\mathbf{p}_N\}$ are defined respectively:

$$\mathbf{v}_i = \dot{\mathbf{r}}_i, \tag{3.83}$$

$$\mathbf{p}_i = m_i\dot{\mathbf{r}}_i = m_i\mathbf{v}_i \qquad i = 1, ..., N. \tag{3.84}$$

Note that $\mathbf{r}_i$, $\mathbf{v}_i$, and $\mathbf{p}_i$ are the time-dependent, d-element vectors, where d is the dimension of the investigated problem. For convenience, the 'classical' coordinates of the atoms will be denoted now as $\{\mathbf{r}_1,...\mathbf{r}_N\}$ instead of notation $\{\mathbf{R}_1,...\mathbf{R}_N\}$ used in Section 2.1. The dot over symbol refers to the derivative with respect to time. The domain Ω, in the MD method called *simulation box*, is usually cubic or rectangular, i.e. a domain with lengths of sides a_1, a_2, a_3 is defined as $\Omega = [0, a_1]\times[0, a_2]\times[0, a_3]$ or $\Omega = [0, a_1]\times[0, a_2]$ in three- or two-dimensional cases, respectively.

The behaviour of the atomic system in the domain Ω can be described by the classical equations of motion. The formulation of Hamilton's equations follows:

$$\dot{\mathbf{r}}_i = \frac{\partial H}{\partial \mathbf{p}_i}, \tag{3.85}$$

$$\dot{\mathbf{p}}_i = -\frac{\partial H}{\partial \mathbf{r}_i}, \quad i = 1, \ldots, N. \tag{3.86}$$

Hamiltonian H is a sum of an atom's kinetic (E_K) and potential energy (V) operators:

$$H = E_K + V. \tag{3.87}$$

The potential energy $V(\mathbf{r}_i, \ldots \mathbf{r}_N)$ is obtained by computing one of the atomic interaction models (e.g. Lennard-Jones potential; see Section 2.1) at the instantaneous positions of the atoms. The kinetic energy can be evaluated by the classical formulation:

$$E_K = \sum_{i=1}^{N} \frac{\mathbf{p}_i^2}{2m_i} = \sum_{i=1}^{N} \frac{m_i \mathbf{v}_i^2}{2}. \tag{3.88}$$

Assuming thermal and mechanical isolation of the considered system, the total energy of such an atomic system is constant, i.e. $dH/dt = 0$; so, the total energy is conserved over time. This leads to the statistical *microcanonical ensemble* – NVE, where a number of atoms N, the volume of the system V, and the total energy E remain constant (see Section 3.2.1.3).

Newton's equations of motion are obtained directly from Hamilton's formulation (3.86):

$$m_i \dot{\mathbf{v}}_i = \mathbf{f}_i, \tag{3.89}$$

or

$$m_i \ddot{\mathbf{r}}_i = \mathbf{f}_i. \tag{3.90}$$

Atomic interaction forces $\mathbf{f}_i$ can be computed as a derivative of the potential energy V with respect to $\mathbf{r}_i$ taken with a minus sign:

$$\mathbf{f}_i = -\frac{\partial V(\mathbf{r}_1, \ldots, \mathbf{r}_N)}{\partial \mathbf{r}_i}. \tag{3.91}$$

Finally, the equations of motion can be written in a similar way, as derived in Section 2.1:

$$m_i \ddot{\mathbf{r}}_i = -\nabla_i V(\mathbf{r}_1, \ldots, \mathbf{r}_N). \tag{3.92}$$

Hamilton's and Newton's equations of motion are mathematically equal. However, the algorithms of solving differential equations commonly used in the MD method are based on Newton's version.

3.2.1.2 Discretization of Equations of Motion

The system of the second-order differential equations, in the general case, cannot be solved analytically. To evaluate the approximate solution of the MD Eq. (3.92), the effective specialized algorithms were developed, and the two most popular of them will be described here in detail. However, certain operations should be performed to prepare the system of equations of motion for numerical solving: the discretization of the simulation's time interval and the derivation of proper approximation schemes for the first and the second derivatives.

The time interval of the simulation t=[t_{start}, t_{end}] has to be discretized into k time steps of length:

$$\Delta t = \frac{t_{end} - t_{start}}{k}. \tag{3.93}$$

As a consequence, a set of time points is created:

$$t_k = n\cdot\Delta t, \quad n = 0, \ldots, k. \tag{3.94}$$

The numerical solution of the equations of motion will be evaluated only at these time points.

The first-order derivative $d\mathbf{r}_i/dt$ at the time point t_n can be approximated using a Taylor expansion of the position variable $\mathbf{r}_i$ of the i-th atom. The previous and the next time points are taken into account:

$$\mathbf{r}_i(t_n + \Delta t) = \mathbf{r}_i(t_n) + \Delta t\dot{\mathbf{r}}_i(t_n) + O(\Delta t^2), \tag{3.95}$$

$$\mathbf{r}_i(t_n - \Delta t) = \mathbf{r}_i(t_n) - \Delta t\dot{\mathbf{r}}_i(t_n) + O(\Delta t^2). \tag{3.96}$$

Subtracting these equations side by side yields the central difference operator:

$$\dot{\mathbf{r}}_i(t_n) = \frac{\mathbf{r}_i(t_n + \Delta t) - \mathbf{r}_i(t_n - \Delta t)}{2\Delta t} + O(\Delta t^2), \tag{3.97}$$

with a discretization error of the order $O(\Delta t^2)$ for the approximation of the first derivative.

The second-order differential operator $d^2\mathbf{r}_i/dt^2$ at time t_n can be evaluated similarly, but the Taylor expansion up to the third-order should be considered:

$$\mathbf{r}_i(t_n + \Delta t) = \mathbf{r}_i(t_n) + \Delta t\dot{\mathbf{r}}_i(t_n) + \frac{1}{2}\Delta t^2\ddot{\mathbf{r}}_i(t_n) + \frac{1}{6}\Delta t^3\dddot{\mathbf{r}}_i(t_n) + O(\Delta t^4), \tag{3.98}$$

$$\mathbf{r}_i(t_n - \Delta t) = \mathbf{r}_i(t_n) - \Delta t\dot{\mathbf{r}}_i(t_n) + \frac{1}{2}\Delta t^2\ddot{\mathbf{r}}_i(t_n) - \frac{1}{6}\Delta t^3\dddot{\mathbf{r}}_i(t_n) + O(\Delta t^4). \tag{3.99}$$

Equations (3.96) and (3.97) added together give an approximation of the second derivative of the $\mathbf{r}_i$, also with an error of the order $O(\Delta t^2)$:

$$\ddot{\mathbf{r}}_i(t_n) = \frac{(\mathbf{r}_i(t_n + \Delta t) - 2\mathbf{r}_i(t_n) + \mathbf{r}_i(t_n - \Delta t))}{\Delta t^2} + O(\Delta t^2). \tag{3.100}$$

One of the most efficient and popular approaches of time discretization of Newton's equations, widely implemented in MD solvers, is the Verlet [115] algorithm and its variants: the velocity-Verlet and leapfrog schemes [36, 51, 54, 55]. All these integration algorithms use finite difference operators (3.97) and (3.100) for computing the first and second derivatives and have an accuracy of the order $O(\Delta t^2)$. However, each of them has a slightly different sensibility to rounding errors and requires specific initial conditions. The Verlet and the velocity-Verlet integration schemes will be presented in detail.

The standard Verlet algorithm can be derived by substitution of the differential term in Newton's Eq. (3.90) by the second-order finite-difference operator (3.100):

$$\frac{m_i}{\Delta t^2}\left(\mathbf{r}_i^{n+1} - 2\mathbf{r}_i^n + \mathbf{r}_i^{n-1}\right) = \mathbf{f}_i^n. \tag{3.101}$$

The abbreviations $\mathbf{r}_i^n, \mathbf{r}_i^{n-1}, \mathbf{r}_i^{n+1}$ refer to the positions of i-th molecule at the current (t_n), previous (t_n–Δt), and next (t_n–Δt) time steps, respectively. The same convention will be used for forces and velocities. The updated positions of the atoms at the next time step $\mathbf{r}_i(t_n+\Delta t)$ can be computed:

$$\mathbf{r}_i^{n+1} = 2\mathbf{r}_i^n - \mathbf{r}_i^{n-1} + \mathbf{f}_i^n\Delta t^2 m_i^{-1}. \tag{3.102}$$

This algorithm of solving equations of motion is not a self-starting method. The initial positions of two sets of atoms $\mathbf{r}^0$ and $\mathbf{r}^1$ at the time points ($t_0 = t_{start}$) and t_1, respectively, are required at the start. All the subsequent trajectories can be estimated using Eq. (3.102). The interatomic forces $\mathbf{f}_i$ can be evaluated analytically or numerically using the formula in Eq. (3.91).

If the velocities of the atoms are needed to compute thermodynamic quantities (e.g. kinetic energy or temperature), they can be obtained using the central difference operator (3.97):

$$\mathbf{v}_i^n = \frac{\mathbf{r}_i^{n+1} - \mathbf{r}_i^{n-1}}{2\Delta t}. \tag{3.103}$$

Unfortunately, the computed velocities are always one time step behind the atom's positions due to such formulation. As a consequence, in the classical Verlet algorithm, the estimation of the kinetic energy is always one step delayed to the estimation of the potential energy. Another drawback of this method is possible rounding errors: in the Eq. (3.102) small term $\mathbf{F}_i^n \Delta t^2 m_i^{-1}$ of the second-order $O(\Delta t^2)$ is added to other, much larger terms of order $O(1)$.

To overcome the previously mentioned problems, the velocity-Verlet scheme was introduced. This variant may be formulated by solving the velocity difference operator (3.103) for $\mathbf{r}_i^{n-1}$:

$$\mathbf{r}_i^{n-1} = \mathbf{r}_i^{n+1} - 2\Delta t \mathbf{v}_i^n, \tag{3.104}$$

and putting Eq. (3.104) into Eq. (3.103). After rearranging, the following algorithm for updating the atomic positions can be obtained:

$$\mathbf{r}_i^{n+1} = \mathbf{r}_i^n - \Delta t \mathbf{v}_i^n + \frac{1}{2}\mathbf{F}_i^n \Delta t^2 m_i^{-1}. \tag{3.105}$$

Inserting Eq. (3.104) into Eq. (3.105) yields the following form:

$$\mathbf{v}_i^n = \left(\mathbf{r}_i^n - \mathbf{r}_i^{n-1}\right)\Delta t^{-1} + \frac{1}{2}\mathbf{F}_i^n \Delta t m_i^{-1}, \tag{3.106}$$

and analogical expression for $\mathbf{v}_i^{n+1}$ can be written:

$$\mathbf{v}_i^{n+1} = \left(\mathbf{r}_i^{n+1} - \mathbf{r}_i^n\right)\Delta t^{-1} + \frac{1}{2}\mathbf{F}_i^{n+1} \Delta t m_i^{-1}. \tag{3.107}$$

After adding these two equations side by side, the schemata for velocities can be developed:

$$\mathbf{v}_i^{n+1} = \mathbf{v}_i^n + \frac{1}{2}\left(\mathbf{F}_i^n + \mathbf{F}_i^{n+1}\right) m_i^{-1} \Delta t. \tag{3.108}$$

Equations (3.105) and (3.108) form the velocity-Verlet algorithm. Although the computer implementation of this scheme requires an extra array to store partial results, the velocity-Verlet algorithm is widely used in the MD method due to its advantages. This variant is not sensitive to the rounding errors, and the atomic positions and velocities are estimated during one time step. Additionally, the necessary initial conditions, atom's initial positions $\mathbf{r}^0$ and velocities $\mathbf{v}^0$ at a time ($t_0 = t_{start}$), are, in practice, more convenient than two subsequent sets of initial positions, required in the standard Verlet method. Note that initial positions and velocities determine the amounts of potential and kinetic energy at the start, respectively.

3.2.1.3 Temperature Controller

Previously, fully isolated atomic systems, where energy conservation condition is satisfied, were taken into consideration. For the Hamiltonian given in the Eq. (3.87), this may be proven:

$$\begin{aligned}\frac{dH}{dt} &= \frac{dE_K}{dr} + \frac{dV}{dt} = \sum_{i=1}^{N} m_i \mathbf{v}_i \dot{\mathbf{v}}_i + \frac{\partial V}{\partial t} + \sum_{i=1}^{N} \frac{\partial V}{\partial \mathbf{r}_i}\frac{\partial \mathbf{r}_i}{\partial t} = \\ &= \sum_{i=1}^{N} m_i \dot{\mathbf{v}}_i \mathbf{v}_i - \sum_{i=1}^{N} \mathbf{f}_i \mathbf{v}_i = 0\end{aligned} \quad . \tag{3.109}$$

In statistical mechanics, such a system represents, as already mentioned, a *microcanonical ensemble – NVE*, where the number of atoms and the volume are fixed, and atoms move on the constant energy hypersurface. However, in many practical problems, the energy or the temperature of the atomic system should be regulated over time. For example, the modelling of a phase transition requires precise temperature control since the proper speed of cooling or melting is crucial and may affect the quality of the simulation. Also, the investigation of some phenomena, such as crack growth or fracture, usually demands maintaining a constant temperature rather than constant energy.

Several methods have been developed for controlling the temperature in the MD method. The simplest approach is based on the *NVE* ensemble and velocity scaling. Other, more sophisticated methods use *canonical ensemble (NVT)* with thermostat. In such an approach, the number of atoms, volume, and temperature are constant, and the atomic system is coupled to a 'heat bath', which imposes the required temperature. Popular thermostats have been proposed by Andersen, Berendsen, and Nosé-Hoover [42, 51]. See original publications [3, 10, 57, 83] for a more detailed description. Another approach is the application of the *NPT* ensemble in which the number of atoms, pressure, and temperature are kept constant, but the simulation domain can change. This section will discuss the velocity scaling algorithm and the NVT ensemble with the Nosé-Hoover thermostat in more detail. An example of MD simulations showing the differences between the NVT and NPT approaches is presented at the end of this section.

For the three-dimensional N-atom system, the relation between kinetic energy E_K and temperature T is given by the well-known equipartition theorem of thermodynamics:

$$E_K = \frac{3N}{2} k_B T. \tag{3.110}$$

Here, k_B is the Boltzmann constant, and the term $3N$ denotes the total number of degrees of freedom in the considered system. Note that this number refers to atomic scaling. In the case of periodic or non-periodic scaling and when additional constraints are applied, three or more degrees of freedom should be subtracted [51].

Taking into account Eq. (3.88), the temperature can be estimated as follows:

$$T = \frac{\sum_{i=1}^{N} m_i \mathbf{v}_i^2}{3N k_B}. \tag{3.111}$$

The simplest algorithm of temperature control in the *NVE* ensemble is the scaling of the velocities of all atoms at certain time intervals by the factor [51]:

$$\beta = \sqrt{\frac{E_K^{Desired}}{E_K}} = \sqrt{\frac{T^{Desired}}{T}}, \tag{3.112}$$

where: $T^{Desired}$ is the target temperature of the system, and T denotes the current temperature, computed using (3.111).

Such an approach is based on the artificial modification of the kinetic energy of the atomic system; therefore, in the computer implementation, velocities are scaled by factor β (i.e. $\mathbf{v}_i^n = \beta \mathbf{v}_i^n$) only every certain number (usually constant) of time steps. Between subsequent scaling procedures, the equations of motion are integrated normally, according to the, e.g. one of the Verlet schemes, so the proper ratio of the potential to the kinetic energy can be restored, and the atomic system can be equilibrated. The scaling procedure is repeated until the temperature reaches the desired value.

However, depending on the values of the current and desired temperature, this algorithm may drastically affect the distribution of the energy and motion of the atomic system. To avoid this, the weighting parameter γ is introduced to the scaling term [10]:

$$\beta = \sqrt{1 + \gamma\left(\frac{T^{Desired}}{T} - 1\right)}. \tag{3.113}$$

For γ=0, the velocity scaling is disabled; for γ=1, the standard formulation (3.112) is obtained. The value of the weighting parameter γ is often chosen as proportional to the time step Δt. In this case, the temperature changes in time proportionally to the difference between T and $T^{Desired}$.

The different, more sophisticated method of controlling temperature, which does not strongly affect the energy and motion of atoms, is the simulation of an atomic system coupled with a heat bath or thermostat. Conceptually, this approach mimics a physical experiment, where a small body immersed in the significantly larger system called a heat bath exchanges heat with it. The heat bath temperature is assumed to be fixed because the influence of the small body on the temperature of the heat bath can be neglected. After some time, the temperature of the small body reaches the temperature of the heat bath, so the heat exchange stops, and the whole system is equilibrated.

In MD, the simulation of the same phenomena can be performed using *canonical ensemble – NVT*. The coupling with a heat bath is introduced by the additional damping term in Newton's equations of motion:

$$m_i \dot{\mathbf{v}}_i = \mathbf{F}_i - \xi(t) m_i \mathbf{v}_i. \tag{3.114}$$

The damping force acting on the i-th atom is proportional to its velocity. The function ξ is time dependent and determines the speed of the changes in the temperature. Due to the above modification, the atomic system is able to lose ($\xi > 0$) or gain ($\xi < 0$) kinetic energy until it reaches the target temperature.

The derivation of the velocity-Verlet algorithm for modified Newton's Eq. (3.114) is the same as in the case of the NVE ensemble and leads to the following integration scheme:

$$\mathbf{r}_i^{n+1} = \mathbf{r}_i^n + \Delta t\left(1 - \frac{1}{2}\Delta t \xi^n\right)\mathbf{v}_i^n + \frac{1}{2}\mathbf{F}_i^n \Delta t^2 m_i^{-1}, \tag{3.115}$$

$$\mathbf{v}_i^{n+1} = \left(1 + \frac{\Delta t}{2}\xi^{n+1}\right)^{-1}\left[\left(1 - \frac{\Delta t}{2}\xi^n\right)\mathbf{v}_i^n + \frac{1}{2}\left(\mathbf{F}_i^n + \mathbf{F}_i^{n+1}\right)\Delta t m_i^{-1}\right]. \tag{3.116}$$

This scheme is also similar to the one described earlier; however, the proper form of the damping function ξ should be chosen. One of the possible formulations is the Nosé-Hoover thermostat [83]. In this approach, the heat bath is integrated with the atomic system and is treated as the next degree of freedom. The differential equation describes damping in the system:

$$\frac{d\xi}{dt} = \left(\sum_{i=1}^{N} m_i \mathbf{v}_i^2 - 3Nk_B T^{Desired}\right) M^{-1}, \tag{3.117}$$

and the value of damping function ξ^{n+1}, in the simplest way, can be approximated as follows:

$$\xi^{n+1} \approx \xi^n + \left(\sum_{i=1}^{N} m_i \mathbf{v}_i^2 - 3Nk_B T^{Desired}\right) \Delta t M^{-1}. \tag{3.118}$$

The factor M refers to the coupling between the atomic system and the heat bath and has to be adjusted carefully. If it is too small, the coupling is strong, and the temperature may fluctuate. In the opposite case, it takes a long time to equilibrate the system at the desired temperature.

The practical difference between the results obtained during MD computations with the canonical and NPT approaches can be clearly demonstrated by the simulation of the tensile test of the two-dimensional material, like graphene or single-layered MoS_2 ($SLMoS_2$; please refer to Section 4.1.1.6 for more information). Generally, the MD simulations of such flat materials can be performed using both the NVE [80, 120] and NPT ensembles [78]. However, the behaviour of atomic lattice in the direction perpendicular to the plane of the material is different [79]. In the case of the NVE dynamics, propagation and reflection of a wave in continuously strained flat material can excite atoms to vibrate in the previously mentioned third dimension (compare it with wave propagation in a pure 2D example presented in Section 3.2.1.7). Application of the NPT ensemble allows to keep the pressure constant by changing the dimensions, thus the volume of the simulation box. Such an approach minimizes the waving effect of the structure, especially when the simulation is performed at higher temperatures. This phenomenon and the rupture of the $SLMoS_2$ sheet under tensile load can be observed in Figure 3.14.

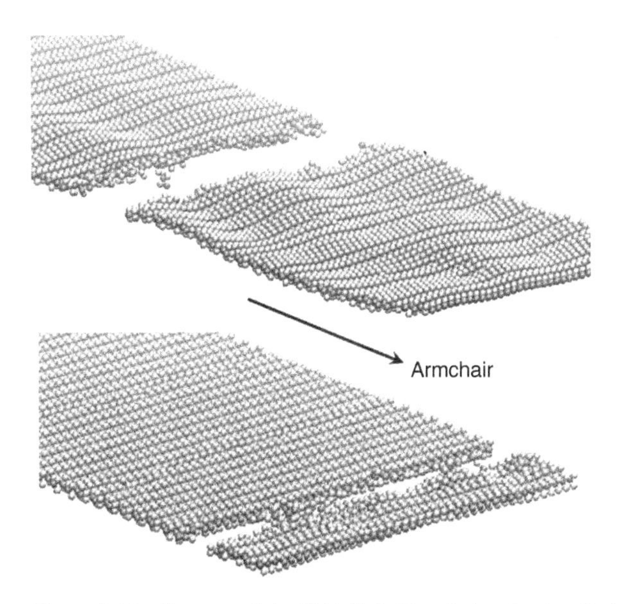

Figure 3.14 Rupture of the $SLMoS_2$ lattice during the tensile test along the armchair direction. Top – NVT, bottom – NPT ensemble [79].

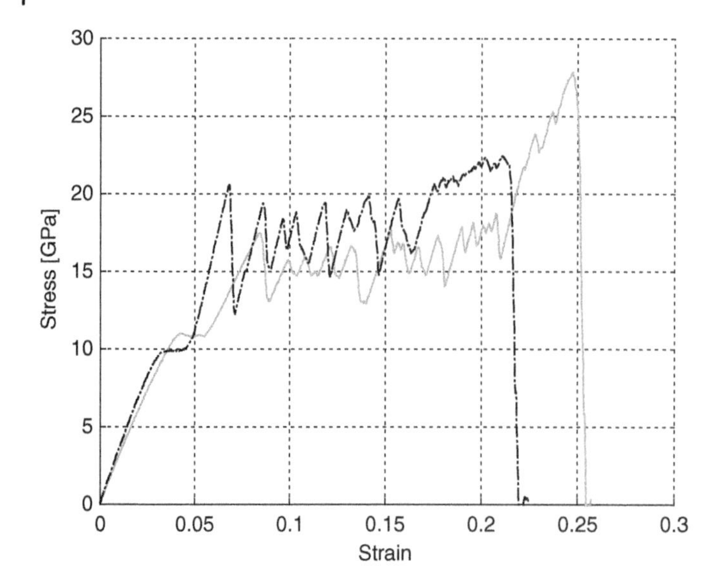

Figure 3.15 Stress–strain relation of the SLMoS_2 lattice obtained in armchair direction: dashed black – NVT; solid green – NPT ensemble.

In both cases, the molecular models and all the conditions of the simulation, including the boundary conditions, temperature, time step, strain rate, and the ReaxFF potential, were the same. The only exception was the pressure damping coefficient in the NPT dynamics. As can be seen from the obtained stress–strain curves (Figure 3.15), the type of the thermodynamic environment practically had no influence on the behaviour of the structure in the linear area, but the slips and the moment of the final failure were accelerated by the vibrations of atoms in the case of the NVT simulation.

3.2.1.4 Evaluation of the Time Step

In the MD simulations, the choice of the optimal time step is a non-trivial problem [24, 36]. The quality of the simulation and the computation cost depends on the value of the time step. Generally, two contradictory conditions have to be considered and satisfied: on the one hand, the time step should be short enough that computed trajectories of atoms mimic the real physical behaviour. On the other hand, the time interval should be as long as possible that the total number of iterations of the integration algorithm will be reduced, and the efficiency of simulation will be increased.

The time step of the MD simulation can be roughly estimated, for a given type of atomic potential, by calculating the Debye frequency. The Debye frequency refers to the theoretical maximum frequency of vibration of the atoms in the crystal's lattice. Chen et al. [23] recommend that the proper time step in the Verlet integration algorithm should be ten to hundred times shorter than the period of the atom's vibrations.

The desired value of the atom's oscillation frequency can be obtained in the following way [36]. Consider two atoms with equal masses, vibrating with the harmonic motion close to the equilibrium distance r_0 in a one-dimensional space. The amount of potential energy of the pair of atoms in harmonic motion can be evaluated using the harmonic (parabolic) potential function:

$$V^H(r_{ij}) = \frac{1}{2}k(r_{ij} - r_0)^2. \tag{3.119}$$

The parameter k is related to the atomic stiffness and can be determined by comparing the second derivative of the harmonic potential (3.119) with the second derivative of specific atomic potential at the equilibrium distances r_0. For example, in the case of Lennard-Jones potential, given in the form shown in Chapter 2 and the equilibrium distance determined by the σ parameter, the atomic stiffness equals:

$$k = \frac{72\varepsilon}{r_0^2}. \tag{3.120}$$

The harmonic force $\mathbf{F}_{ij}^H = -k\mathbf{r}_{ij}$ is obtained as a derivative of the potential with respect to the distance, and the following system of Newtonian equations of the harmonic motion for two atoms can be written:

$$\begin{cases} m\ddot{\mathbf{r}}_i = \mathbf{F}_{ij}^H \\ m\ddot{\mathbf{r}}_j = -\mathbf{F}_{ij}^H \end{cases}. \tag{3.121}$$

These are classical differential equations of the harmonic oscillator. The motion of each atom is thus described by the general solution, given in the form:

$$r_{ij} = A\cos(2\pi ft + \varphi), \tag{3.122}$$

where: the amplitude A and phase φ can be obtained from the initial conditions, and the period of the atom's vibrations T^{Deb} may be determined as:

$$T^{Deb} = 2\pi\sqrt{\frac{m}{k}} = 2\pi\sqrt{\frac{mr_0^2}{72\varepsilon}}. \tag{3.123}$$

According to [23], the time step Δt in the Verlet algorithm should be no longer than:

$$\Delta t = 0.01 \div 0.1 \cdot T^{Deb}. \tag{3.124}$$

For the Lennard-Jones potential, e.g. with the given set of parameters [111] fitted to the properties of the bulk aluminium: $\varepsilon = 0.1699$ nN•nm, $\sigma = 0.2575$ nm, and atomic mass $m_{al} = 26.982$ u, the Debye's period is equal to the 0.11 ps. Taking into account the recommendation in Eq. (3.124), the proper time step of the simulation should be of the order of several femtoseconds.

3.2.1.5 Cutoff Radius and Nearest-Neighbour Lists

One of the most time-consuming tasks in MD simulations is the computation of atomic forces. Taking into account all the pair interactions, i.e. interactions of each atom with all others, atoms in the N-atom system results with $(N^2-N)/2$ evaluations of the atomic forces in each step of integration. Such a simple approach makes a bottleneck, even for small atomic systems. Note that the same number of computations is needed when total potential energy has to be estimated. The computational cost of the force evaluations may be significantly reduced by introducing cutoff radius.

In the case of short-range potentials, which decay rapidly with distance (see Section 2.1), there is no need to take into account all the interactions. In practice, atomic forces acting on the i-th atom can be computed only with its nearest neighbours in a certain circular (or

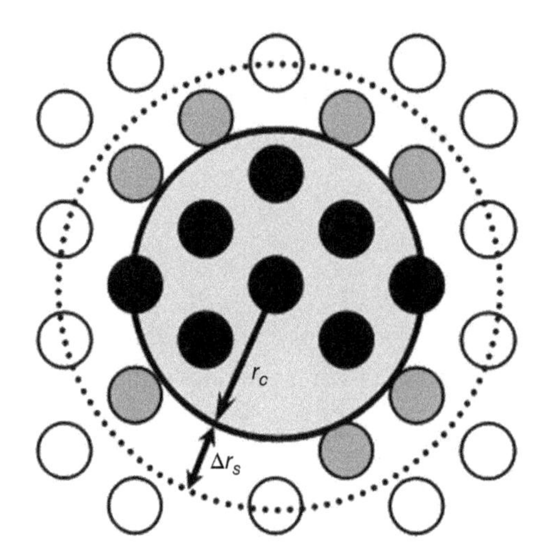

Figure 3.16 Illustration of the cutoff radius and the 'skin' parameter.

spherical) area determined by the cutoff radius r_c (Figure 3.16). Interactions with atoms placed further than r_c are neglected. The modified potential function has the following form:

$$V_c = \begin{cases} V(r_{ij}) & r_{ij} \le r_c \\ 0 & r_{ij} > r_c \end{cases}. \tag{3.125}$$

This procedure significantly reduces the number of computations of forces to the level of $(Nn)/2$, where n is the number of atoms placed in the cutoff area. However, an error may be introduced to the estimation of the atomic forces if the cutoff radius chosen is too short. Additionally, the total energy of the atomic system may not be conserved due to the discontinuity of the truncated potential function. Different techniques were developed to overcome this phenomenon. Most of them use smoothing or shifting routines to make potential and force functions continuous or even differentiable again [11]. However, usually a large enough cutoff radius provides that the fluctuations of the energy in the truncation area are extremely small and may be neglected.

Further reduction of the computation cost can be obtained using neighbour lists. Such a list is created for each atom and contains all its neighbours in the area slightly larger than the cutoff radius. In this approach, it is not necessary to evaluate the distances between each pair of atoms in the simulated system, and it is also not required to search for nearest neighbours in every step of integration. Lists are constructed only once and need to be refreshed after a certain number of iterations, depending on the character of the simulated problem. This method is especially effective in the simulations of solids in low temperatures where the atoms do not move too much during two subsequent time steps. Naturally, simulations of liquids and gases require frequent refreshing of the neighbour lists.

The frequency of updating the neighbour lists can be adjusted by 'skin' parameter Δr_s, which corresponds to the thickness of the cutoff boundary (Figure 3.16).

If the value Δr_s is too small, it results in frequent rebuilds of neighbour lists; on the other hand, thick 'skin' results in longer lists and additional computational effort. In practice, typical values of Δr_s are in the range of 1–2 Å.

3.2.1.6 Boundary Conditions

The MD method, based on the velocity-Verlet integrations algorithm, needs two sets of the initial conditions: the set of initial positions of the particles and the set of their initial velocities. The preparation of the initial solutions will be described in Section 4.1.

As was mentioned previously, atoms are placed in a certain domain called the *simulation box*. Note that the MD method does not need any definition of the specific boundary conditions to work. However, such unrestricted simulation is suitable only for a rather small class of specialized problems, such as forming of the atomic clusters.

Depending on the considered problem, certain conditions are introduced and applied on the boundary of the simulation's domain. The periodic, reflective, and rigid boundary conditions will be discussed in this section.

The MD simulations are usually performed on a separated, very small part of the real, macroscopic system (e.g. crack tip). In such an approach, a large number of atoms is placed on the boundary (or surface) of the simulation box. These atoms have fewer neighbours than atoms lying inside the domain, and as a consequence, some negative surface effects can arise and disturb the quality of the simulation. The introduction of periodic boundary conditions can overcome this problem to a certain extent because the small, finite atomic system can be considered as an infinite array of identical simulation domains. The main idea of periodic boundary conditions is shown in Figure 3.17.

Note that the atom that leaves the simulation box crosses a certain boundary immediately and approaches from the opposite side. This mechanism ensures that the total number of atoms in the domain of the simulation remains constant. Additionally, atoms placed near the boundary may interact with the atoms in the neighbouring copies of simulation boxes (or, equivalently, with the atoms lying close to the opposite boundary). Due to this feature, some surface effects can be removed, but problems with handling long-range interactions can also occur [99]. Evidently, periodic boundary conditions are imposed on the boundaries of the natural periodic systems like crystals. However, this kind of boundary condition may introduce artificial periodicity to the simulated system. Such periodicity is usually difficult to isolate from other effects. According to [69], it is discussable whether a small atomic region with

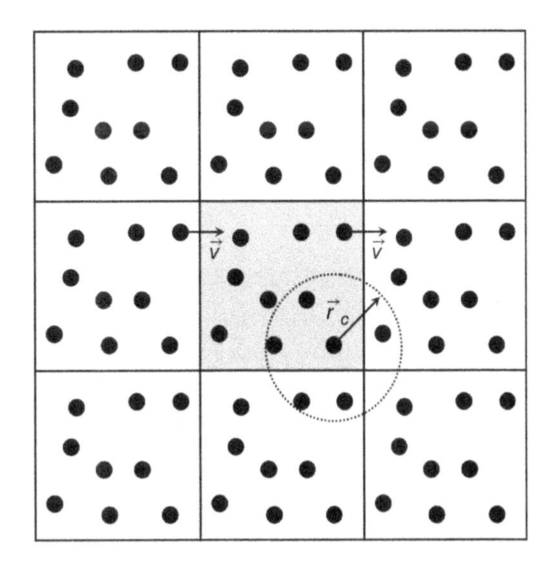

Figure 3.17 Illustration of the periodic boundary conditions.

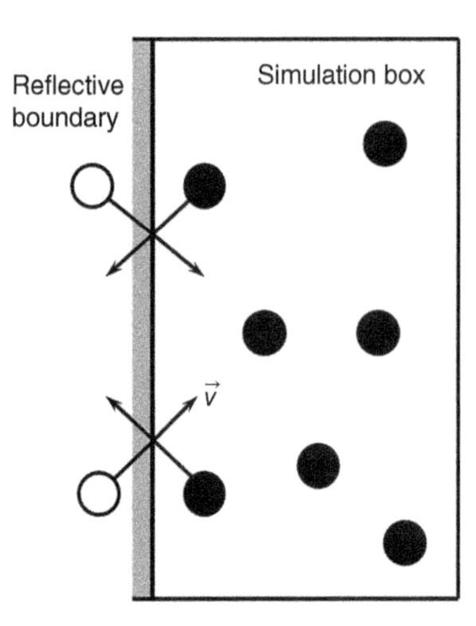

Figure 3.18 The reflecting wall.

periodic boundary conditions represents a real macroscopic system to a satisfactory degree. In the general approach, some tests with increasing simulation domain sizes should be performed.

The reflective and rigid boundary conditions are applied to model solid walls (moving or fixed) at the edge of the simulation domain, which avoids atoms leaving the simulation box. Reflective boundary conditions can be realized in many different ways, such as introducing repulsive force acting on the atoms that move close to the walls or flipping appropriate components of the atom's velocity vectors. In the first case, the magnitude of the repulsive force is equal to the force of the additional artificial particle, placed at the mirrored position (on the other side of the boundary, outside of the simulation box), acting on the real atom, which has to be reflected from the wall. Details are shown in Figure 3.18.

The artificial particle has the same mass and usually the same potential parameters as the real atom. The interaction force is computed only when the distance between two mirrored atoms is smaller than doubled equilibrium distance length; thus, only the repulsive part of the potential is taken into account. Note that artificial reflecting particles may also be located directly on the boundary. Alternatively, reflective walls can be created using a uniformly distributed force field, acting on the atoms in a direction perpendicular to the boundary. The reflective wall can be implemented as a fixed boundary condition or a moving one. Setting the proper velocity to this kind of barrier gives an opportunity to simulate, e.g. compression of the molecular structure (see Section: 4.1: Creation of the polycrystalline structures).

The rigid boundary conditions are constructed from layers of frozen atoms, which do not change their positions due to interaction with other atoms. It can be simply achieved by setting the force exerted by all unconstrained particles on frozen atoms to zero. Additionally, the velocity of the fixed atoms can be set to a specified value, which allows the modelling of moving boundaries. Since each unconstrained atom placed near the wall should have a full set of neighbours, the width of the rigid boundaries must be greater than the cutoff radius used in the simulation.

3.2.1.7 Size of the Atomistic Domain – Limitations of the Molecular Simulations

The size of the simulated atomistic system should be discussed. As was mentioned, molecular simulations, especially in mechanical engineering, are performed on the small representative volume elements of the real macroscopic system. In the case of homogenous materials or materials with periodic structure, the negative surface effects can be significantly reduced by the application of appropriate boundary conditions. However, simulations of some mechanical phenomena like crack propagation or dislocation motion require additional discussion because insufficient size and boundaries of the molecular domain can disturb all the simulations. This process will be presented in the following example of a simple, two-dimensional simulation of fracture. A rectangular plate with a horizontal crack (Figure 3.19) is taken under consideration. The lower and upper regions of the molecular lattice are treated as rigid boundaries and allowed to move in opposite directions with constant linear velocity. The vertical edges remain free. Such an arrangement results in the generation of tensile load and then in the opening of the crack introduced in the middle of the plate.

The results of the simulation are presented in Figure 3.20. An elastic wave propagated to form the crack edge (a) cannot spread away freely and cannot be dissipated to the environment, so the wave has been reflected from the boundary of the molecular domain that is too small (b). The reflected wave starts to propagate in the reverse direction, interfering and perturbing all the fracture dynamics (c).

Achieved results are in good agreement with simulations presented by [70] and lead to the conclusion that the atomic simulation domain should be at least of an order of magnitude larger than imperfection (crack, dislocation, grain boundary, etc.). Otherwise, the quality of performed simulation is questionable due to interference of the boundary effects. Only large domains with many millions of atoms can satisfy this condition. However, in many cases it is unnecessary to maintain real atomistic resolution far from the investigated nano-defect, where only elastic deformations occur. Thus, multiscale techniques can effectively handle these problems, where a small discrete molecular domain is coupled with a large continuum domain modelled using the FEM or BEM.

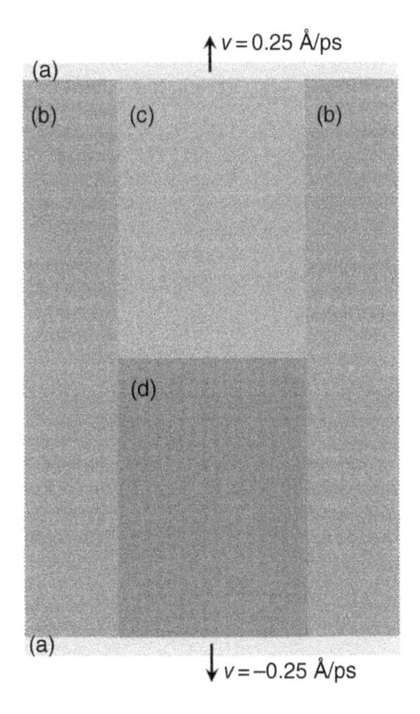

Figure 3.19 Rectangular plate with a crack under tensile load. A constant linear velocity was applied to the rigid boundaries (a). Areas marked by (b), (c), and (d) are made from the HCP lattice of unconstrained atoms. Interactions between the regions (c) and (d) were excluded from computations to form the cleavage.

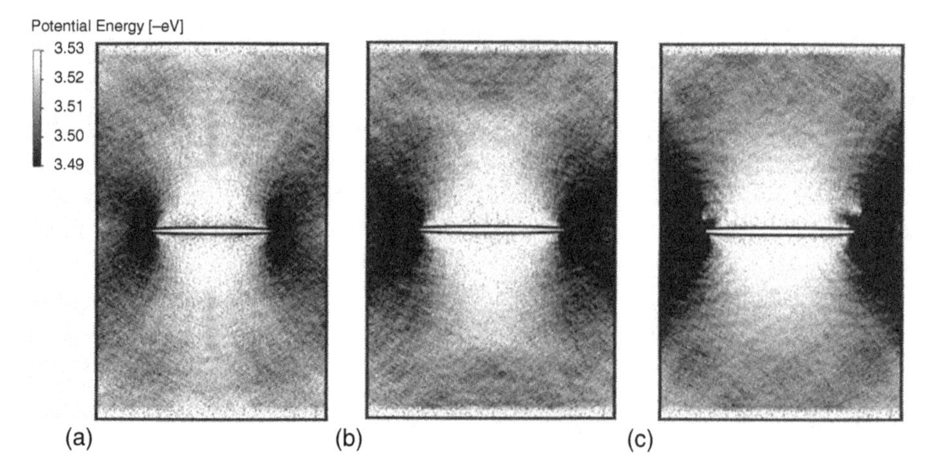

Figure 3.20 Results (a), (b), and (c) after, respectively, 6, 7, 8 ps of simulation (see text for details).

3.2.2 Molecular Statics

3.2.2.1 Equilibrium of Interatomic Forces

In this section, we consider a molecular system with N atoms placed in the domain Ω. Now, a static problem is taken under consideration (compare with Section 3.2.1.1), i.e. one assumes that the atom's velocities are zeroed, and all atoms are characterized only by the set of their coordinates $\{r_1,...r_N\}$. According to Eq. (3.91) and Figure 3.21, the interaction forces between each pair of atoms equals:

$$\mathbf{f}_{ij} = -\frac{\partial V(r_{ij})}{\partial r_{ij}}\mathbf{n}_{r_{ij}} \quad \text{and} \quad \mathbf{f}_{ji} = -\mathbf{f}_{ij}, \tag{3.126}$$

where the unit directional vector is defined as:

$$\mathbf{n}_{ij} = \frac{\mathbf{r}_{ij}}{r_{ij}}. \tag{3.127}$$

Consider the homogeneous deformation of an infinite representative atomistic lattice (Figure 3.22). The kinematic relation is given by the first order Cauchy-Born [14, 111] rule:

$$\mathbf{r}_{ij} = \mathbf{K} \times \mathbf{R}_{ij}, \tag{3.128}$$

where: symbol $\mathbf{K}$ denotes the first-order deformation gradient, which defines a linear tangent map, given by the tensor:

$$\mathbf{K} = \nabla\varphi. \tag{3.129}$$

The deformation map $\varphi(\mathbf{X})$ renders the placement $\mathbf{X}$ in the material configuration into the placement $\mathbf{x} = \varphi(\mathbf{X})$ in the spatial configuration (Figure 3.22).

The operator $\nabla\varphi$ is defined as:

$$\nabla\varphi = \mathbf{I} + \nabla\mathbf{u}, \tag{3.130}$$

where:

$$\nabla\mathbf{u}_{ij} = \frac{\Delta\mathbf{u}_{ij}}{\mathrm{R}_{ij}}. \tag{3.131}$$

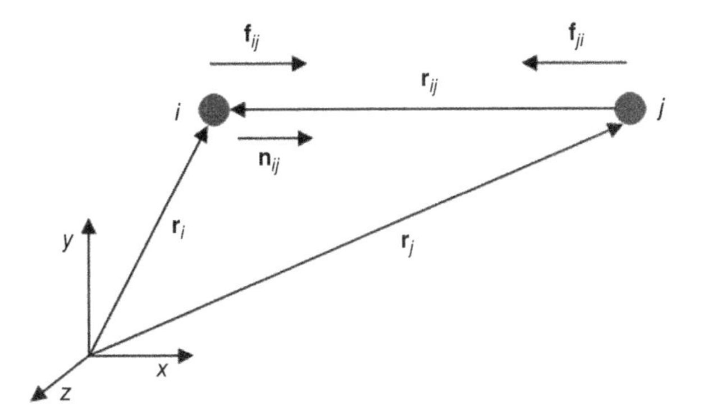

Figure 3.21 Interaction forces between the pair of atoms.

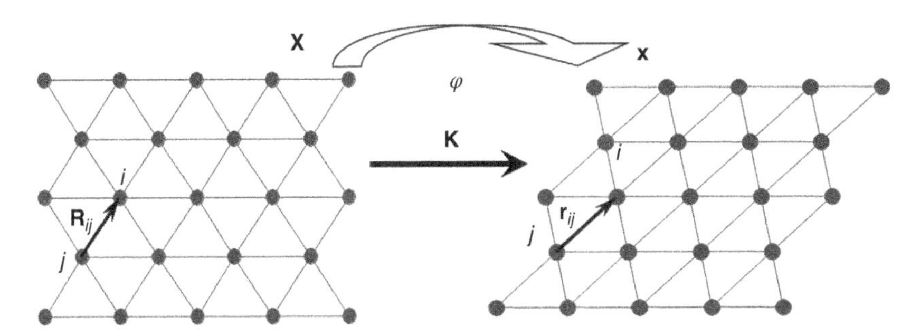

Figure 3.22 Homogenous deformation of the atomic lattice.

Substitution of Eqs. (3.129)–(3.131) to Eq. (3.128) leads to the following vector relations:

$$\mathbf{r}_{ij} = \mathbf{R}_{ij} + \Delta\mathbf{u}_{ij}, \tag{3.132}$$

$$\Delta\mathbf{u}_{ij} = \mathbf{u}_j - \mathbf{u}_i. \tag{3.133}$$

Graphical interpretation of Eq. (3.132) is presented in Figure 3.23.

In Figure 3.23, $\mathbf{R}_{ij}$ denotes an initial distance between atoms i and j; $\mathbf{u}_i$ and $\mathbf{u}_j$ are vectors of displacements of these atoms, respectively; and $\mathbf{r}_{ij}$ is the resultant distance vector. The force acting between atoms in a displaced position now can be written as:

$$\mathbf{f}_{ij}\left(r_{ij}\right) = f_{ij}\left(r_{ij}\right)\frac{\mathbf{r}_{ij}}{r_{ij}}. \tag{3.134}$$

Putting Eq. (3.132) into Eq. (3.134) yields:

$$\mathbf{f}_{ij}\left(r_{ij}\right) = f_{ij}\left(r_{ij}\right)\frac{\mathbf{R}_{ij}}{r_{ij}} + f_{ij}\left(r_{ij}\right)\frac{\Delta\mathbf{u}_{ij}}{r_{ij}}. \tag{3.135}$$

In the equilibrium state, the forces cancel each other, thus the equilibrium equations of the atomic forces can be formulated:

$$\mathbf{f}_{ij}\left(r_{ij}\right) = 0, \tag{3.136}$$

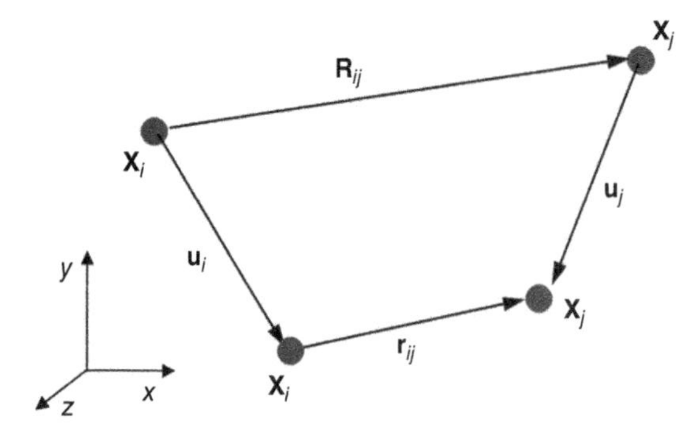

Figure 3.23 Initial ($\mathbf{X}_i$, $\mathbf{X}_j$) and final ($\mathbf{x}_i$, $\mathbf{x}_j$) positions of the pair of atoms.

or

$$f_{ij}(r_{ij})\frac{\mathbf{R}_{ij}}{r_{ij}}+f_{ij}(r_{ij})\frac{\Delta\mathbf{u}_{ij}}{r_{ij}}=0. \tag{3.137}$$

3.2.2.2 Solution of the Molecular Statics Problem

From the numerical point of view, equilibrium Eq. (3.137) should be rewritten in the more convenient matrix form. For the two-dimensional case, one can be obtained:

$$\begin{bmatrix} \kappa & 0 & -\kappa & 0 \\ 0 & k & 0 & -\kappa \\ -\kappa & 0 & \kappa & 0 \\ 0 & -\kappa & 0 & \kappa \end{bmatrix} \begin{Bmatrix} u_{ix} \\ u_{iy} \\ u_{jx} \\ u_{jy} \end{Bmatrix} = \begin{Bmatrix} f_{ix} \\ f_{iy} \\ f_{jx} \\ f_{jy} \end{Bmatrix}, \tag{3.138}$$

where κ is the atomistic stiffness:

$$\kappa=\frac{f_{ij}(r_{ij})}{r_{ij}}, \tag{3.139}$$

and the appropriate components of the forces are formulated in this way:

$$\begin{aligned} f_{ix} &= -f_{jx} = \kappa(X_i-X_j) \\ f_{iy} &= -f_{jy} = \kappa(Y_i-Y_j) \end{aligned}. \tag{3.140}$$

This system of equations describes one bonding between two atoms. Symbols X_i and X_j, Y_i, and Y_j denote initial coordinates of i-th and j-th atom, respectively. The considered system of equations is nonlinear and must be transformed into the form, which can be solved using an iterative method.

System of Eq. (3.138) can be transformed to the form:

$$\begin{bmatrix} \kappa & 0 & -\kappa & 0 \\ 0 & \kappa & 0 & -\kappa \\ -\kappa & 0 & \kappa & 0 \\ 0 & -\kappa & 0 & \kappa \end{bmatrix} \begin{Bmatrix} u_{ix} \\ u_{iy} \\ u_{jx} \\ u_{jy} \end{Bmatrix} - \begin{Bmatrix} \kappa(X_i-X_j) \\ \kappa(Y_i-Y_j) \\ \kappa(X_j-X_i) \\ \kappa(Y_j-Y_i) \end{Bmatrix} = \begin{Bmatrix} 0 \\ 0 \\ 0 \\ 0 \end{Bmatrix}, \tag{3.141}$$

or

$$\mathbf{W}(\mathbf{u}) = 0, \tag{3.142}$$

where

$$\mathbf{W}(\mathbf{u}) = \begin{Bmatrix} W_1 \\ W_2 \\ W_3 \\ W_4 \end{Bmatrix} = \kappa \begin{Bmatrix} (u_{ix} - u_{jx}) - (X_i - X_j) \\ (u_{iy} - u_{jy}) - (Y_i - Y_j) \\ (u_{jx} - u_{ix}) - (X_j - X_i) \\ (u_{jy} - u_{iy}) - (Y_j - Y_i) \end{Bmatrix}. \tag{3.143}$$

Such a system of nonlinear Eq. (3.142) can be solved iteratively using the Newton-Raphson method. The general scheme goes like this:

$$\mathbf{u}^{it+1} = \mathbf{u}^{it} - \left[\mathbf{J}(\mathbf{u}^{it})\right]^{-1} \mathbf{W}(\mathbf{u}^{it}), \tag{3.144}$$

and Jacobian matrix J contains partial derivatives of the elements of the vector (3.143):

$$\mathbf{J}(\mathbf{u}) = \begin{bmatrix} \frac{\partial W_1}{\partial u_{ix}} & \frac{\partial W_1}{\partial u_{iy}} & \frac{\partial W_1}{\partial u_{jx}} & \frac{\partial W_1}{\partial u_{jy}} \\ \frac{\partial W_2}{\partial u_{ix}} & \frac{\partial W_2}{\partial u_{iy}} & \frac{\partial W_2}{\partial u_{jx}} & \frac{\partial W_2}{\partial u_{jy}} \\ \frac{\partial W_3}{\partial u_{ix}} & \frac{\partial W_3}{\partial u_{iy}} & \frac{\partial W_3}{\partial u_{jx}} & \frac{\partial W_3}{\partial u_{jy}} \\ \frac{\partial W_4}{\partial u_{ix}} & \frac{\partial W_4}{\partial u_{iy}} & \frac{\partial W_4}{\partial u_{jx}} & \frac{\partial W_4}{\partial u_{jy}} \end{bmatrix}. \tag{3.145}$$

The index *it* in the equation denotes the counter of the interactions.

For the whole atomic system, local vectors **W** (3.143) and local Jacobian matrices **J** (3.145) have to be computed for all atoms, which interact with others in a certain area, determined by the cutoff radius (compare Figures 3.16, 3.17, and 3.24). In the next step, all determined local components are aggregated into a global nonlinear system of equations. Aggregation of the local matrices to the global ones is shown in Figure 3.24. This procedure is similar to the aggregation of the stiffness matrices and force vectors of the truss model in the FEM.

If necessary, certain atoms can be fixed on their initial positions by the simple elimination method, i.e. by removing the verse, column, and elements corresponding to the appropriate component of the constrained atom's displacement. Additionally, the initial displacements of atoms can also be introduced in a similar way: by adding values of displacements to the corresponding initial coordinates of the 'frozen' atoms.

A fully assembled system of equations is solved using an algorithm (3.144). To improve stability and convergence of the Newton-Raphson scheme, especially when the initial guess is not sufficiently close to the root or more complex potential functions are used (such as EAM), the backtracking algorithm should be applied [91]. The iterations are stopped when the following condition is satisfied:

$$\|\mathbf{W}(\mathbf{u})\| < \omega, \tag{3.146}$$

i.e. the Euclidean norm of the vector **W**(**u**) will be less than a certain admissible value. The ω parameter should be adjusted individually to the solved problem.

However, in the case of large initial displacements (e.g. greater than half of the equilibrium distance) or deformed, far-from-equilibrium initial spatial configurations, some problems

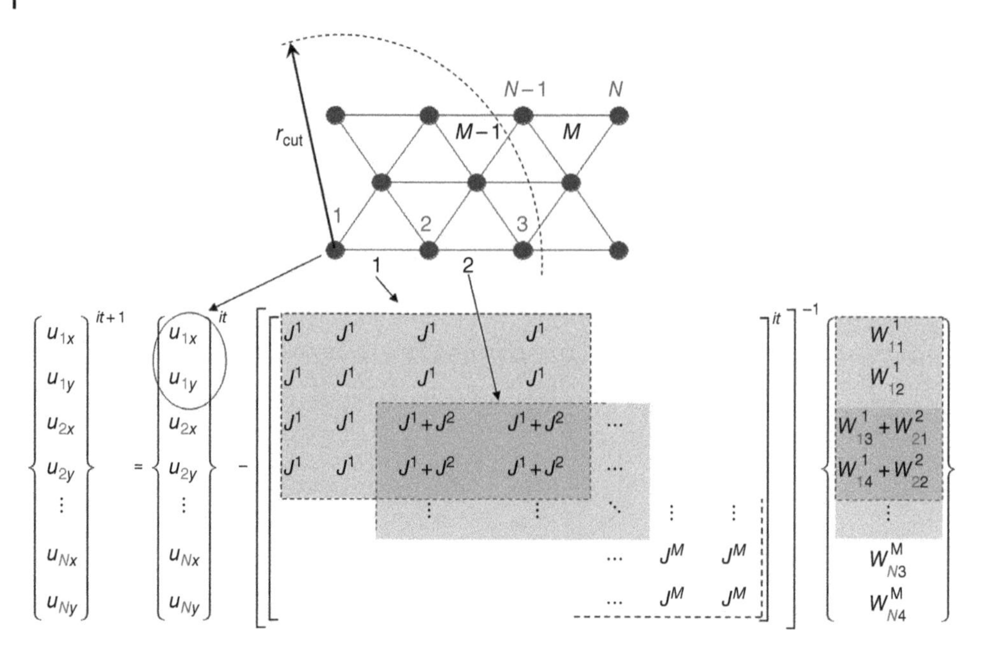

Figure 3.24 Assembly of the global system of equations: M and N denote bonding and atom numbers, respectively. r_{cut} is the cutoff radius.

with the convergence of the solution may occur. In such cases, initial displacements should be divided into the n_s parts and applied incrementally until they reach the desired value. In each step, the initial displacements are increased by the value:

$$\Delta \mathbf{u}_0 = \frac{1}{n_s} \mathbf{u}_0, \tag{3.147}$$

and then the atomistic system is equilibrated, solving a system of Eq. (3.101). The complete algorithm is presented in Figure 3.25.

The proposed approach allows simulation of the deformations of the atomistic lattice under static loads. It is suitable to investigate the basics of the atomic lattice's mechanics or determine elastic constants in the zero temperature (see Section 4.1.1.6). Note that from the computational point of view, static analysis is usually much quicker than dynamic analysis.

3.2.2.3 Numerical Example of the Molecular Statics

This simple example shows the ability of the presented algorithm to investigate slips in the hexagonal close-packed (HCP) atomic lattice. Shearing of the plate with a rectangular notch (Figure 3.26a) is considered. The left region of the molecular lattice is fixed at the initial position, and the right side is displaced vertically by the value of $u_0 = 0.05$ Å in each step of the algorithm presented in Figure 3.26. The whole model consists of 819 atoms, and Lennard-Jones' potential is applied to describe atomic interactions.

The results of the simulation for increasing values of displacement are presented in Figure 3.26. Slips in the preferred system $\{0001\}\langle 11\bar{2}0\rangle$ and opening of the crack at the corner of the notch can be observed.

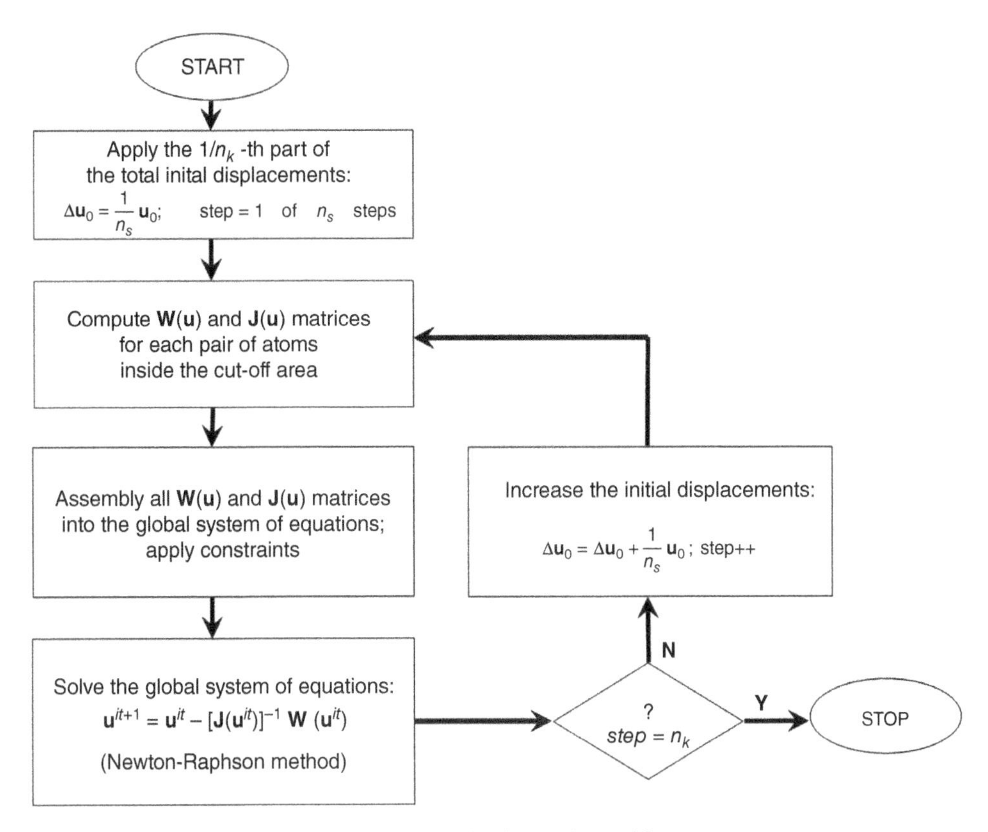

Figure 3.25 The algorithm of solving the molecular statics problem.

3.2.3 Cellular Automata

The cellular automata technique was originally developed in the 1950s by Janos Von Neumann [116] to simulate the behaviour of discrete and complex systems. Initially, the limited capabilities of computers were the main obstacle in the development of this method and its wider applications. However, this technique has been popular for the last two decades due to continuously increasing computational power of computers [25, 119]. Nowadays, the CA method is used in different scientific areas, from biology and chemistry to solid-phase physics and electronics [37]. During the last few years, the CA method has also been applied to model material behaviour during thermomechanical processing [5]. This includes modelling of static, dynamic, and metadynamic recrystallization behaviour, precipitate coarsening, grain boundary migration, corrosion, grain fragmentation, or solidification [8, 17, 26–28, 30–32, 43, 44, 46, 49, 50, 52, 53, 56, 61, 62, 66, 72, 73, 75, 76, 82, 93–97, 101, 103, 107, 112–114, 118, 121, 124].

3.2.3.1 Cellular Automata Definitions

The cellular automata method is based on four major components that have to be precisely defined during the development of the particular model of the physical phenomenon. The first component is the cellular automata space, then the set of cell variables, the

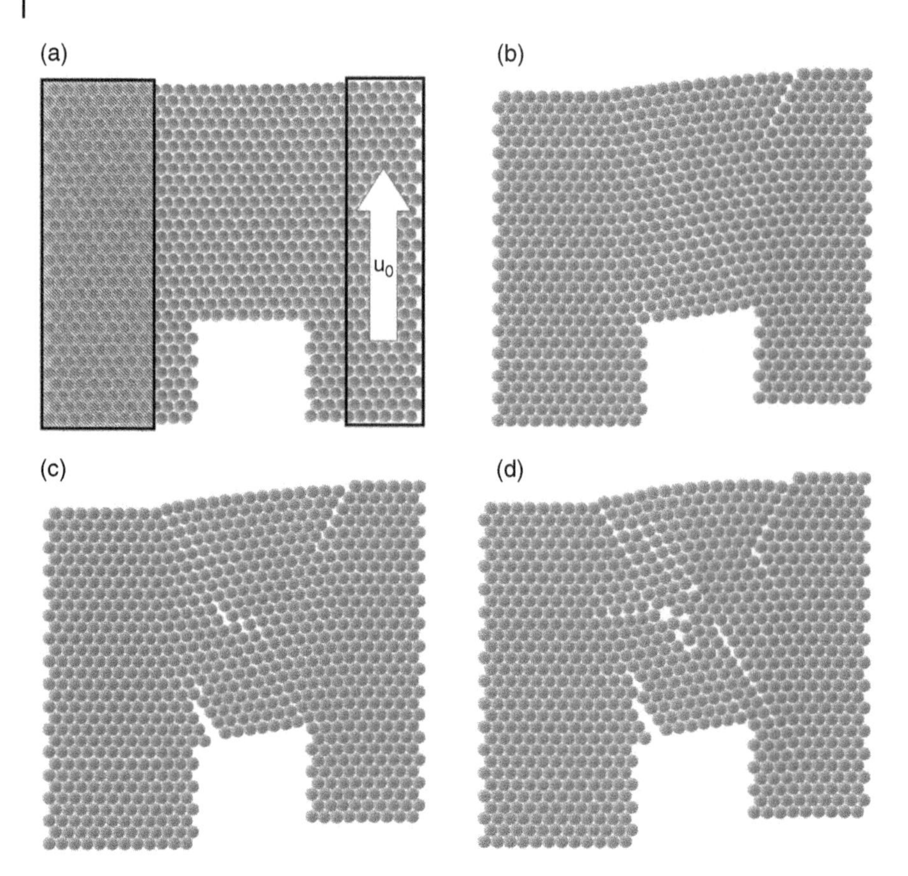

Figure 3.26 Shearing of the HCP lattice: (a) is the initial state; (b), (c), and (d) are subsequent results for $\mathbf{u}_0$=5.25 Å, 6 Å, and 6.8 Å, respectively.

neighbourhood, and finally, the fourth component is the set of transition rules. Therefore, the cellular automata model is a quadrupole:

$$CA = (\omega, \mathbf{Y}, \mathbf{N}, f), \tag{3.148}$$

where: ω – cellular automata space, $\mathbf{Y}$ – set of cell variables, $\mathbf{N}$ – set of neighbours, f – transition rules.

The main idea of the cellular automata technique introduces a computational domain composed of lattices of finite cells. In practical applications, mainly 2D and 3D CA spaces are investigated. That is why the 1D CA space concept is not discussed in this book; for more information, please refer to [104]. Each cell in the computational domain is called a cellular automaton, and the entire set of the cells creates the cellular automata space ω. In most cases, the CA space represents the regular geometrical region and is composed of a set of regular cellular automatons. In such a case, the 2D CA space can be composed of squares, triangles, hexagons, etc. as seen Figure 3.27. The same concept applies to the 3D computational domains.

The CA space is always based on the square division in the present work to obtain the square $n \times n$ or rectangular $n \times m$ shape of cellular automata space, as seen in Figure 3.28.

Another very important aspect of the CA method is the definition of the appropriate CA neighbourhood type. A large variety of different neighbourhood types is now available in the

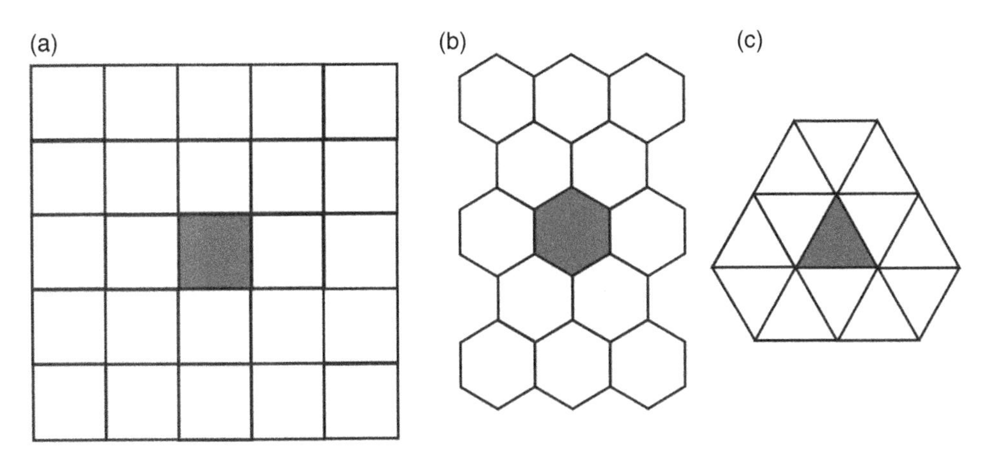

Figure 3.27 Examples of cellular automata space division into (a) squares, (b) triangles, and (c) hexagons.

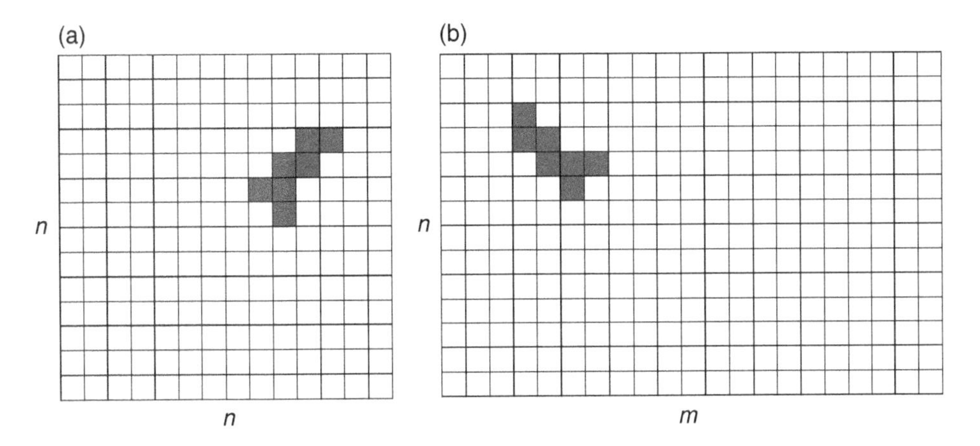

Figure 3.28 Examples of (a) $n \times n$ and (b) $n \times m$ CA space composed of square cellular automatons.

scientific literature. The most commonly used CA neighbourhoods in 2D space are the von Neumann (four neighbours surround the cell) and Moore (eight neighbours surround the cell). Graphical representation for these types of neighbourhoods is presented in Figure 3.29.

As mentioned, various modifications of these two classical approaches were proposed during the years, including extended von Moore, reversed von Neumann, or random hexagonal were developed as seen in Figure 3.30.

Similar examples of 3D CA neighbourhoods are presented in Figure 3.31.

In the case of neighbourhoods with stochastic components, more possible solutions occur in 3D, as seen in Figure 3.32.

Finally, it has to be pointed out that other non-standard neighbourhoods can also be found in the scientific literature [122]. Especially interesting is a fully random neighbourhood operating on randomly distributed cellular automatons (Figure 3.33). This class of the CA approaches, called the random cellular automata method, is often coupled with finite element solutions [74, 88].

Additionally, a set of cell variables characterizes each cellular automaton in the defined CA space. When cell variables contain information about possible states of the CA cells, they are

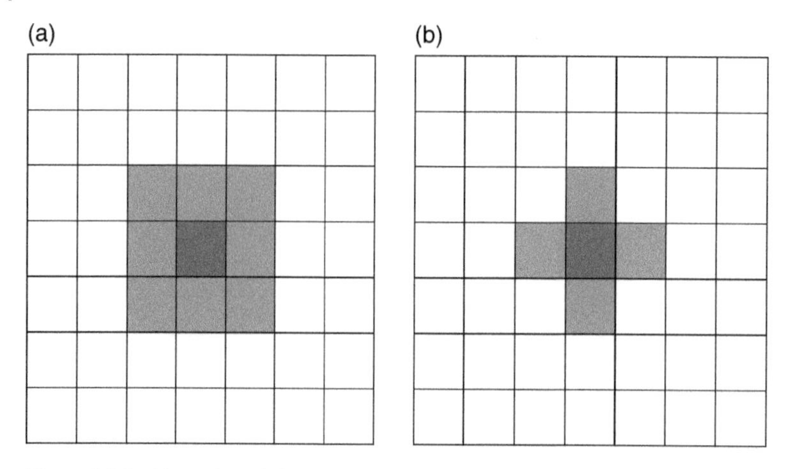

Figure 3.29 Examples of CA neighbourhoods in the 2D space: (a) Moore and (b) von Neumann.

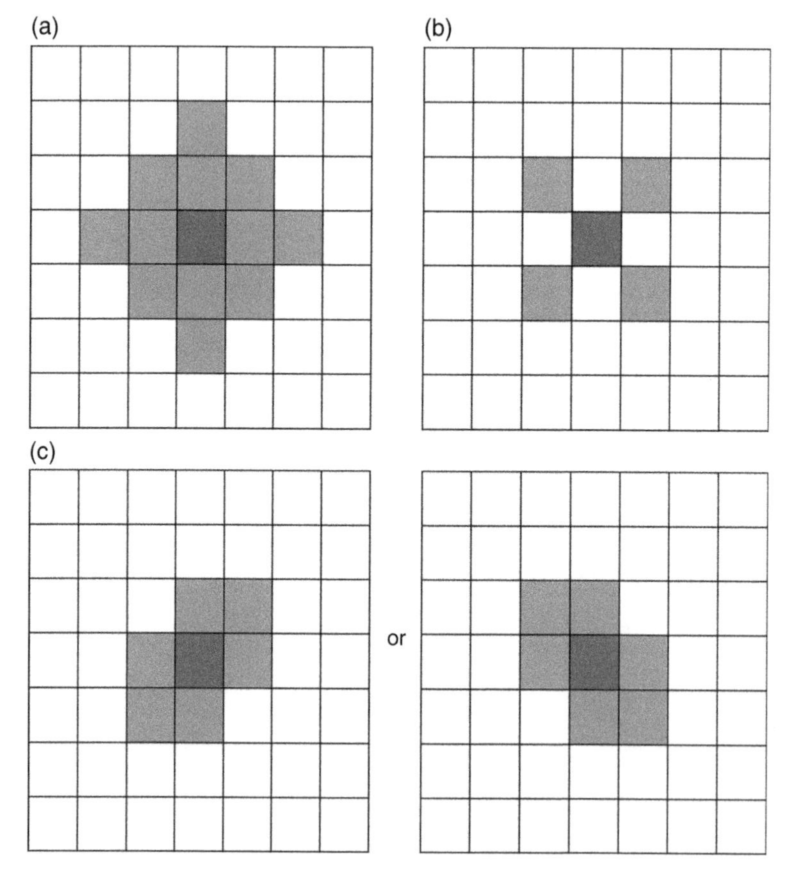

Figure 3.30 Examples of other types of CA neighbourhoods, (a) extended Moore, (b) reversed von Neumann, and (c) random hexagonal.

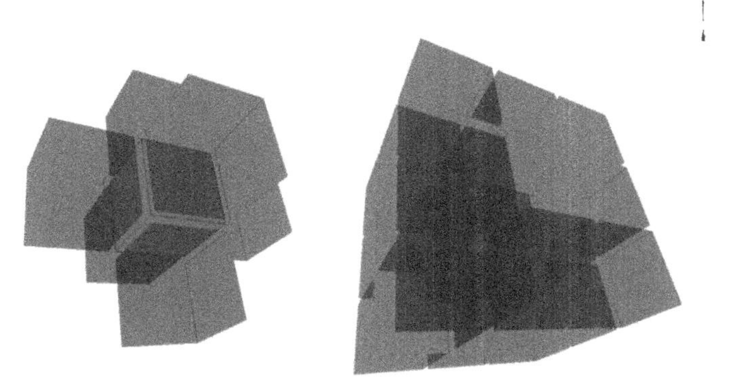

Figure 3.31 Examples of CA neighbourhoods in the 3D space: von Neumann and Moore.

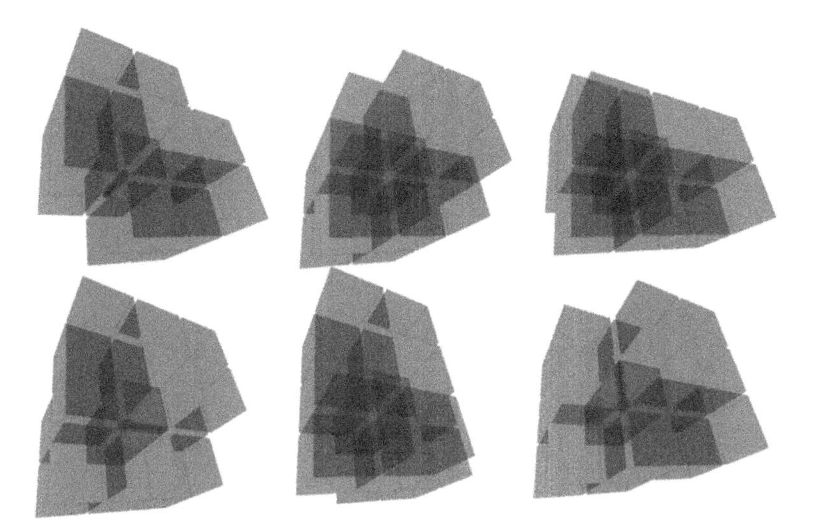

Figure 3.32 A 3D representation of random hexagonal neighbourhood.

usually called the set of state variables. On the other hand, when they contain information about the physical parameters of the investigated phenomenon, they are referred to as internal variables. However, both types of cell variables are crucial during the definition of the transition rules in the CA method. These rules are used to control the evolution of the cellular automata model and replicate the physics of the investigated phenomenon. Therefore, the transition rules are created on the basis of the available knowledge on the particular phenomenon. Transition rules *f* are evaluated in each time step and assume that the state of each cell in the CA space is a function of the previous states of its neighbours and the cell itself:

$$Y_i^{t+1} = f\left(Y_j^t\right) \quad where \quad j \in N(i). \tag{3.149}$$

In this case, the state of the CA cell can remain the same, or it can change to a new state if a certain logical function Λ is fulfilled:

$$Y_i^{t+1} = \begin{cases} new\_state \Leftrightarrow if\ \Lambda \\ Y_i^t \end{cases} \quad where \quad j \in N(i), \tag{3.150}$$

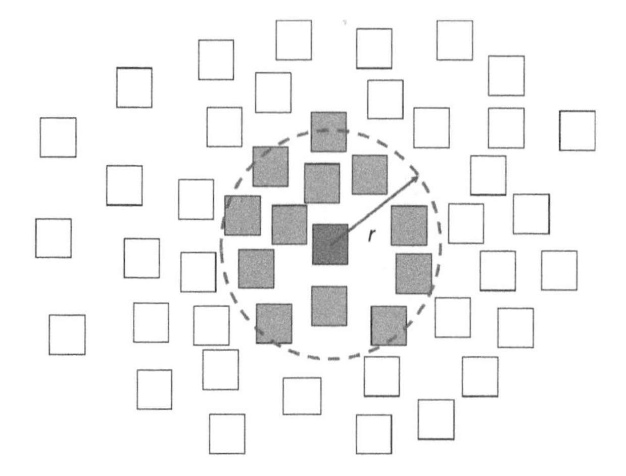

Figure 3.33 Cellular automata space is a random cellular automata method.

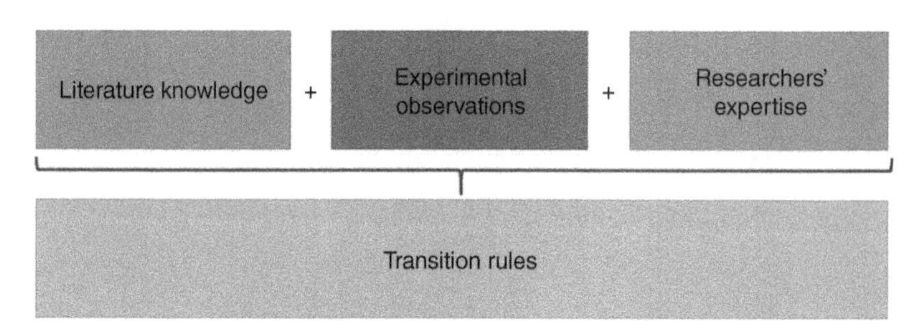

Figure 3.34 Source of information for the definition of transition rules.

where: $N(i)$ – surrounding of the i-th cell, Y_i – state of the i-th cell, t – time step, $\Lambda = \Lambda\,(p, q, Y_i, Y_j)$ – logical function, based on the transition rules, which depends on the internal q and external p variables and which describes a condition for the cell to change the state.

Since the transition rules control the behaviour of the cell during calculations (i.e. during the recrystallization of phase transformation), the proper definition of these rules in the process of designing a CA model directly affects the accuracy of this approach. The transition rules are created on the basis of literature knowledge, experimental observation, and researchers' knowledge regarding mechanisms leading to the development and propagation of particular physical phenomenon (Figure 3.34)

These four major components (cellular automata space, set of cell variables, the neighbourhood, and transition rules) of the CA method have to be defined to create a robust numerical model capable of replicating investigated phenomenon. However, to deal with different kinds of physical phenomena, the appropriate boundary conditions must also be selected. One can choose from a set of different approaches, namely:

- absorbing boundary conditions – the state of cells located on the edges of the CA space are adequately fixed with a specific state to absorb moving quantities,
- reflective boundary conditions – the state of cells located on the edges of the CA, with a specific state to reflect moving quantities, or

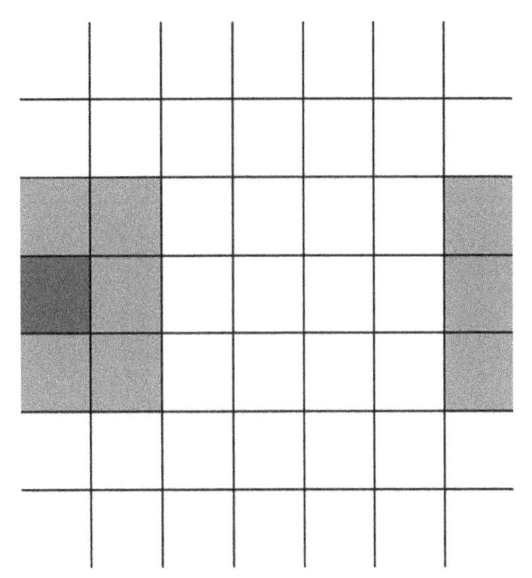

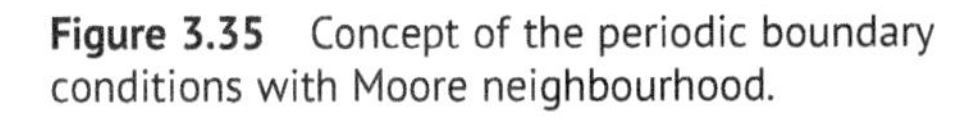

Figure 3.35 Concept of the periodic boundary conditions with Moore neighbourhood.

- periodic boundary conditions – takes into account cells located on subsequent edges of the CA space and considers them as neighbours. As a result, a CA space continuity is obtained, as seen in Figure 3.35. This type of boundary condition is commonly used in many physical simulations as it assures the fulfilment of the principle of conservation.

Capabilities, as well as examples of the CA method applications, will be shown in Chapters 4 and 5 based on microstructure evolution models. Another discrete numerical method presenting similar advantages is the Monte Carlo (MC) group of methods.

3.2.4 Monte Carlo Methods

The MC methods do not represent a single numerical algorithm. It is instead a group of algorithms characterized by common assumptions [68]. The first practical applications of this method are dated to 1930 when Enrico Fermi used a statistical numerical approach to model neutron diffusion. It was intensively developed during the second world war in Los Alamos Laboratories. It was applied by Stanisław Ulam and Janos Von Neumann to model nuclear physics phenomena.

Over the years, this class of methods has been intensively developed, and these days MC methods are applied in various areas of physics, engineering, or medicine.

The similarities between algorithms are mainly based on statistical evaluation of various investigated mathematical functions, formulas, or physical systems using random solution space sampling. In this case, an efficient and reliable random numbers generator plays a significant role in obtaining accurate results. Due to the stochastic nature, in every case, there is an error associated with the solution. However, the larger the set of random samples is, the more accurate result can be obtained. This group of algorithms is mainly designed for problems that are difficult to solve with deterministic algorithms.

As a result, MC is mainly applied in modelling fluid dynamics, disordered materials, or cellular structures. They are also commonly applied in modelling systems with various uncertainties in input data, e.g. stock exchange predictions.

MC algorithms vary between each other depending on the specific application. Nevertheless, four general steps are similar and composed of:

- definition of a domain of possible input parameters for the model,
- generation of defined input parameters in a random matter over the investigated domain/space,
- realization of required deterministic calculations using generated input parameters, and
- aggregation of obtained solutions.

MC methods, like auxiliary field Monte Carlo, dynamic Monte Carlo, kinetic Monte Carlo, quantum Monte Carlo, quasi-Monte Carlo, Makarov Chains, etc. are based on four mentioned steps. One of the most commonly used MC approaches based on the Pots model is often used in material science applications [13, 41, 123]. The Pots model originated as a model for modelling the behaviour of spins that are interacting along the crystalline lattice. Now, the model is widely used to investigate the cooperative behaviour of cellular structures similar to the CA method described earlier. Actually, the MC Pots model can be treated as a generalized cellular automata approach based on similar definitions. The difference lies in the definition of transition rules that in the MC approach have purely probabilistic character.

In the Monte Carlo Pots approach, the whole computational domain (lattice) is discredited in the form of a finite number of cells (lattice sites). Each of the cells gets a random value Q_i which determines the actual state of the investigated cell (Figure 3.36). As a result, a set of states is defined:

$$\Omega = \{Q_0, \ldots, Q_{N-1}\}, \tag{3.151}$$

where: N is the number of available states.

The physical meaning of the particular state is defined by the user depending on the investigated phenomenon, e.g. recrystallization, phase transformation, or fracture.

Similar to the CA space, an appropriate neighbourhood of a selected cell has to be defined prior to calculation. In the MC approaches, the Moore neighbourhood with eight neighbouring cells is commonly used (Figure 3.29a).

The MC Pots algorithm is then composed of three main steps [90]:

- The first is a random selection of a lattice cell from the entire available domain. Then, the initial energy for that site is computed using the Hamiltonian:

$$E_i = -J \sum_{j=1}^{Z} \left(\delta_{S_i S_j} - 1\right), \tag{3.152}$$

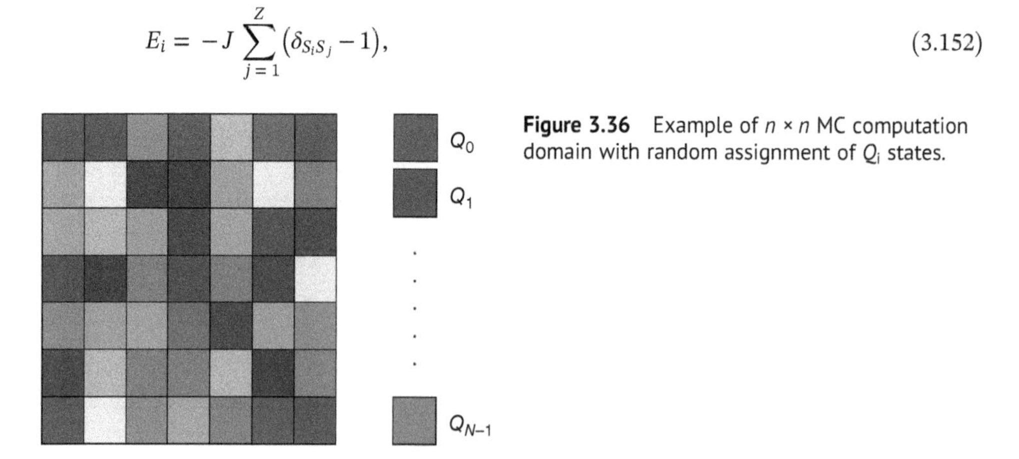

Figure 3.36 Example of $n \times n$ MC computation domain with random assignment of Q_i states.

where: J – energy between two lattice sites, Z – number of neighbours adjacent to the site, δ – Kronecker's delta with the value of 1 for the lattice sites that belong to the same state ($S_i = S_j$) and the value of 0 for the lattice sites that belong to different grains ($S_i \neq S_j$).

- The second is responsible for the change in the lattice state. The new state for investigated lattice is chosen randomly from the entire available set of states, and the energy is computed again using Eq. (3.152). This is the main difference between CA and MC approaches, as in the CA transition rules are precisely defined by the researcher.
- In the third step, the difference in the energy ΔE is determined, and the new site state is accepted when the new value of energy is no higher than the old one. This criterion can be modified according to model requirements; however, it should always lead to the reduction of the energy of the system.

The MC Pots model will be described in detail in Chapter 4 when the applications to modelling microstructure evolution are discussed.

3.3 Methods of Optimization

3.3.1 Optimization Problem Formulation

The optimization problem formulation contains at least a definition of the objective function, design variables, and in most cases, constraints. The design variables describe the properties of a numerical model, experimental setup, part of the results, and form design variables vector $\mathbf{x}$ [39, 48, 110]. The objective function value depends on the design variables:

$$\min f(\mathbf{x}). \tag{3.153}$$

The problem is called constrained when constraints on the design variable exist. There are two groups of constraints, linear $\mathbf{g}(\mathbf{x})$ and nonlinear $\mathbf{h}(\mathbf{x})$:

$$\mathbf{g}(\mathbf{x}) = 0, \tag{3.154}$$

$$\mathbf{h}(\mathbf{x}) \leq 0. \tag{3.155}$$

The optimization goal is to determine design vector values for the minimum or maximum of the objective function. The constraints limit the design variable space, and their presence leads to methods of objective function modification or algorithms allowing their presence. The optimization problem can be unimodal with only one extremum, but in many cases, often in engineering problems, one has to deal with problems with many minima. Figure 3.37 shows the typical objective functions for one design variable, the unimodal, multimodal, and noncontinuous objectives functions.

The choice of proper optimization method is important, and the modality, constraints, and continuity of objective function should be taken into account. The methods of choice of a proper optimization algorithm can be found in [64]. The following sections describe chosen optimization methods.

3.3.2 Methods of Conventional Optimization

The conventional methods can be divided into two types, gradient and non-gradient. The first group of methods can be used if the continuous objective function with gradient is available.

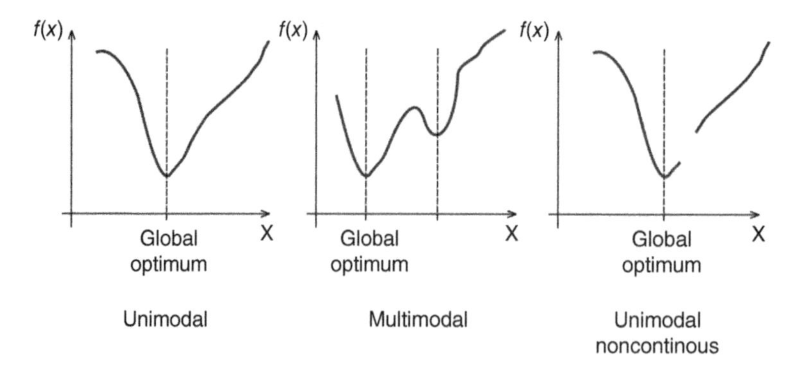

Figure 3.37 The examples of objective functions.

One of the simplest representatives of gradient-based optimization algorithms is descent methods [110]. The methods can be used for unconstrained minimization problems with the possibility of gradient vector determination for each set of design variable values. The optimization algorithm is presented in Figure 3.38.

The starting design vector must be chosen in the first step. The starting point in design space influences the number of iterations needed to converge to the optimum. If there are many optimums, the starting point is crucial because the algorithm will, in most cases, converge to the nearest local optimum. The search direction of improving the objective function is defined as:

$$\mathbf{d}(\mathbf{x}) = -\nabla f(\mathbf{x}). \tag{3.156}$$

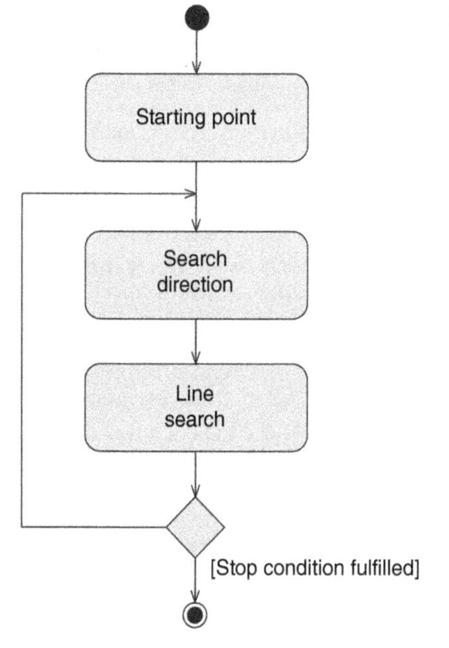

Figure 3.38 The flowchart of descent algorithm.

The change of design vector values depends on gradient values and the length of step in the direction of objective improvement. The line search algorithms are used for step length determination. There are many approaches, one of the most known is the steepest descent method using gradients values for step length determination.

One of the most commonly used gradient-based methods is sequential quadratic programming (SQP). The method allows solving an optimization problem with constraints and the continuous, smooth objective function. SQP and similar methods are implemented in many engineering software like Matlab and Octave. The methods, as the name suggests, use the iteratively quadratic programming method to solve a reformulated optimization problem with constraints. The quadratic programming method uses an approach with Lagrange multipliers for introducing the information about constraints into an optimization problem; the method also uses gradients and hessians of the modified function to obtain the location of optimum.

The problem arises when the objective function is multimodal, gradient-based methods will converge to the nearest local optimum. The global optimum may be localized using multi-start, gradient-based methods – the optimization is repeated for different starting points, such an approach does not guarantee to find global optimum but allows better search in design space.

The non-gradient algorithms typically converge slower due to a lack of precise information of descent direction. One of the well-known non-gradient algorithms is random search optimization. The cost of this method is very high; the algorithm is simple: iteratively choose a random point in design space and check the objective function value, keep information about the point if the objective is better than previously known, and iterate until stop condition formulated as a maximum number of iterations is fulfilled. The method seems to be inefficient, and many other methods are based on random search-guided algorithms where some mechanisms of the direction of search are introduced, and the cost of the method is significantly reduced.

3.3.3 Methods of Nonconventional Optimization

The nonconventional methods are, in many cases, bioinspired algorithms that have emerged in the 1970s. Most of the methods are based on randomness, but there are random typical search algorithms. The design space is searched for the global optimum by changing the design vector variables in a guided, random way. The guiding methods are based on phenomena observed in biology, mostly behaviours of systems and species. There are dozens of bioinspired algorithms nowadays [21, 85], from the first attempts based on the evolution of species – genetic algorithms – to methods based on bat behaviour, the way the weed grows etc. The following subsections are devoted to three representatives: evolutionary algorithms, artificial immune systems, and particle swarm optimization. These algorithms are based on different biology phenomena, but all of them can operate with floating point numbers as design variables without the need for introducing additional coding, and all are robust and well tested.

3.3.3.1 Evolutionary Algorithm

The evolutionary algorithm is based on the mechanism present in the evolution of species [6, 77]. The genetic algorithms operate on design variables represented by binary values. The names of genetic operators are very similar, and the design variable is called a gene. Genes create chromosomes, as in biology. The binary values of genes were used to be close to biological representation with four bases in DNA; however, such an approach is less efficient than using floating point numbers (for problems where the floating point number are natural – like properties of structures and material, dimensions of inclusions etc.). Most evolutionary and genetic algorithms use individuals with only one chromosome, and it is equivalent to design variables vector. The optimization algorithm is iterative and uses many individuals in each step. The individuals are modified with the use of genetic operators and are 'promoted' to the next iteration using the selection mechanism. The genetic operators are based on real life, and the names – mutations and crossovers – are taken directly from biology. A huge number of operators can be found in the scientific literature; they work on gene representation (binary, integer, floating point numbers, graphs, etc.) and also on a mechanism which can improve the speed or accuracy of the algorithm.

The problem is, in most cases, formulated as maximization of the objective function and the objective function is part of the fitness function. Sometimes the fitness function contains additional terms, e.g. some penalty functions, to fulfil constraints. The value of the fitness function says how the individual is close to the environment, and the goal of the evolutionary algorithm is to find the best-fitted individual.

The individual with one chromosome containing floating point number genes is shown in Figure 3.39a. The box constraints on each design variable – gene – are imposed in the algorithm:

$$g_{i_{\min}} \leq g_i \leq g_{i_{\max}}, \tag{3.157}$$

where: gene i minimum value is $g_{i_{\min}}$ and maximum is $g_{i_{\max}}$.

The evolutionary algorithm schema operating on a set of individuals is presented in Figure 3.39b. The individuals are created in the first step. The gene values can be created in a random way, but in some cases, other methods are used; sometimes the individual locations in the design space are set manually. Next, the fitness function for each individual should be computed. The individuals undergo a selection process. The idea behind the selection is to create a new set of individuals. The selection process in the evolution of species promotes the best-fitted species, but also the worst one sometimes survives. The new set of individuals is modified using evolutionary operators, and the process is repeated until

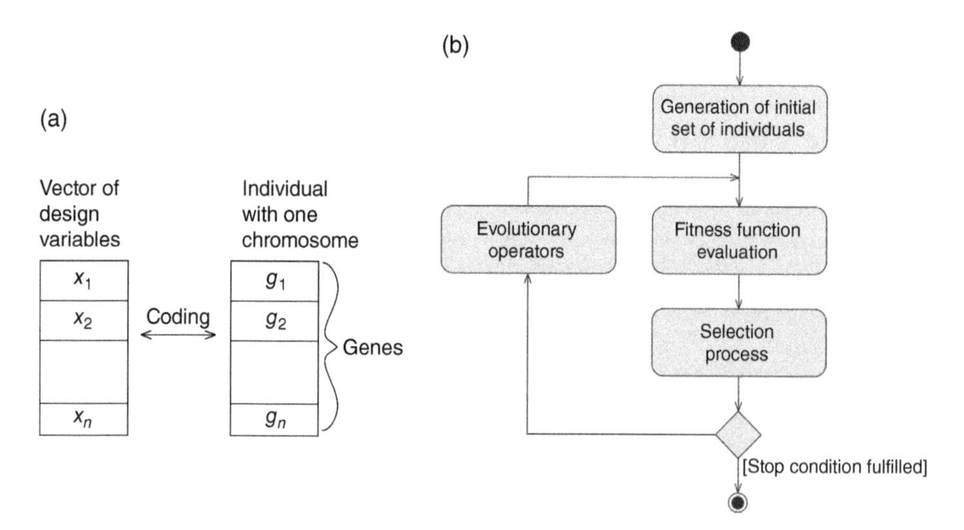

Figure 3.39 (a) Individual genes; (b) the flowchart of evolutionary algorithm.

the stop condition is fulfilled. The typical stop conditions are a maximum number of iterations, the value of fitness function for the best individual, or the decrease rate of the fitness function value.

The evolutionary operators based on biological mutation change values of genes, the random genes are chosen, and their values are altered. The number of individuals with modified genes by mutations is a parameter of evolutionary algorithm sometimes given as a probability of mutation. The probability is used during making a decision if an individual will undergo mutation; so in each iteration, the number of mutated individuals may be slightly different. The commonly used mutations are uniform and Gaussian. The uniform mutation changes the gene values by adding a random number generated using uniform distribution; consequently, Gaussian mutation uses normal distribution. The range of random numbers can be another parameter of the algorithm or can be set on the base of box constraints for the mutated gene.

The crossovers need two parent individuals exchanging information based on gene values. The simple crossover is applied to part of an individual based on the crossover probability parameter. The two individuals are randomly chosen. Next, the position of cut is randomly determined, and new offspring, individuals, are created (Figure 3.40). The floating point individuals can also be modified with arithmetic crossovers where the genes after crossover contain average values from two parent individuals. There exist many other crossover operators, e.g. similar to simple crossover but with many cut lines; also in some algorithms, crossovers are used together with mutation. An example of such an approach is a simple crossover with mutation where two parent individuals undergo simple crossover and offspring are modified using Gaussian mutation. The individuals can be modified by these two types of operators in the same iteration in such an approach.

The selection process can also be performed in many ways; the main idea is the fittest individual has the biggest chance to survive – become part of next iteration set (population) of individuals. The oldest methods are based only on fitness function value; the example of such an approach is roulette wheel selection. The method assigns part of the virtual roulette wheel according to the fitness function related to the sum of fitness of every individual. The individuals taking part in the next steps of the algorithm are chosen randomly from the roulette wheel. The individual with the highest fitness also has the highest probability to be present, but the worst individual may also be chosen. The problem with roulette wheel selection is the need for only positive fitness function values and maintaining a low difference between fitness function values of individuals (if one of them has, for example, fitness function value of 1000 and all others of around 0.01, it is very hard to not promote the best individual). The

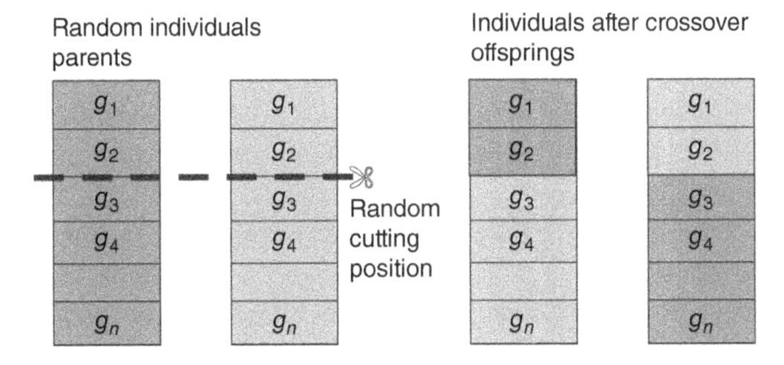

Figure 3.40 Mechanism of simple crossover operator.

remedy for such situations is in ranking (or rank) selection. These methods introduce additional rank functions based on individual numbers sorted on the basis of the fitness function list. Another popular method is tournament selection. The population of individuals is divided into groups, and each group performs a tournament. The tournament goal is to choose the best individual within the group and promote it to the next steps of the algorithm. The selection process should promote different individuals because the diversity in a set of individuals evolving during optimization is very important and allows the discovery of the basin of attraction of the global optimum.

The evolutionary algorithms have many parameters, and the performance is measured as distance to the global optimum and cost. A number of fitness function evaluations depend on parameter values but also on the optimized function. The best set of parameters cannot be decided a priori, so in most cases, the parameters used for similar problems are used in practical applications because the cost of determining parameters would be too high.

There are also popular modifications of evolutionary algorithms called distributed or multi-island versions where the population of individuals is divided into many subpopulations or islands. The evolution is performed for each part of the individual separately except migration – exchange of individuals between subpopulation occurring after some predefined number of iterations [63].

3.3.3.2 Artificial Immune System

The immune system of mammals is a complex biological system containing specialized cells, barriers, and mechanisms. The artificial immune system is based on the cellular part of the biological version. There are many algorithms used for optimization and learning, data partition etc. The optimization algorithms called clonal selection algorithms [22] are based on mechanisms observed in some of the lymphocyte B cells. The B cell can recognize the pathogen and become activated, and later some of them may become memory cells. The B cell is present in the organism for some period of time and is responsible for a primary response to the presence of pathogens. The memory cell plays a crucial role during the secondary response (if the same or a very similar pathogen is present in an organism) is fast response, the proliferation (cloning) of B cells similar to the memory cell. This biological mechanism is, of course, much more complicated. It involves many other types of cells, mechanisms, and chemical signals. The artificial immune system uses only some features of the biological system. The design variable vectors are stored as B cells and memory cells. The B and memory cells play a similar role to individual chromosomes in evolutionary algorithms. The optimization algorithm is shown in Figure 3.41.

The first step is like in an evolutionary algorithm, the values of design variables stored in memory cells are generated randomly or using other methods, and the objective function for each B cell is computed. The proliferation occurs in the following step. The proliferation is a process where the B cells are created as copies of memory cells. The number of created B cells from each memory cell depends on the objective function value. The better the memory cell is, the more B cells are created containing the same vector of design variables. The B cells undergo a hypermutation process; this is similar to the biological immune system. Hypermutation means that the changes of the genetic material recognizing pathogens can be huge. In the case of an artificial immune system, similar operations to evolutionary algorithm mutation can be used. It is typical that crossover mechanisms are not present in the algorithm. The objective function is computed for modified B cells, and some of the B cells may exchange memory cells. The process of removing memory cells is based on their objective function value, B cell value, and distance between the memory cell and B cell in design space. The

Figure 3.41 Flowchart of the clonal algorithm, which is one of the artificial immune systems.

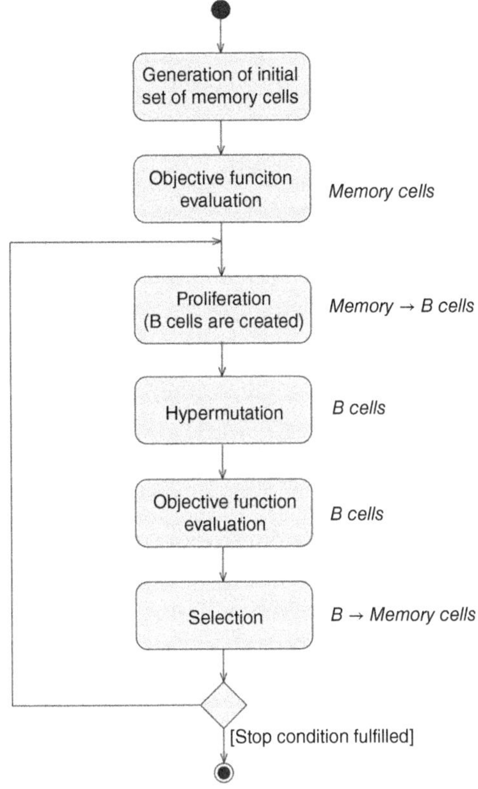

comparison is made for the memory cell and all B cells present in the algorithm. If the better (with higher objective function value) B cell exists near the memory cell, the B cell becomes a memory cell, and the old memory cell is removed. The often-used mechanism in clonal selection algorithms is crowding. The presence of many memory cells near the same basin of attraction of the objective value lowers the resistance of the algorithm to local minima. The crowding mechanism checks distances between memory cells in a design space and removes some of them if they are too close to each other (and the new B cell is introduced in the place of the memory cell). The algorithm iterates until the end of the computing condition is fulfilled like in evolutionary algorithms.

3.3.3.3 Particle Swarm Optimization

Swarms of animals in nature seem to be moving in a coordinated, complex way; however, if one looks closer, a few typical rules for each member of the swarm can be distinguished. The animals move in a similar direction avoiding collisions, and there are leaders followed by rest of the swarm members. The particle swarm optimization algorithm [58] took a few rules out of the swarms found in nature. The position of each member particle is a vector of design variables describing a location in the design space. Each particle moves through the design space with velocity described by velocity vector in time – iterations of the algorithm. The particle positions are updated on the basis of the current velocity of each of them, and the velocity is the main factor allowed to go through the design space during a search for the best solution of an optimization problem. The impact of the velocity of a particle is on the particles in the

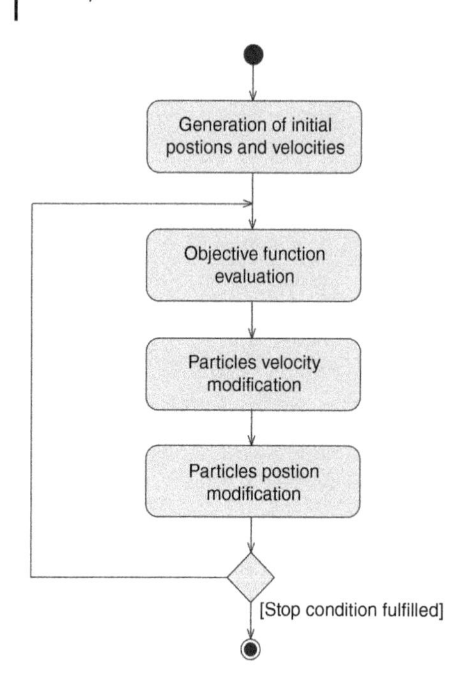

Figure 3.42 Flowchart of particle swarm optimization.

neighbourhood (so we avoid collision and go in a similar direction) but also the velocity of the leader of the swarm – a particle with the highest objective function value. The velocity may also be influenced by the best particle's velocity in the neighbourhood and some relatively small, random changes. The design variables are changed by two equations determining position and velocities:

$$x_i^{\text{next iteration}} = x_i + v_i \Delta t \tag{3.158}$$

$$v_i^{\text{next iteration}} = v_i + c_1 v_i^{\text{leader}} + c_2 v_i^{\text{local neighbourhood}} + c_3 v_i^{\text{random}}, \tag{3.159}$$

where: $x_i^{\text{next iteration}}$ is a position in the design space in next iteration, design variable i, x_i is current position, design variable i, v_i is a velocity in direction i, Δt is a time step used for position computing on the basis of velocity, $v_i^{\text{next iteration}}$ is the velocity in next iteration in direction i, c_j is weighting coefficients for velocity components, v_i^{leader} is the velocity of the leader of the swarm in direction i, $v_i^{\text{local neighbourhood}}$ is the average velocity of the local neighbourhood of the particle, v_i^{random} is a random modifier of the velocity.

The optimization algorithm flowchart is shown in Figure 3.42. The first step is to determine initial positions and velocities for particles. This step is similar to the evolutionary algorithm. Next, the objective function for all particles must be evaluated. The velocities and positions are updated, and the algorithm iterates until the end condition is satisfied.

3.3.3.4 Hybrid Optimization Algorithms

Hybrid optimization algorithms are often built with global and local algorithms. It is a common procedure to perform in the first stage of optimization using for example evolutionary algorithm. Later a gradient-based algorithm for the best-found solution is initiated to improve results. The hybridization can also be carried out by introducing operators based on gradient algorithms together with evolutionary operators [87]. The hybrid algorithm may also be built on top of global algorithms working similar to multi-island evolutionary algorithms where each 'island' is performing an optimization process using different algorithms exchanging results every few iterations. There are also concepts using different types of bioinspired algorithms one after the other to improve the result of optimization and lower the number of objective function evaluations [105]. The goal of hybridization is to use all advantages of algorithms while reducing its disadvantages.

References

1 Alder, B.J. and Wainwright, T.E. (1957). Phase transition for a hard sphere system. *Journal of Chemical Physics* 275: 1208–1209.

2 Alder, B.J. and Wainwright, T.E. (1959). Studies in molecular dynamics. I. General method. *Journal of Chemical Physics* 31 (2): 459–466.

3 Andersen, H. (1980). Molecular dynamics simulation at constant pressure and/or temperature. *Journal of Chemical Physics* 72: 2384–2393.

4 Armero, F. and Garikipati, K. (1995). Recent Advances in the analysis and numerical simulation of Strain localization in inelastic solids. Proceedings Conference COMPLAS95, (eds. Owen, D.R.J. and Onate, E.), Barcelona, 547–561.

5 Arun Babua, K., Prithiv, T.S., Gupta, A., and Mandal, S. (2021). Modeling and Simulation of dynamic recrystallization in super austenitic stainless steel employing combined cellular automaton, artificial neural network and finite element method. *Computational Materials Science* 195: 110482.

6 Back, T., Fogel, D.B., and Michalewicz, Z. (1997). *Handbook of Evolutionary Computation.* Bristol: IOP Publishing Ltd.

7 Banerjee, P.K. (1994). *The Boundary Element Method in Engineering*. London: McGraw-Hill Book Company.

8 Baudin, T., Penelle, R., and Liu, Y. (1996). Simulation of normal grain growth by cellular automata. *Scripta Materialia* 34: 1679–1683.

9 Beer, G. (1983). Finite element, boundary element and coupled analysis of unbounded problems in elastostastics. *International Journal for Numerical Methods in Engineering* 19: 567–580.

10 Berendsen, H., Postma, J., van Gunsteren, W. et al. (1984). Molecular dynamics with coupling to an external bath. *Journal of Chemical Physics* 81: 3684–3690.

11 Bishop, M., Kalos, M.H., and Frisch, H.L. (1979). Molecular dynamics of polymeric systems. *Journal of Chemical Physics* 70: 1299–1304.

12 Bitzek, E., Koskinen, P., Gähler, F. et al. (2006). Structural relaxation made simple. *Physical Review Letters* 97: 170201.

13 Blikstein, P. and Tschiptschin, A.P. (1999). Monte Carlo simulation of grain growth. *Materials Research* 2: 133–137.

14 Born, M. and Huang, K. (1954). Elasticity and stability. In: *Dynamical Theory of Crystal Lattices* (ed. D.H. Wilkinson and W. Marshall). Oxford: Clarendon Press.

15 Brebbia, C.A. and Dominiguez, J. (1989). *Boundary Elements: An Introductory Course.* New York: Computational Mechanics Publications.

16 Brebbia, C.A., Telles, J.C.F., and Wrobel, L.C. (1984). *Boundary Element Techniques.* Berlin: Springer–Verlag.

17 Burbelko, A. (2004). *Mezomodelowanie krystalizacji metodą automatu komórkowego*. Wydawnictwo AGH (in polish).

18 Burczyński, T. (1995). *The Boundary Element Method in Mechanics*. Warsaw: WNT.

19 Burczyński, T. (ed.) (2001). *Advanced Mathematical and Computational Mechanics Aspects of the Boundary Element Method*. Dordrecht: Kluwer Publishers.

20 Burczyński, T. and Grabacki, J. (1998). The boundary element method, part IV. In: *Handbook of Computational Solid Mechanics* (ed. M. Kleiber). Berlin: Springer-Verlag.

21 Burczyński, T., Kuś, W., Beluch, W. et al. (2020). *Intelligent Computing in Optimal Design*. Springer.

22 de Castro, L.N. and von Zuben, F.J. (2002). Learning and optimization using the clonal selection principle. *IEEE Transaction on Evolutionary Computation Special Issue Artificial Immune System* 6 (3): 239–251.

23 Chen, S.P., Egami, T., and Vitek, V. (1988). Local fluctuations and ordering in liquid and amorphous metals. *Physical Review B* 37: 2440–2448.

24 Choe, J.I. and Byungchul, K. (2000). Determination of proper time step for molecular dynamics simulation. *Bulletin of the Korean Chemical Society* 21: 419–424.

25 Conway, J.H. (1976). *On Numbers and Games*. New York: Academic Press.

26 Davies, C.H.J. (1995). The effect of neighbourhood on the kinetics of a cellular automation recrystallisation model. *Scripta Metallurgica et Materialia* 33: 1139–1143.

27 Davies, C.H.J. (1997). Growth of nuclei in a cellular automaton simulation of recrystalization. *Scripta Materialia* 36: 35–40.

28 Davies, C.H.J. and Hong, L. (1999). The cellular automaton simulation of static recrystallization in cold-rolled AA1050. *Scripta Materialia* 40: 1145–1150.

29 Daw, M.S. and Baskes, M.I. (1984). Embedded-atom method: derivation and application to impurities, surfaces, and other defects in metals. *Physical Review B* 29: 6443–6453.

30 Ding, R. and Guo, Z. (2001). Coupled quantitative simulation of microstructural evolution and plastic flow during dynamic recrystallization. *Acta Materialia* 49: 3163–3175.

31 Ding, R. and Guo, Z. (2002). Microstructural modelling of dynamic recrystallisation using an extended cellular automaton approach. *Computational Material Science* 23: 209–218.

32 Ding, R. and Guo, Z. (2004). Microstructural evolution of a Ti–6Al–4V alloy during β-phase processing, experimental and simulative investigations. *Material Science and Engineering, A* 365: 172–179.

33 Długosz, A. and Burczyński, T. (2013). Identification in multiscale thermoelastic problems. *Computer Assisted Methods in Engineering and Science* 20: 325–336.

34 Dłużewski, P. and Traczykowski, P. (2003). Numerical simulation of atomic positions in quantum dot by means of molecular statics. *Archives of Mechanics* 55: 501–514.

35 Dominguez, J. (1993). *Boundary Elements in Dynamics*. Southampton-Boston, London-New York: Computational Mechanics Publications, Elsevier Applied Science.

36 Elizondo, A. (2007). Horizontal coupling in continuum atomistics. PhD thesis.

37 Elsayeda, W.M., Elmogy, M., and El-Desouky, B.S. (2021). DNA sequence reconstruction based on innovated hybridization technique of probabilistic cellular automata and particle swarm optimization. *Information Sciences* 547: 828–840.

38 Fish, J. (2014). *Practical Multiscaling*. Wiley.

39 Fletcher, R. (2000). *Practical Methods for Optimization*. Wiley.

40 Florindo, J.B. and Metze, K. (2021). A cellular automata approach to local patterns for texture recognition. *Expert Systems with Applications* 179: 115027.

41 Francis, A., Roberts, C.G., Cao, Y. et al. (2007). Monte Carlo simulations and experimental observations of templated grain growth in thin platinum films. *Acta Materialia* 55: 6159–6169.

42 Frenkel, D. and Smit, B. (2002). *Understanding Molecular Simulation, from Algorithms to Applications*. Academis Press.

43 Gandin, C.A. and Rappaz, M. (1994). A coupled finite element – cellular automaton model for the prediction of dendritic grain structures in solidification processes. *Acta Metallurgica* 42: 2233–2246.

44 Gandin, C., Desbiolles, J., Rappaz, M., and Thevoz, P. (1999). A three-dimensional cellular automaton–finite element model for the prediction of solidification grain structures. *Metallurgical and Materials Transactions A* 30A: 3153–3165.

45 Gaul, L., Kögl, M., and Wagner, M. (2003). *Boundary Element Methods for Engineers and Scientists: An Introductory Course with Advanced Topics*. Springer-Verlag.

46 Gawąd, J., Maciol, P., and Pietrzyk, M. (2005). Multiscale modeling of microstructure and macroscopic properties in thixoforming process using cellular automaton technique. *Archives of Metallurgy and Materials* 50: 549–562.

47 Geers, G.D., Kouznetsova, V.G., Matous, K., and Yvonnet, J. (2017). *Homogenisation Methods and Multiscale Modelling: Nonlinear Problems*, Encyclopedia of Computational Mechanics. Wiley.

48 Gill, P.E., Murray, W., and Wright, M.H. (1997). *Practical Optimization*. Academic Press.

49 Goetz, R. (2005). Particle stimulated nucleation during dynamic recrystallization using a cellular automata model. *Scripta Materialia* 52: 851–856.

50 Goetz, R. and Seetharaman, V. (1998). Static recrystallization kinetics with homogeneous and heterogeneous nucleation using a cellular automata model. *Metallurgical and Materials Transactions A* 29: 2307–2321.

51 Griebel, M., Knapek, S., and Zumbusch, G. (2007). *Numerical Simulation in Molecular Dynamics: Numerics, Algorithms, Parallelization, Applications*, Texts in Computational Science and Engineering, 5. Springer.

52 Guillemot, G., Gandin, C.A., Combeau, H., and Heringer, R. (2004). A new cellular automaton – finite element coupling scheme for alloy solidification. *Modelling and Simulation in Materials Science and Engineering* 12: 545–556.

53 Guiso, S., Caprio, D., de Lamare, J., and Gwinnera, B. (2021). Influence of the grid cell geometry on 3D cellular automata behavior in intergranular corrosion. *Journal of Computational Science* 53: 101322.

54 Haile, J.M. (1992). *Molecular Dynamics Simulation, Elementary Methods*. USA: Wiley.

55 Heermann, D.W. (1990). *Computer Simulation Methods in Theoretical Physics*. Berlin Heidelberg: Springer-Verlag.

56 Hesselbarth, H. and Göbel, I. (1991). Simulation of recrystallization by cellular automata. *Acta Metallurgica* 39: 2135–2143.

57 Hoover, W. (1985). Canonical dynamics: equilibrium phase-space distributions. *Physical Review A* 31: 1695–1697.

58 Kennedy, J. and Eberhart, R.C. (2001). *Swarm Intelligence*. Morgan Kauffman.

59 Kobayashi, S., Oh, S.I., and Altan, T. (1989). *Metal Forming and the Finite Element Method*. New York, Oxford: Oxford University Press.

60 Kouznetsova, V.G. (2002). Computational homogenisation for the multi-scale analysis of multi-phase materials. PhD Thesis. TU Eindhoven.

61 Kroc, J. (2001). Simulation of dynamic recrystallization by cellular automata. PhD Thesis. Charles University, Prague. Czech Republic.

62 Kühbach, M., Barrales-Mora, L.A., and Gottstein, G. (2014). A massively parallel cellular automaton for the Simulation of recrystallization. *Modelling and Simulation in Materials Science and Engineering* 22: 075016.

63 Kuś, W. (2007). Grid-enabled evolutionary algorithm application in the mechanical optimization problems. *Engineering Applications of Artificial Intelligence* 20 (5): 629–636.

64 Kuś, W. and Mucha, W. (2014). Idea of the optimization strategy for industrial processes. *Computer methods in Material Science* 14: 13–19.

65 Kwon, Y.W. and Jung, S.H. (2003). Discrete atomic and smeared continuum modelling for static analysis. *Engineering Computations* 20: 964–978.

66 Lach, L., Nowak, J., and Svyetlichnyy, D. (2018). The evolution of the microstructure in AISI 304L stainless steel during the flat rolling – modeling by frontal cellular automata and verification. *Journal of Materials Processing Technology* 255: 488–499.

67 Lenard, J.G., Pietrzyk, M., and Cser, L. (1999). *Mathematical and Physical Simulation of the Properties of Hot Rolled Products*. Amsterdam: Elsevier.

68 Liu, J.S. (2008). *Monte Carlo Strategies in Scientific Computing*. Springer Series in Statistics.

69 Liu, W.K., Jun, S., and Qian, D. (2005). Multiscale modelling of materials mechanics. In: *Computational Nanomechanics of Materials, Handbook of Theoretical and Computational Nanotechnology* (ed. M. Rieth and W. Schommers). Stevenson Ranch: American Scientific Publishers.

70 Liu, W.K., Karpov, E.G., and Park, H.S. (2006). *Nano Mechanics and Materials: Theory Multiscale Methods and Applications*. Wiley.

71 Lo, S.H. (2013). *Finite Element Mesh Generation*. CRC Press.

72 Madej, L., Hodgson, P., Gawad, J., and Pietrzyk, M. (2004), Modeling of rheological behavior and microstructure evolution using cellular automaton technique. Proceedings ESAFORM 2004, (ed. Støren, S.), Trondheim, 143–146.

73 Madej, L., Sitko, M., and Pietrzyk, M. (2016). Perceptive comparison of mean and full field dynamic recrystallization models. *Archives of Civil and Mechanical Engineering* 16: 569–589.

74 Madej, L., Sitko, M., Legwand, A. et al. (2018). Development and evaluation of data transfer protocols in the fully coupled random cellular automata finite element model of dynamic recrystallization. *Journal of Computational Science* 26: 66–77.

75 Majta, J., Madej, L., Svyetlichnyy, D.S. et al. (2016). Modeling of the inhomogeneity of grain refinement during combined metal forming process by finite element and cellular automata methods. *Materials Science and Engineering: A* 671: 204–213.

76 Marx, V., Reher, F.R., and Gottstein, G. (1999). Simulation of primary recrystallization using a modified three-dimensional cellular automaton. *Acta Materialia* 47: 1219–1230.

77 Michalewicz, Z. (1996). *Genetic Algorithms + Data Structures = Evolution Programs*. Berlin: Springer.

78 Mortazavi, B., Ostadhossein, A., Rabczuk, T., and van Duin, A.C.T. (2016). Mechanical response of all-MoS2 single-layer heterostructures: a ReaxFF investigation. *Physical Chemistry Chemical Physics* 18 (34): 23695–23701.

79 Mrozek, A. (2019). Basic mechanical properties of 2H and 1T single-layer molybdenum disulfide polymorphs: a short comparison of various atomic potentials. *International Journal for Multiscale Computational Engineering* 17 (3): 339–359.

80 Mrozek, A. and Burczyński, T. (2013). Examination of mechanical properties of graphene allotropes by means of computer simulation. *Computer Assisted Methods in Engineering and Science* 20 (4): 309–323.

81 Mrozek, A., Kuś, W., and Burczyński, T. (2007). Application of the coupled boundary element method with atomic model in the static analysis. *Computer Methods in Materials Science* 7: 284–288.

82 Mukhopadhyay, P., Loeck, M., and Gottstein, G. (2004). Simulation of microstructure evolution during recrystallization using a high-resolution three-dimensional cellular automaton. *Journal de Physique IV* 120: 225–230.

83 Nosé, S. (1984). A molecular dynamics method for simulations in the canonical ensemble. *Molecular Physics* 53: 255–268.

84 Ogierman, W. and Kokot, G. (2017). Homogenization of inelastic composites with misaligned inclusions by using the optimal pseudo-grain discretization. *International Journal of Solids and Structures* 113/114 (9): 230–240.

85 Okwu, M.O. and Tartibu, L.K. (2021). *Metaheuristic Optimization: Nature-Inspired Algorithms Swarm and Computational Intelligence, Theory and Applications, Studies in Computational Intelligence 927*. Springer.

86 Olivier, J. (1995). Continuum modelling of strong discontinuities in solid mechanics. Proceedings Conference COMPLAS95, (eds. Owen, D.R.J. and Onate, E.), Barcelona, 455–479.

87 Orantek, P. (2004). Hybrid evolutionary algorithm in optimization of structures under dynamical loads. In: IUTAM Symposium on Evolutionary Methods in Mechanics. Proceedings of the IUTAM Symposium held in Cracow, Poland, 24–27 September 2002, Springer, pp. 297–308.

88 Perzynski, K. and Madej, L. (2016). Fracture modelling in dual phase steel grades based on the discrete/continuum random cellular automata finite element RCAFE approach. *Simulation* 92: 195–207.

89 Pietrzyk, M. (2000). Finite element simulation of large plastic deformation. *Journal of Materials Processing Technology* 106: 223–229.

90 Pietrzyk, M., Madej, L., Rauch, L., and Szeliga, D. (2015). *Computational Materials Engineering: Achieving High Accuracy and Efficiency in Metals Processing Simulations*. Butterworth Heinemann Elsevier.

91 Press, W.H., Teukolsky, S.A., Vetterling, W.T., and Flannery, B.P. (2007). *Numerical Recipes*, The Art of Scientific Computing, 3e. Cambridge University Press.

92 Ptaszny, J. (2015). Accuracy of the fast multipole boundary element method with quadratic elements in the analysis of 3D porous structures. *Computational Mechanics* 56 (3): 477–490.

93 Qian, M. and Guo, Z. (2004). Cellular automata simulation of microstructural evolution during dynamic recrystallization of an HY-100 steel. *Material Science and Engineering A* 365: 180–185.

94 Raabe, D. (1998). *Computational Materials Science*, The Simulation of materials microstructures and properties. Weinheim, New York: Wiley-VCH.

95 Raabe, D. (2002). Cellular automata in materials science with particular reference to recrystallization simulation. *Annual Review of Materials Research* 32: 53–76.

96 Raabe, D. (2004). Mesoscale Simulation of spherubibe growth during polymer crystallization by use of a cellular automaton. *Acta Materialia* 52: 2653–2664.

97 Raabe, D. and Hantcherli, L. (2005). 2D cellular automaton simulation of the recrystallization texture of an if sheet steel under consideration of Zener pinning. *Computational Material Science* 34: 299–313.

98 Rahman, A. (1964). Correlations in the motion of atoms in liquid Argon. *Physical Review* 136: 405–411.

99 Rapaport, D. (2004). *The Art of Molecular Dynamics Simulation*. UK: Cambridge University Press.

100 Reddy, J.N. (2006). *An Introduction to the Finite Element Method*, 3e. McGraw Hill.

101 Reuther, K. and Rettenmayr, M. (2014). Perspectives for cellular automata for the simulation of dendritic solidification – A review. *Computational Materials Science* 95: 213–220.

102 Rojek, J., Zienkiewicz, O.C., Onate, E., and Postek, E. (2001). Advances in FE explicit formulation for simulation of metalforming processes. *Journal of Materials Processing Technology* 119: 41–47.

103 Rollett, A. and Raabe, D. (2001). A hybrid model for mesoscopic simulation of recrystallization. *Computational Material Science* 21: 69–78.

104 Roth, T.O. (2021). About the robustness of 1d cellular automata revising their temporal entropy. *Physica D: Nonlinear Phenomena* 425: 132953.

105 Sebastjan, P. and Kuś, W. (2021). Optimization of material distribution for the forged automotive component using hybrid optimization techniques. *Computer Methods in Material Science* 21: 63–74.

106 Sheppard, D., Terrell, R., and Henkelman, G. (2008). Optimization methods for finding minimum energy paths. *The Journal of Chemical Physics* 128: 134106.

107 Sitko, M., Pietrzyk, M., and Madej, L. Time and length scale issues in numerical modelling of dynamic recrystallization based on the multi space cellular automata method. *Journal of Computational Science* 16: 98–113.

108 Sladek, V. and Sladek, J. (1983). Boundary integral equation method in thermoelasticity, part I: general analiysis. *Applied Mathematical Modelling* 7: 241–253.

109 Sluzalec, A. (1992). *Introduction to Nonlinear Thermodynamics*. London: Springer Verlag.

110 Snyman, J. (2005). *Practical Mathematical Optimization, An Introduction to Basic Optimization Theory and Classical and New Gradient-Based Algorithms*. Springer.

111 Sunyk, R. and Steinmann, P. (2002). On higher gradients in continuum-atomistic modeling. *International Journal of Solids and Structures* 40: 6877–6896.

112 Svyetlichnyy, D. and Milenin, A.(2005). Modelowanie procesów rekrystalizacji za pomocą automatów komórkowych, (eds. Piela, A., Lisok, J., and Grosman, F.), Proceedings KomPlasTech05, Ustroń, (in Polish).

113 Svyetlichnyy, D.S., Muszka, K., and Majta, J. (2015). Three-dimensional frontal cellular automata modeling of the grain refinement during severe plastic deformation of microalloyed steel. *Computational Materials Science* 102: 159–166.

114 Traka, K., Sedighiani, K., Bos, C. et al. (2021). Topological aspects responsible for recrystallization evolution in an IF-steel sheet – Investigation with cellular-automaton simulations. *Computational Materials Science* 198: 110643.

115 Verlet, L. (1967). Computer 'experiments' on classical fluids, I. Thermodynamical properties of Lennard–Jones molecules. *Physical Review* 159: 98–103.

116 Von Neumann, J. (1966). Fourth lecture. In: *Theory of Self Reproducing Automata* (ed. A.W. Bamk), 64–87. Urbana: University of Illinois.

117 Wagoner, R.H. and Chenot, J.L. (1997). *Fundamentals of Metal Forming*. New York: Wiley.

118 Wetterich, C. (2021). Probabilistic cellular automata for interacting fermionic quantum field theories. *Nuclear Physics B* 963: 115296.

119 Wolfram, S. (1994). Universality and complexity in cellular automata. *Physica, D* 10: 1–35.

120 Xiong, S. and Cao, G. (2015). Molecular dynamics simulations of mechanical properties of monolayer MoS2. *Nanotechnology* 10: 185705.

121 Yan, F., Zhang, W., Pan, P.-Z., and Li, S.-J. (2021). Dynamic crack propagation analysis combined the stable scheme and continuous-discontinuous cellular automaton. *Engineering Fracture Mechanics* 241: 107390.

122 Yazdipour, N., Davis, C.H.J., and Hodgson, P.D. (2007). Simulation of dynamic recrystalization using random grid cellular automata. *Computer Methods in Materials Science* 7: 168–174.

123 Yu, Q. and Esche, S.K. (2003). A Monte Carlo algorithm for single phase normal grain growth with improved accuracy and efficiency. *Computational Materials Science* 27: 259–270.

124 Yu, Q. and Esche, S.K. (2005). A multi-scale approach for microstructure prediction in thermo-mechanical processing of metals. *Journal of Material Processing Technology* 169: 493–502.

125 Zienkiewicz, O.C. and Taylor, R.L. (2000). *The Finite Element Method*. London: Butterworth Heinemann.

126 Zienkiewicz, O.C., Taylor, R.L., and Zhu, J.Z. (2013). *The Finite Element Method: Its Basis and Fundamentals*, 7e. Elsevier.

4

Preparation of Material Representation

This chapter is devoted to innovative approaches to numerical modelling, taking into account different nanoscale or microscale features explicitly during simulation with the previously described numerical methods. Different approaches to the generation of nanostructures and microstructures for subsequent numerical calculations are detailed in the following sections.

4.1 Generation of Nanostructures

4.1.1 Modelling of Polycrystals and Material Defects

Atomistic simulations based on the molecular dynamics (MD) or molecular statics (MS) methods (see Section 3.2) need initial solutions – sets of the atom's coordinates that serve as starting positions. Depending on the type and aim of the simulation, the initial positions of atoms can be generated randomly or formed into a desired, unique spatial structure called the digital representation of the material.

In the simplest molecular models, atoms are placed in the nodes of the regular crystal lattice, generated based on crystallography. The crystal lattice is built from the unit cell, replicated infinitely in all of the crystallographic directions. The four of the most common types of unit cells, the simple cubic cell (SC), the body-centred cell (BCC), the face-centred cell (FCC), and the hexagonal close-packed cell (HCP), are shown in Figure 4.1 [8].

However, such models have anisotropic mechanical properties, distinctive for monocrystals. Preparation of a proper nanoscale material representation that properly mimics features well known from the macroscopic world is a real challenge because, in addition to the right type of atomic interactions, the spatial configuration of atoms and nanodefects (dislocations, vacancies, inclusions) is crucial to the overall macroscopic properties. This is a complex problem related to dislocation's mechanics and generally is beyond the scope of this book. In this chapter, methods of building atomistic polycrystalline structures with randomly orientated monocrystal grains and the role of point defects in mechanical properties of two-dimensional materials are described.

Several methods of creation of polycrystalline structures have already been developed [32, 33, 50, 52]. Such techniques are usually based on the three different approaches:

- geometrical constructions,
- digital image processing, and
- MD simulations.

Multiscale Modelling and Optimisation of Materials and Structures, First Edition. Tadeusz Burczyński, Maciej Pietrzyk, Wacław Kuś, Łukasz Madej, Adam Mrozek, and Łukasz Rauch.

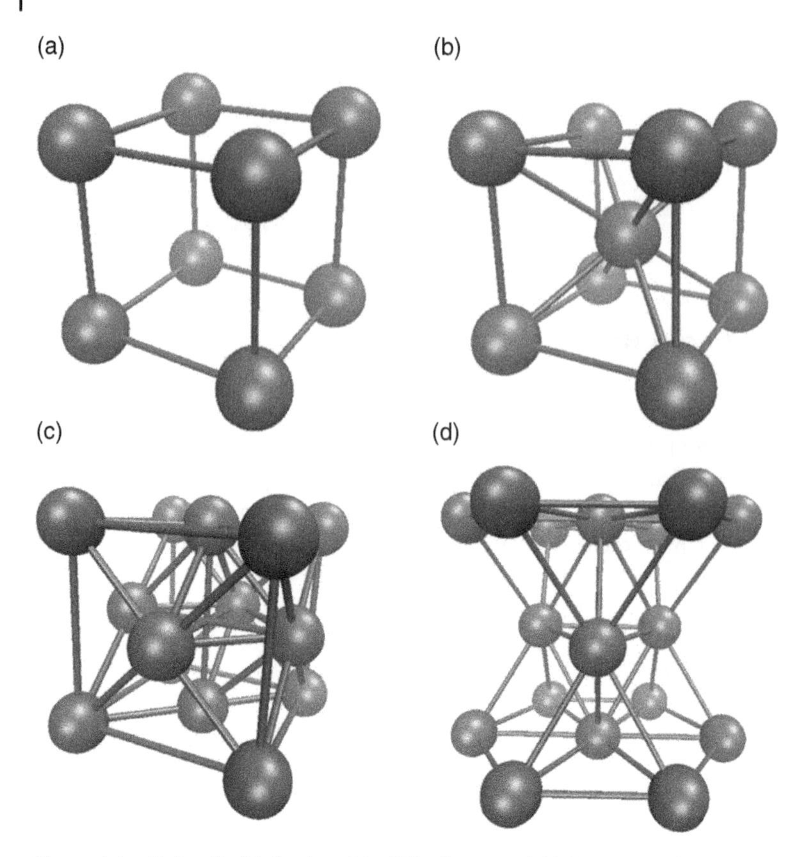

Figure 4.1 Unit cells: (a) simple cubic, (b) body-centred, (c) face-centred, and (d) hexagonal close-packed.

The geometrical methods (e.g. Voronoi tessellation [14]) divide the simulation domain into a set of polyhedra. In each polyhedron, the regular atomic lattice is created with random crystallographic orientation. Algorithms of this kind are fast and don't require huge computational powers. The main drawbacks of these methods are problems with obtaining natural, realistic shapes with a proper random distribution of the artificially generated nanocrystals. Additionally, molecular models created in this way need extra MD run due to equilibration of the whole structure, especially the grain boundaries. The equilibration should be performed with the atomic potential and at the temperature appropriate for further simulations. The geometrical methods, grain-growth approaches, as well as image processing are described in the subsequent chapter related to the models of the microstructures.

Herein, methods of creation of the polycrystalline materials based on the MD simulations will be presented with examples and discussed in detail. The essentials of the MD method have already been described and explained in Section 3.2. Unlike the purely geometrical approaches, in the MD simulations, the nanocrystalline structure is created as a result of the atom's motion according to Newton's equations of motion under specified thermodynamic conditions and interaction models. This obviously implies higher computational cost; however, the algorithms are simpler and more intuitive than ones used in geometrical constructions or image processing and result in models of materials created on a physical basis. Additionally, polycrystalline structures obtained in this way are ready for use in further simulations because they usually utilize the same atomic potentials and final temperatures.

4.1.1.1 Controlled Cooling

The most common MD-based approach to creating polycrystalline structures is the controlled melting and cooling of the molecular system. Like in thermal processing of real metallic materials, grains of different sizes and shapes can be obtained by adjusting the speeds of the melting and cooling and initial and final temperatures. This method has many practical variants, e.g.:

- melting of the ideal monocrystal and controlled cooling,
- starting from unstable spatial configuration at low temperatures, and
- starting from high temperature at an unstable spatial configuration and subsequently cooling.

During the process of melting of the regular atomic lattice, the temperature, thus the kinetic energy of the thermostated system, increases. Atoms start to move with high velocities like in real boiling metallic liquid and take random positions with proper interatomic distances determined by features of the applied potential in a certain time step. This is an important thing because in the case of purely randomly generated coordinates, distances between the atoms much shorter than equilibrium bond length result in high potential energies and may cause MD simulation unstable. On the other hand, distances that are too long may isolate atoms from interaction with others in the considered system.

A subsequent decrease in the temperature reduces the kinetic energy of atoms, which start to solidify, forming a polycrystalline structure. The melting temperature and the speed of the cooling are critical and determine the size and shape of the created, randomly orientated polycrystalline grains. Values of these parameters depend on the type (mass) of atoms, potential model (pair-wise, many-body), and size of the molecular system and should be tuned experimentally. This method can be used with any common type of boundary conditions (i.e. periodic or reflecting walls) in two- or three-dimensional simulations.

Alternatively, the melting part may be replaced with a specially prepared set of coordinates of atoms that define the initial solution for MD solver: a type of regular atomic lattice that does not correspond to the applied atomic potential [25, 32, 50, 52]. An attempt to determine the equilibrium state of such a system forces spontaneous rearrangement of atoms, which forms another stable spatial configuration with a nanocrystalline structure. For example, in the molecular modelling of FCC metals like Ni or Al, the unstable initial solution can be built of regular SC or BCC lattice, generated with appropriate (for a given atom's type) lattice constant.

This process can also be boosted by the high kinetic energy of the particles (i.e. high starting temperature). It can be noted that skipping the melting part decreases the total number of integration steps, and thus reduces the cost of the computations.

The capabilities and drawbacks of this method were presented in the subsequent numerical examples and discussed. The MD simulation (NVT ensemble) of the controlled cooling of the aluminium cluster is shown in Figure 4.2. The initial solution was formed from approximately 257000 atoms, arranged into regular BCC lattice with the 4 Å spacings (Figure 4.2a), and randomly generated velocities, seeded to the temperature of 1800 K. The embedded atom method (EAM) [12, 13] fitted for Al [26] was applied as a many-body interaction model for the monoatomic FCC metals [19]. The integration time step was equal to 1 fs. All the atoms can move freely inside the simulation domain constrained by reflecting walls.

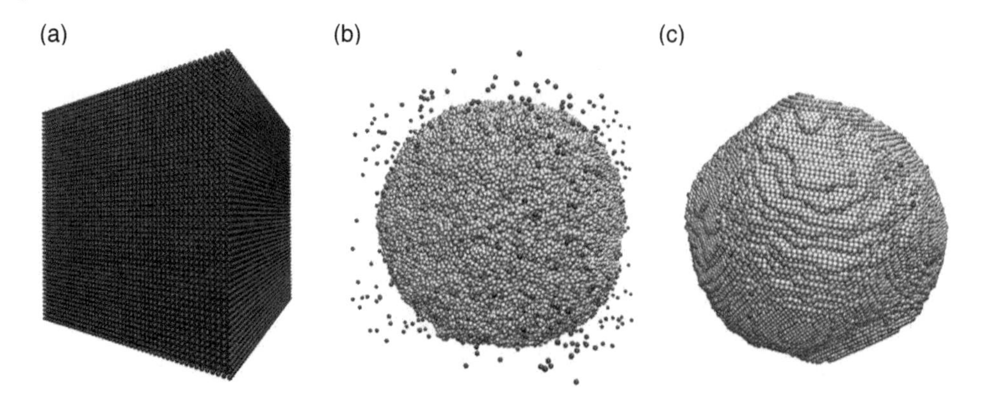

Figure 4.2 Controlled cooling of the Al cluster: (a) unstable initial configuration, (b) boiling state, and (c) solidified structure.

The unstable spatial configuration and high kinetic energy force atoms to boil (Figure 4.2b). Subsequent reduction of the temperature (to 300 K in this case) decreases the speed of atoms and starts the process of solidification. The resultant structure is stable, fully equilibrated at final temperature (Figure 4.2c), and may serve as a numerical model of polycrystalline material in further MD simulations. Obtained results for cooling times equal to 200 ps, 1ns, and 5ns are presented in Figure 4.3a–c, respectively. In each case, the drop-like structure with randomly orientated FCC nanograins was created, and the longer cooling times result in larger nanocrystals.

Polycrystalline structures created with a cooling method and the simplest pair-wise Lennard-Jones potential (ε =1 eV, σ = 0.25 nm [75]) are shown in Figure 4.4a and b. The simple cube lattice (Figure 4.1a) with 2.85 Å interatomic distances serves as a starting configuration. The domain of the simulation was surrounded by reflecting walls. Initial velocities were generated randomly and correspond to the temperature of 4000 K. The atomic system was cooled down to 300 K with an integration time step of 1 fs. Results for cooling times equal to 100 ps and 50 ns are shown in Figure 4.4a and b, respectively. Obtaining the polycrystals with larger grains is also possible; however, it requires longer cooling times, i.e. longer runs of MD algorithm, which dramatically raises the simulation cost, even on a computer cluster. The method

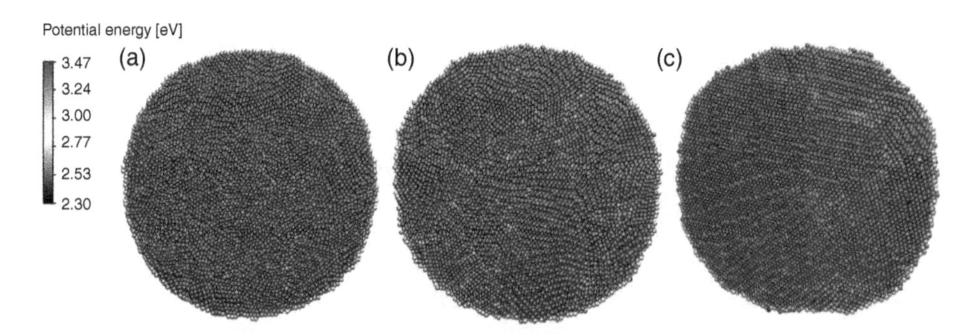

Figure 4.3 3D nanocrystalline structures of diameter approximately 20 nm obtained using embedded atom method, during various cooling times: (a) 200 ps, (b) 1 ns, (c) 5 ns.

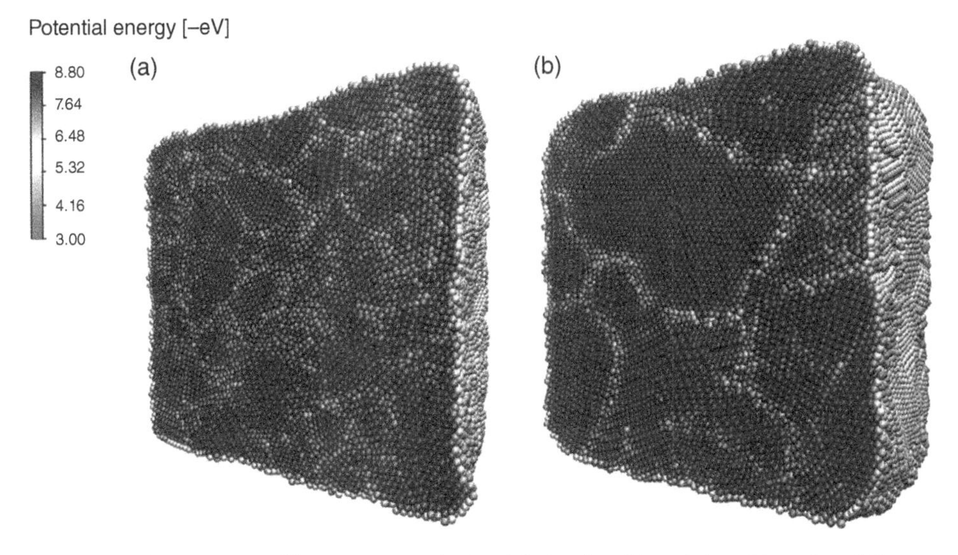

Figure 4.4 3D nanocrystalline structures obtained during fast (a) and slow (b) cooling. Cube dimensions: 19 nm × 19 nm × 19 nm, Lennard-Jones potential.

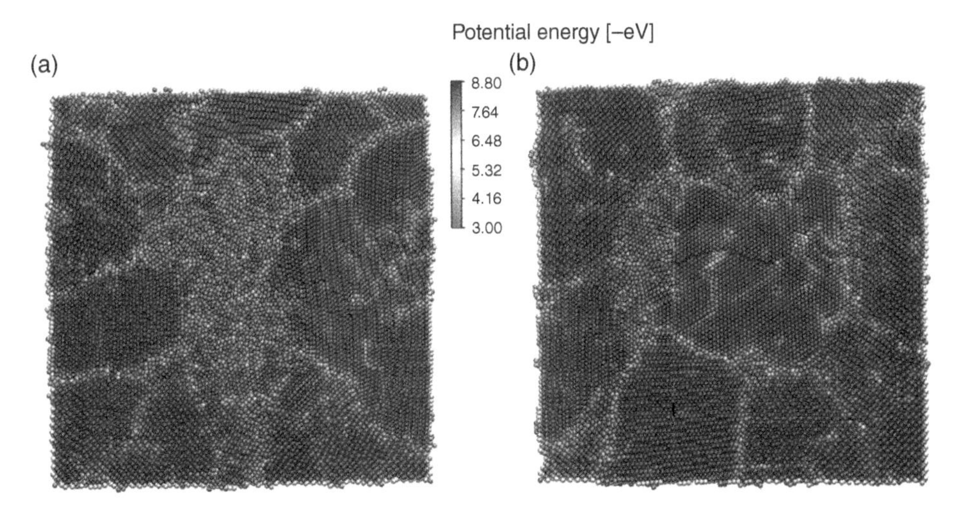

Figure 4.5 Effects of fast and proper cooling: (a) amorphous phase and (b) HCP grain in the centre of the polycrystalline structure.

may be somehow accelerated by sequential incensement of the integration time step at lower temperatures when the particles start to slow down and solidify.

Rapid cooling of the molecular structure should be discussed. The short cooling times may result in structures with small grains and amorphous, uncrystallized phases that may remain in the solidified structure. This phenomenon often occurs in the centre of the large molecular structures and is clearly visible in Figure 4.5a. This structure is a larger version (1 million atoms, 23 nm × 23 nm × 23 nm) of the previously described sample and was cooled down from 3000–300 K at 50 ps. Such a treatment results in the amorphous areas, mainly in the centre of the molecular model, while the slower cooling rates (from 2000–300 K at 400 ps) guarantee the creation of the fully crystallized structures (Figure 4.5b).

4.1.1.2 Adjustable Range of Atomic Interactions

The second method presented combines energetically unstable initial configuration, cooling, and the features of different shapes of the atomic potential functions. This algorithm is similar to the previously described one; however, the cooling process is performed at constant speed, and the size of the grains and the type of structure are controlled by changing the effective interaction range of the potential. After equilibrating at the desired final temperature, the potential function can be replaced with another one.

The Morse potential (see Section 2.1 for details) can be easily applied in this method due to a very convenient mathematical formulation: the shape of the function, thus the range of the atomic interactions, depends directly on the scaling parameter α. This feature is illustrated in Figure 4.6.

The 2D nanocrystalline structures obtained using the previously described technique are shown in Figure 4.7a–c. The initial solution was built of 251 000 atoms formed into an energetically unstable simple square lattice (a two-dimensional case of unit cell presented in Figure 4.1a) with spacing equal to 2.86 Å. The reflecting boundaries were imposed. The set of randomly distributed initial velocities seeded to the temperature of 3000 K was generated, and the Morse potential with $D_e = 1$ ev and $r_0 = 2.86$ Å was applied. The molecular system was cooled down to 300 K at 200 ps time with an integration step of 1 fs.

The simulation results are presented in Figure 4.7a–c for the values of the α parameter equal to 8, 3, and 2, respectively. They show that simulations with 'wide' potential functions, i.e. longer interactions, produce structures with large grains.

The approximate side dimensions of the final structures from Figure 4.7a–c are equal to 140, 137, and 136 nm, respectively, and depend on the size of grains and their boundaries as well as other defects. Each structure was created after the same number of integration steps (i.e. constant cooling speed); however, this fact doesn't correspond to the constant time of the

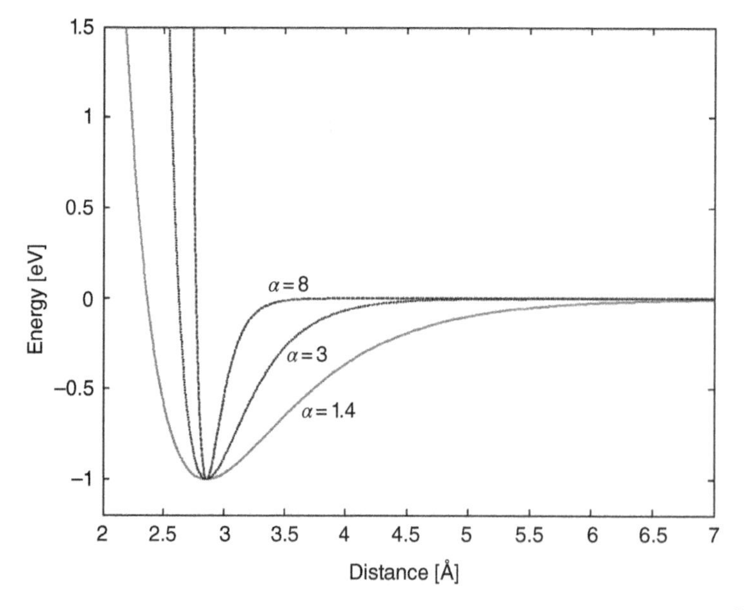

Figure 4.6 The shape of the Morse potential functions for $D_e = 1$ ev, $r_0 = 2.86$ Å, and various values of α parameters.

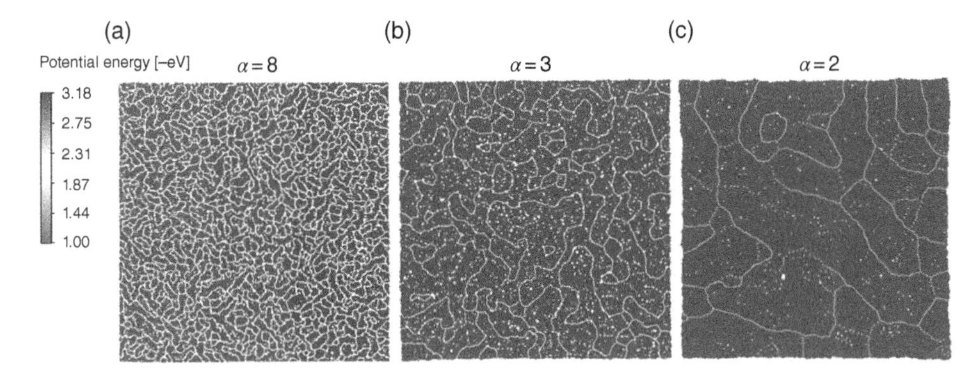

Figure 4.7 2D nanocrystalline structures obtained for: (a) short ($\alpha = 8$), (b) medium ($\alpha = 3$), (c) long ($\alpha = 2$) range of Morse potential.

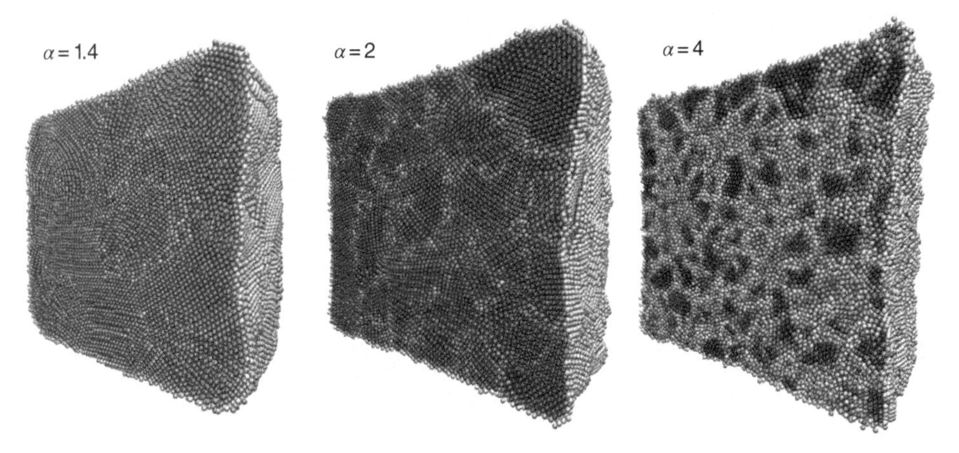

Figure 4.8 3D nanocrystalline structures obtained for various ranges of interactions. Approximately 357 000 atoms, dimensions: 19 nm × 19 nm × 19 nm.

computation since the longer interaction range needs defining the larger cut-off radius and neighbour lists (see Section 3.2.1 for details).

The following examples show the capabilities of this method to create 3D polycrystalline models of various sizes. All the assumptions and parameters remain unchanged, except the dimension of the problem and cooling time of 2 ns. Created polycrystalline samples with various sizes of grains, as well as corresponding values of α parameters, are shown in Figures 4.8 and 4.9.

4.1.1.3 Squeezing of the Nanoparticles

The idea of this method is self-explanatory. The set of small, separated molecular structures of arbitrary shape and size is being squeezed in a controllable way and form a large polycrystalline structure. In contrast to previous methods, this process can be carried out at low temperatures, even near absolute zero. The effects of the squeezing method were presented on the 2D and 3D models, with various initial configurations.

In the first example, the initial solution was prepared from the 36 separate, square polycrystalline structures obtained earlier by using one of the cooling methods (Figure 4.10a).

Figure 4.9 3D nanocrystalline structures obtained for various ranges of interactions. Approximately 1.7 million atoms, dimensions: 33 nm × 33 nm × 33 nm.

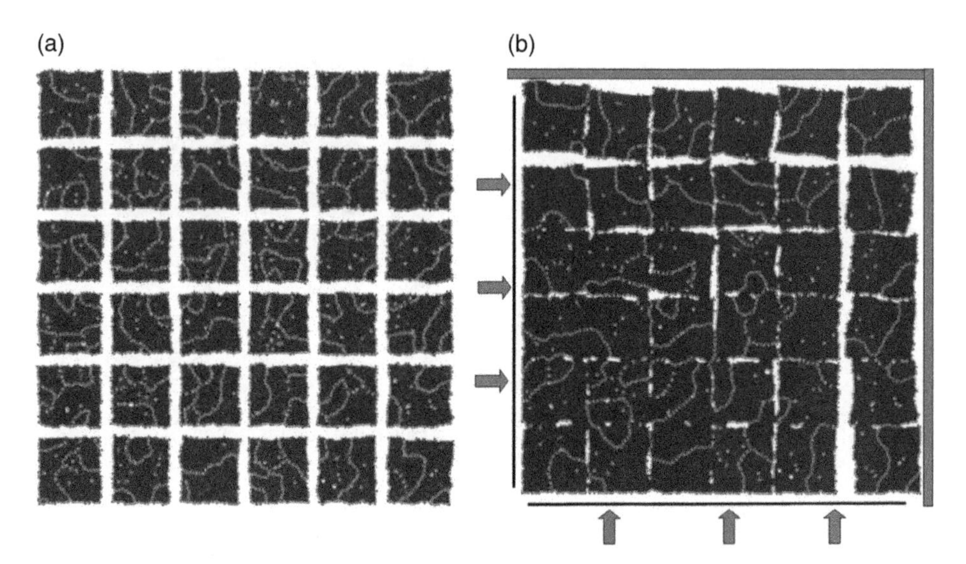

Figure 4.10 Squeezing of the set of 2D nanocrystalline structures: (a) initial solution, (b) moving reflective boundaries.

The MD simulation was performed using the NVT ensemble at the temperature of 300 K with the Lennard-Jones potential and time step of 1 fs. Four reflecting walls surrounded the whole molecular system: two of them were fixed, and two were moving with a constant velocity of the order of 1 Å/ps towards the fixed ones (Figure 4.10b).

As the result of squeezing, all the particles were merged into one big polycrystalline structure, shown in Figure 4.11a. In this method, the most critical parameter is the speed of retraction of the moving walls, i.e. the speed of release of the load exerted on the molecular model. It should not be greater than the speed of squeezing. A rapid removal of the reflecting walls results in rupture of the created polycrystalline. This phenomenon is shown in Figure 4.11b.

This method can also be used to create models of the polycrystalline materials with predefined size and crystallographic orientation of the grains and generate a structure with

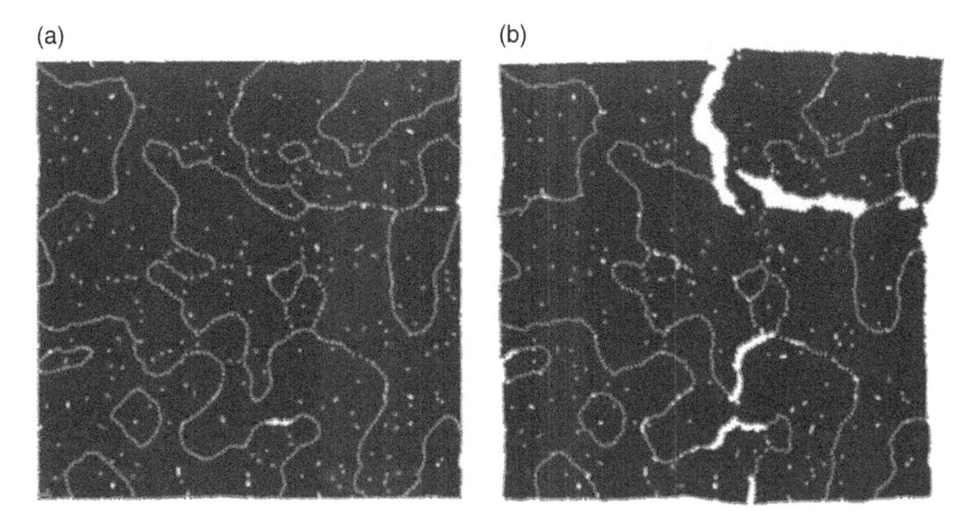

Figure 4.11 2D structures obtained during squeezing: (a) proper polycrystal; (b) cracked one due to rapid removal of the constraints.

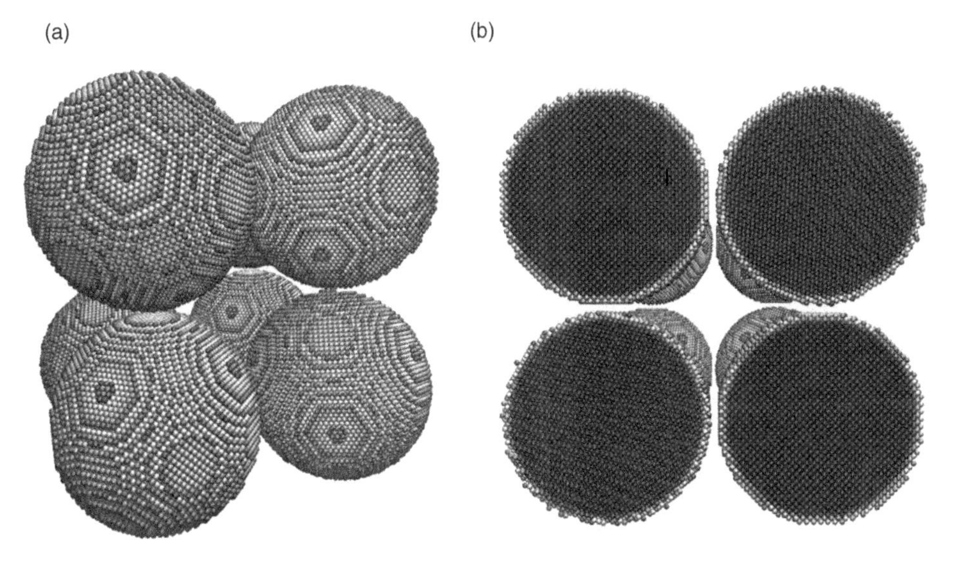

Figure 4.12 The set of eight FCC balls: (a) general view; (b) cross-section view.

controllable porosity [32]. The squeezing of the ball-shaped particles will be considered to reveal this feature.

Eight balls with a diameter of 12 nm each were made of regular FCC lattice (Figure 4.12a) with various crystallographic orientations, shown in the cross-section view in Figure 4.12b. The Morse potential with the parameters $D_e = 0.2703$ eV, $r_0 = 3.253$ Å, and $\alpha = 1.1646$ fitted to the properties of the bulk aluminium [20] was used. Particles were squeezed symmetrically by reflecting walls with a constant velocity of 1 Å/ps at a low temperature of 60 K.

Results of the simulation are presented in Figure 4.13a and b. The size of the grains of the created cube-like structure corresponds to the size of the squeezed particles. Additionally, the initial orientation of the FCC lattices is also preserved.

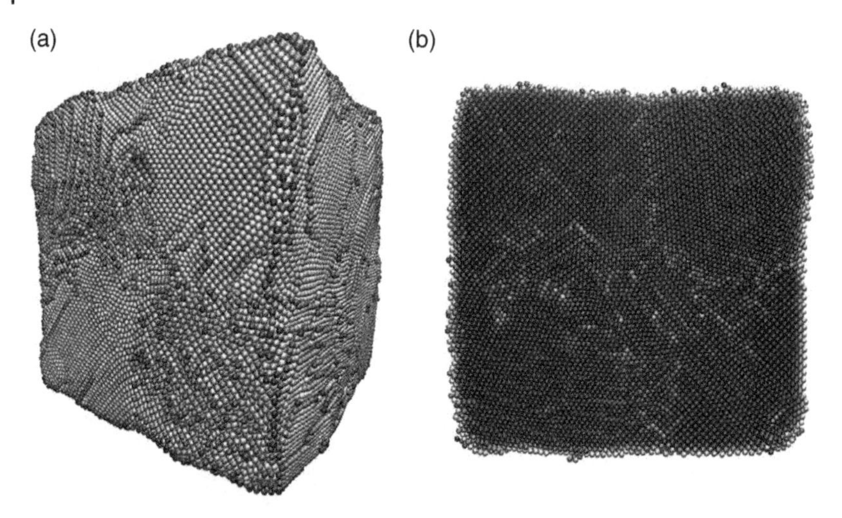

Figure 4.13 Obtained polycrystalline structure: (a) general view; (b) cross-section view.

Initial configurations for this method can be generated with geometrical or grain-growth algorithms, presented in Section 4.2. Such an approach makes the control of the size, shape, and orientation of the grains easy, while the MD-based squeezing produces equilibrated, ready to use, polycrystalline models.

4.1.1.4 Modelling of Structures with Voids

In conjunction with the periodic boundary conditions, the cooling method can be used to create molecular models with voids of the controllable size. The main difference between this and previous methods is that atoms in the 'boiling state' are not reflected by the walls of the simulation box, but may cross the periodic boundaries and spread uniformly over all domains. Assuming the constant volume of the simulation box, the number of particles, and the constant cooling rate during MD run, the size, and number of the voids depend on the initial density of atoms that fill the simulation domain. The initial solution with controllable atomic density can be easily created by choosing the type and adjusting the lattice constant of the regular atomic network.

Effects of this technique were presented in two examples. In the first case, the initial solution was arranged from FCC lattice with slightly larger than equilibrium interatomic distances (2.95 Å instead of 2.89 Å, over 500 000 atoms). In the second simulation, the initial configuration was made of SC lattice with equilibrium spacings (approximately 320 000 atoms). In both cases, the dimensions of the cubic simulation box were equal to 20 nm × 20 nm × 20 nm. The structures were cooled down from 2300 to 300 K in 80 ps with a 1 fs step, and the Lennard-Jones potential was applied ($\varepsilon = 0.44$ eV, $\sigma = 0.2575$ nm [16]). The results – fully equilibrated structures with vacancies and voids – are presented in Figure 4.14a and b. It can be noted that the size of the voids and the ratio between densities of the obtained structures corresponds to the ratio of atomic packing factors of the SC (0.52) and FCC (0.74) lattices.

Other methods of creation of porous materials, especially rocks and concrete, called 'condensation techniques' are described and illustrated in [32, 33].

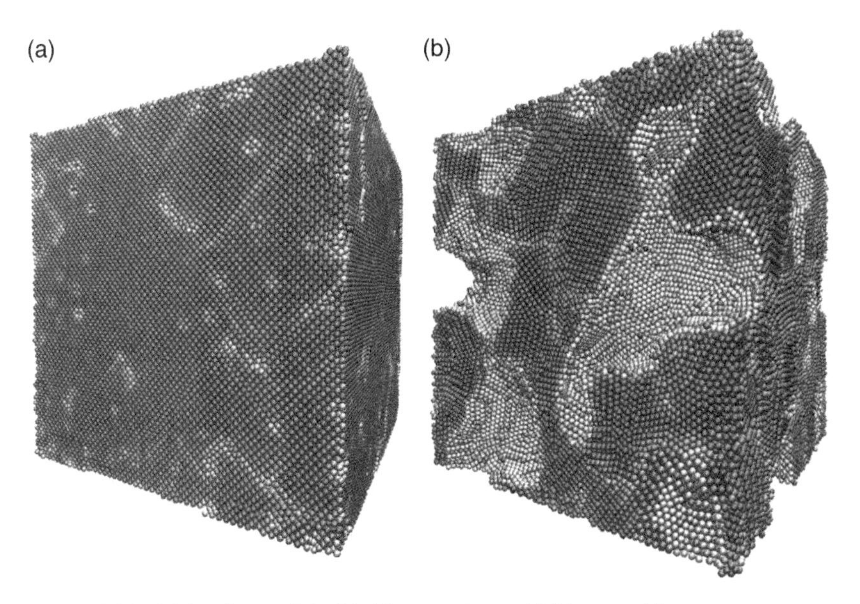

Figure 4.14 Molecular model with (a) vacancies and (b) large voids.

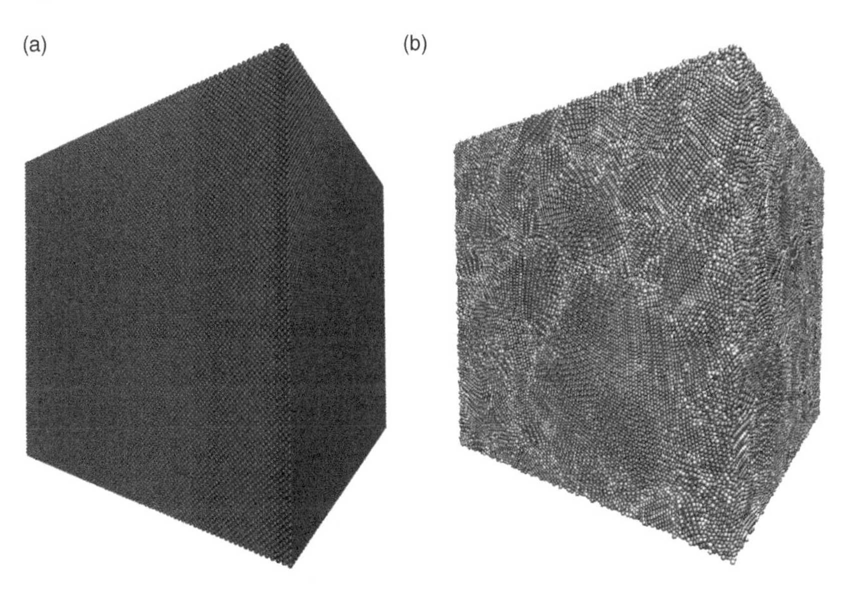

Figure 4.15 Molecular models used in tensile and shear tests: (a) ideal FCC monocrystal; (b) polycrystalline.

4.1.1.5 Material Properties of the Nanostructures

In many methods of multi-scale modelling, like concurrent and upscaling approaches, numerical homogenization, etc., the evaluation of the effective macroscopic mechanical properties is necessary. This knowledge also allows validation of the created numerical models of materials. The simulations of the tensile and shearing tests on the molecular level at finite temperature are presented in this chapter. The mechanical properties and stress–strain characteristics of the ideal FCC monocrystal (Figure 4.15a) and nanocrystalline structure (Figure 4.15b) were computed and compared.

Both cubes have dimensions of approximately 32 nm × 32 nm × 32 nm and contain over 2 million Al atoms. Both models were stretched from 0 to 35% with a constant strain rate corresponding to the velocity 10 m/s at the temperature of 10 K and with a time step of 1 fs. In the shearing test, the strain was applied from 0 to 20° with a rate equal to 0.1 deg/ps. Periodic boundary conditions were imposed on the simulation box. The atomistic stress tensor was calculated using the virial theorem [71]:

$$\boldsymbol{\sigma} = \frac{1}{\Omega}\sum_{i}^{N}\left[-m_i\mathbf{v}_i\otimes\mathbf{v}_i + \frac{1}{2}\sum_{j\neq i}^{N}\mathbf{r}_{ij}\otimes\mathbf{f}_{ij}\right], \tag{4.1}$$

where: i and j are atom's indices, $\mathbf{v}_i$ is the velocity of the i-th particle, and $\mathbf{f}_{ij}$ and $\mathbf{r}_{ij}$ are the interaction force and distance between two particles, respectively. Summation is performed over all atoms occupying volume Ω,. The virial stress theorem has two parts, kinetic and potential ones, and generally is not an equivalent of the Cauchy stress or any other macroscopic mechanical stress tensor. However, when the kinetic part of the virial equation is neglected, and the values obtained in each integration step are averaged over time and geometry, the atomistic stress may be reduced to the Cauchy stress with physical meaning [71, 85]. Apart from these considerations, the kinetic energy contribution to the atomistic stress is quite small due to the low temperature of the simulation. All quantities necessary to obtain strains and appropriate components of the stress tensor (4.1) were averaged every 500 time steps.

The continuum theory of linear elasticity describes the behaviour of aluminium materials under small deformations. In the 3D case, the generalized Hooke law has the form:

$$\begin{aligned}
\varepsilon_x &= \frac{1}{E}\left[\sigma_x - \nu\left(\sigma_y + \sigma_z\right)\right], & \gamma_{xy} &= \frac{\tau_{xy}}{G},\\
\varepsilon_y &= \frac{1}{E}\left[\sigma_y - \nu\left(\sigma_x + \sigma_z\right)\right], & \gamma_{yz} &= \frac{\tau_{yz}}{G},\\
\varepsilon_z &= \frac{1}{E}\left[\sigma_z - \nu\left(\sigma_x + \sigma_y\right)\right], & \gamma_{zx} &= \frac{\tau_{zx}}{G}.
\end{aligned} \tag{4.2}$$

Under the uniaxial strain, e.g. along x axis, the relations $\sigma_x >> \sigma_y$ and $\sigma_x >> \sigma_z$ occur, and Eq. (4.2) can be simplified and written in an inverse manner:

$$\sigma_x = E\varepsilon_x, \tag{4.3}$$

and for the other axes:

$$\begin{aligned}
\sigma_y &= E\varepsilon_y,\\
\sigma_z &= E\varepsilon_z.
\end{aligned} \tag{4.4}$$

The Young moduli in the small deformation range can be easily determined from Eqs. (4.2) and (4.3) by computing the slope coefficient of the linear approximation of the stress–strain curve.

Obtained stress–strain curves, as well as deformed mono- and polycrystals, are presented in Figures 4.16–4.18. The determined mechanical properties for the small strains are as follows: Young's moduli are equal to 105 and 98 GPa and the shear moduli are 45 and 12.5 GPa, respectively, for ideal monocrystal and polycrystalline structure. The difference between shearing characteristics of the investigated molecular models is caused by the movement

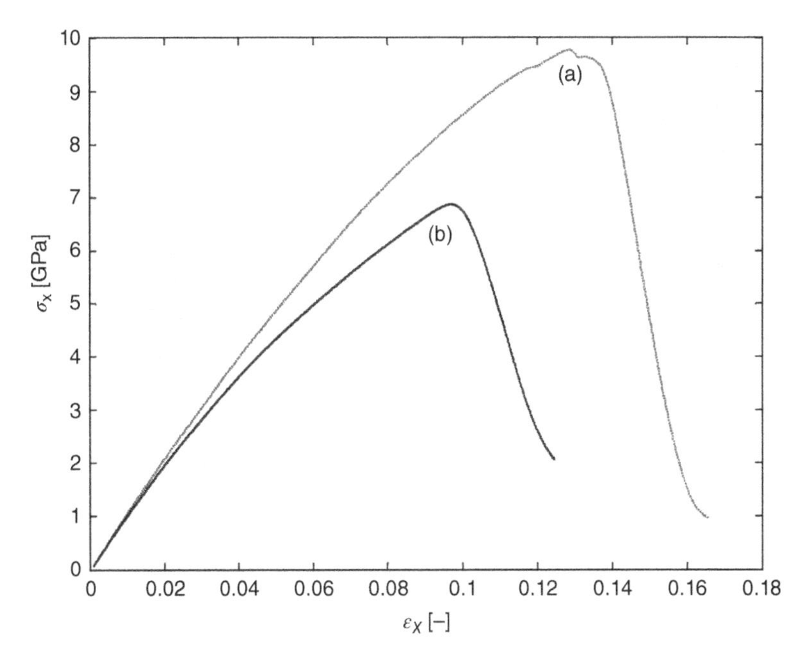

Figure 4.16 Stress–strain curves – tensile test: (a) ideal FCC monocrystal; (b) polycrystalline.

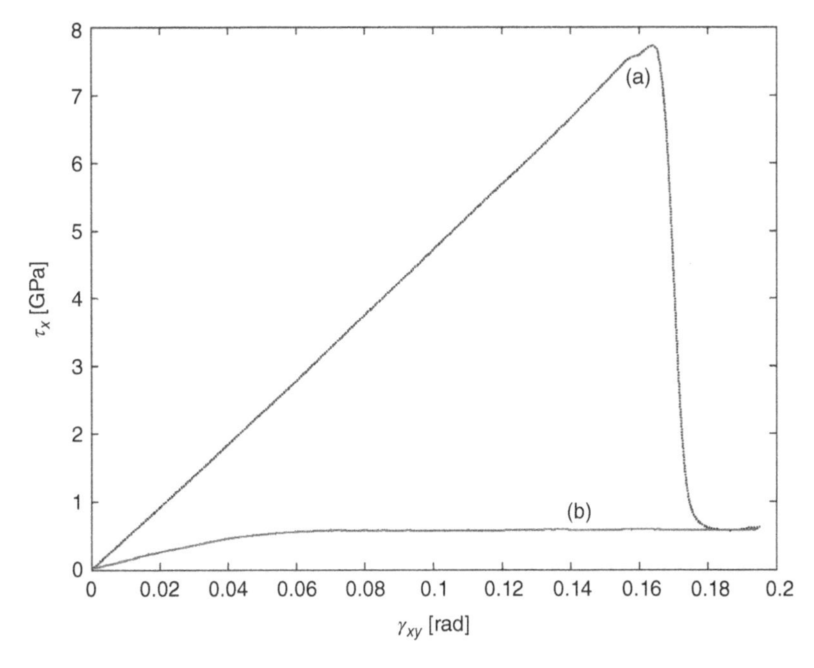

Figure 4.17 Stress–strain curves – shear test: (a) ideal FCC monocrystal; (b) polycrystalline.

of the internal defects, which contribute to the overall plasticity and strength of the polycrystalline structure. For comparison, typical macroscopic values for the various aluminium-based materials or alloys oscillate between 70–90 GPa and 24–26 GPa for the Young's and shear moduli, respectively.

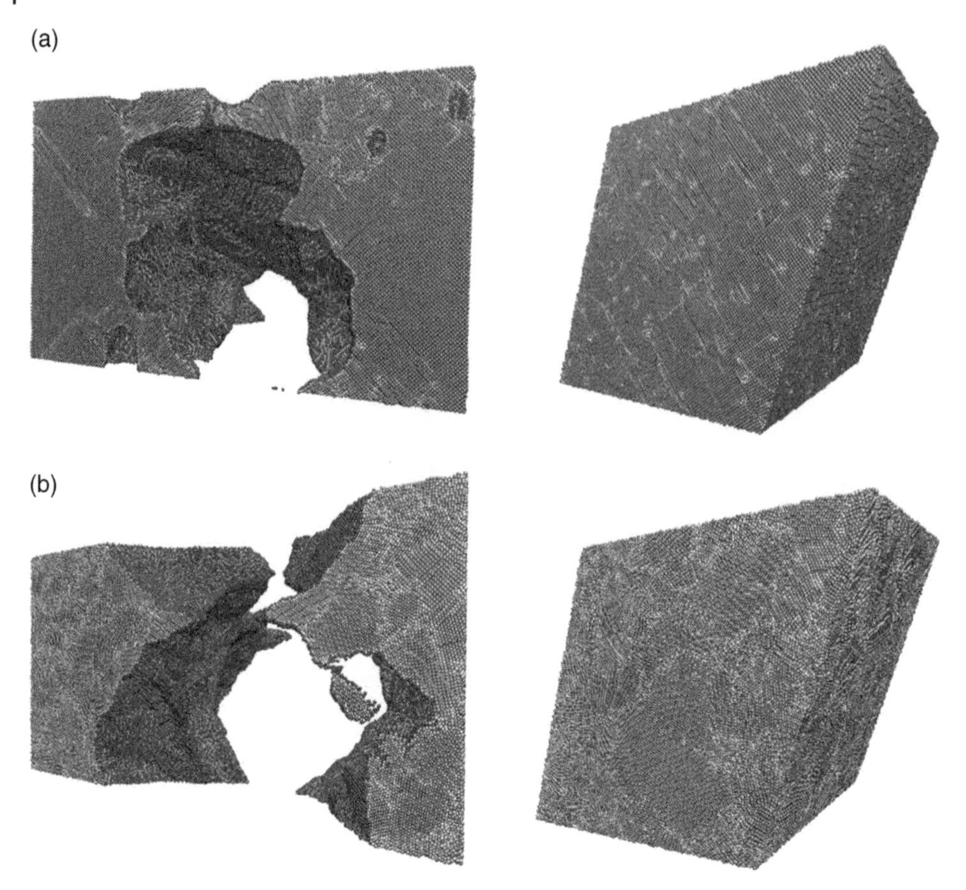

Figure 4.18 Damaged structures during tensile and shear tests: (a) ideal FCC monocrystal; (b) polycrystalline.

4.1.1.6 Models and Mechanical Properties of 2D Materials with Point Defects

In recent years, two-dimensional (2D) materials play an important role in modern material science [10, 17, 27, 51, 54, 79, 82]. This is caused mainly by the unique electronic, thermal, and mechanical properties of such flat structures [31, 81, 86] and the fact that such 2D materials can be used to create other, more complex nanostructures like nanotubes [7, 10], sensors [57], actuators [58], optoelectronic devices [31], or transistors [23, 65]. Most of the research was focused on the family of graphene-like materials. From the technical point of view, 2D materials are periodic flat lattices made of stable configurations of atoms.

The structure, molecular models, and the impact of the vacancies on mechanical properties of one of the most important 2D materials, the single-layered molybdenum disulfide ($SLMoS_2$) [28, 49, 61, 84], are presented and discussed in this chapter.

Similar to graphene, the most common stable 2H phase [48, 49] of the $SLMoS_2$ has a honeycomb-like hexagonal lattice. Although the $SLMoS_2$ is considered as a 2D material, it actually is made up of three layers of atoms: single central layer of Mo atoms is symmetrically covered by two layers of S atoms, with stacking sequence A-B-A, as seen in Figure 4.19 (right part).

Structural parameters and mechanical properties of the pristine $SLMoS_2$ 2H lattice depend on the method of investigation (usually DFT calculations, MD and MS simulations with proper atomic potential, or experimental approaches). For example, the distance between

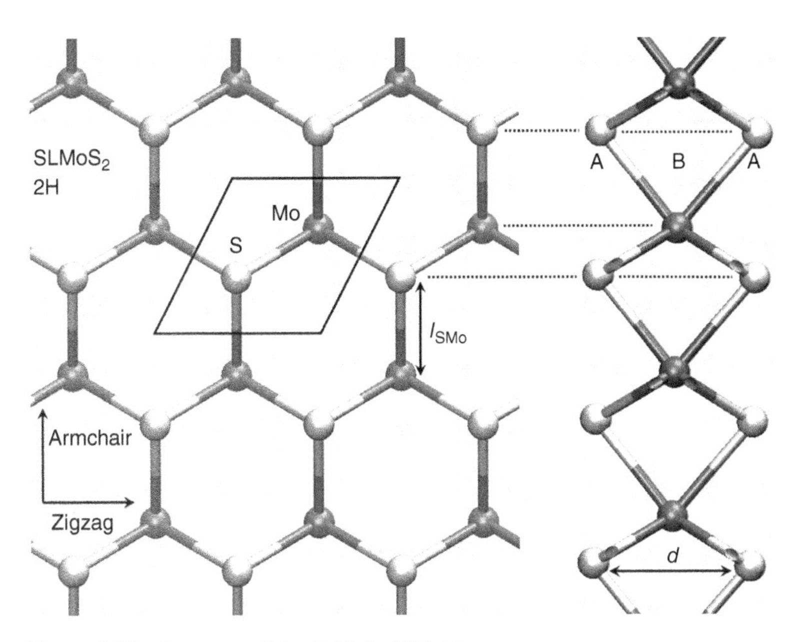

Figure 4.19 Structure of the $SLMoS_2$ 2H lattice.

neighbouring S and Mo atoms l_{SMo}, obtained during MD simulation, varies from 2.398 Å (Stillinger-Weber potential) to 2.444 Å (ReaxFF interaction model), while the same parameter computed using one of the DFT methods is equal to 2.408 Å [82]. Analogically, the distance between S layers d varies from 3.114 Å (DFT) to 3.242 Å (ReaxFF). A comprehensive juxtaposition of structural parameters of the $SLMoS_2$ can be found in [49].

In 2D materials, values of the mechanical properties like elastic constants and Young moduli are often expressed in the force per length [N/m] units, rather than more common force per area [GPa] units. It is due to problems with the evaluation of the effective thickness of the particular single atomic lattice, especially when a certain class of 2D materials, like graphene, has many allotropes [17, 51, 53]. In such cases, the forces acting between two or more layers of 2D materials – thus the resultant distance between them – depends on their type and mutual spatial orientation [55].

The pristine $SLMoS_2$ 2H phase reveals isotropic properties. The average Young modulus usually is in the range of 130 ± 30 N/m [49]. Assuming the effective thickness of the $SLMoS_2$ 2H lattice equal to half of the vertical lattice constant (6.092 Å [28]), the Young moduli, expressed in common stress units, varies in the range of 213 ± 50 GPa. Analogically as in the case of geometrical parameters, an appropriate model of atomic interactions plays a crucial role in determining the motion of the atoms, bond forming, and the accuracy of the behaviour of the whole model, especially when the atomic lattice is subjected to large strain. This is clearly visible on the stress–strain graphs presented in Figure 4.20.

Simulation of the tensile test on rectangular $SLMoS_2$ 2H lattice with dimensions of 280 Å × 130 Å with approximately 12 000 atoms was carried out in a similar way to the previous one described in Section 4.1.1.4. Periodic boundary conditions were imposed along the zigzag and armchair directions. The tensile load was applied in the direction parallel to the longer side of the model. Such geometry helps to propagate the wave along the tensile direction and

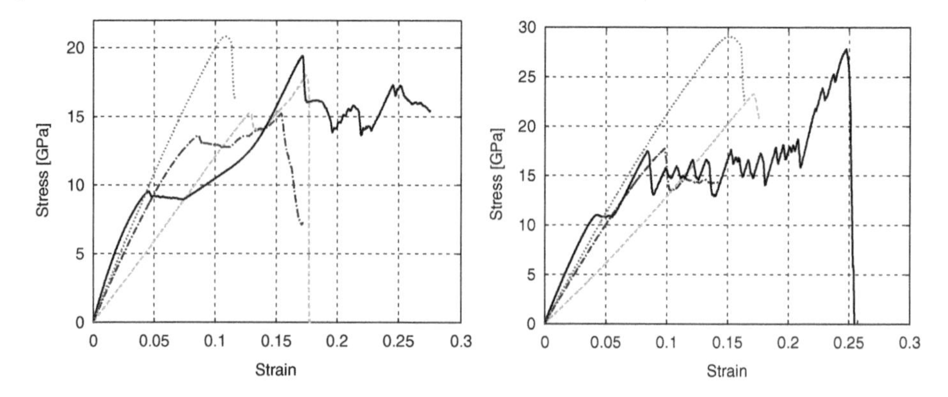

Figure 4.20 Stress–strain relations of the ideal $SLMoS_2$ 2H lattice – (a) zigzag; (b) armchair direction. The colours of the plots denote applied atomic potential: black – ReaxFF, red – REBO, green and blue – Stillinger-Weber [28] and [30], respectively. *Source:* Based on Jiang et al. [28], Kandemir et al. [30].

minimizes the effect of artificial periodicity (see Section 3.2.1.6). During the tensile test, an atomic model was stretched from 0 to 100%, with the strain rate corresponding to the effective velocity of 10 m/s. The canonical (NVT) ensemble and Nose-Hoover thermostat with the temperature damping factor equal to 50 fs was applied. The temperature was kept at a constant level of 30 K, and the time step was set to 0.25 fs. The tensile test has been repeated four times with the following empirical atomic potentials: two variants of the Stillinger-Weber model [28, 30], REBO [36, 37], and ReaxFF [60].

Three models of atomic interactions (except Stillinger-Weber [28]) reveal similar, proper behaviour in the linear range – up to several percentage of strain. However, after exceeding the linear region, each type of atomic potential acts differently. Such a problem may have a purely technical nature, e.g. due to the choice of the cutoff radii, applied REBO potential allows only simulation of up to 11% of strain. Both Stillinger-Weber potentials tend to form the BCC lattices above linear area, while in the case of ReaxFF the further decrease of the distance between two neighbouring sulphur atoms results in the creation of the new bonds between them.

Such phenomenon, as shown in Figure 4.21, was also noticed and discussed in [48, 49]. Described phase transitions that occur during large stains are a good starting point for further research and validation with more accurate *ab initio* or DFT-based methods.

Similar to the previously presented 3D polycrystalline structures, the existence of defects in the lattice of the 2D material impacts the mechanical and electronic properties. Research on the defected graphene sheets is presented in [9, 56]. Theoretical investigations of MoS_2 with imperfections using MD simulations with REBO potential, DFT, and first-principles calculations have been published in [35, 74, 81], respectively.

Besides Stone-Wales defects, one of the most common imperfections in the atomic lattices of 2D materials are point defects. The mechanical properties of $SLMoS_2$ with varying concentrations of randomly distributed vacancies are presented in the following example.

The atomic model of pristine $SLMoS_2$ lattice initially contained 25 000 atoms [1]. Vacancies were created by randomly removing atoms from the ideal lattice using a random number generator with uniform distribution. The percentage of defect concentration varies from 1 to 15%. The periodic boundary conditions in all directions were imposed along the armchair

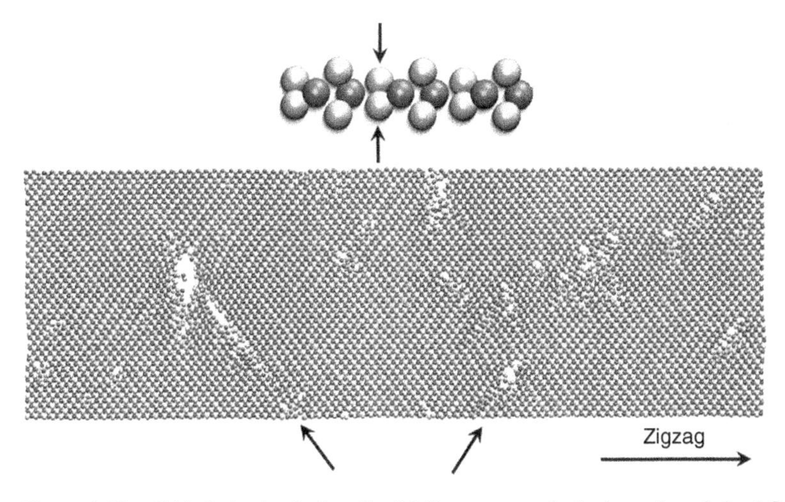

Figure 4.21 $SLMoS_2$ lattice before final failure: arrows indicate series of short S-S bonds along easy slip directions (ReaxFF, NPT, temperature=30 K).

Table 4.1 Number of boundary elements of the continuum models. The elastic constants of MoS2 H2 lattice with different defect fractions.

Vacancies [%]	C_{11} [N/m]	C_{22} [N/m]	C_{21} [N/m]
Ideal lattice	149.42	149.42	52.29
1%	149.01	148.85	52.15
2%	140.70	141.51	48.52
5%	128.89	128.89	43.25
10%	108.30	107.94	33.70
15%	94.21	93.47	28.75

and zigzag directions. Minimization of the total potential energy, based on the Polak-Ribiere Conjugate-Gradient (CG) method (built-in Lammps software [34]), has been applied twice: once for relaxation of the unstrained atomistic model, and once for computation of the new positions of atoms after application of finite, small strain at the chosen direction. As an alternative to the CG routine, the Molecular Statics solver, as described in Section 3.2.2, can be applied. Components of the micro-stress tensor of the strained model were computed using Eq. (4.1). Averaged elastic constants obtained for a series of atomic models with different number of randomly distributed vacancies are gathered in Table 4.1. Two examples of imperfect $SLMoS_2$ lattices are presented in Figure 4.22.

It can be clearly seen that $SLMoS_2$ material with randomly distributed defects maintains isotropic properties, and stiffness decreases faster with an increasing number of vacancies. Presented values of elastic constants for the ideal $SLMoS_2$ 2H phase are in good agreement with results published in [35] (148.4 N/m) and [58] (130.4 N/m).

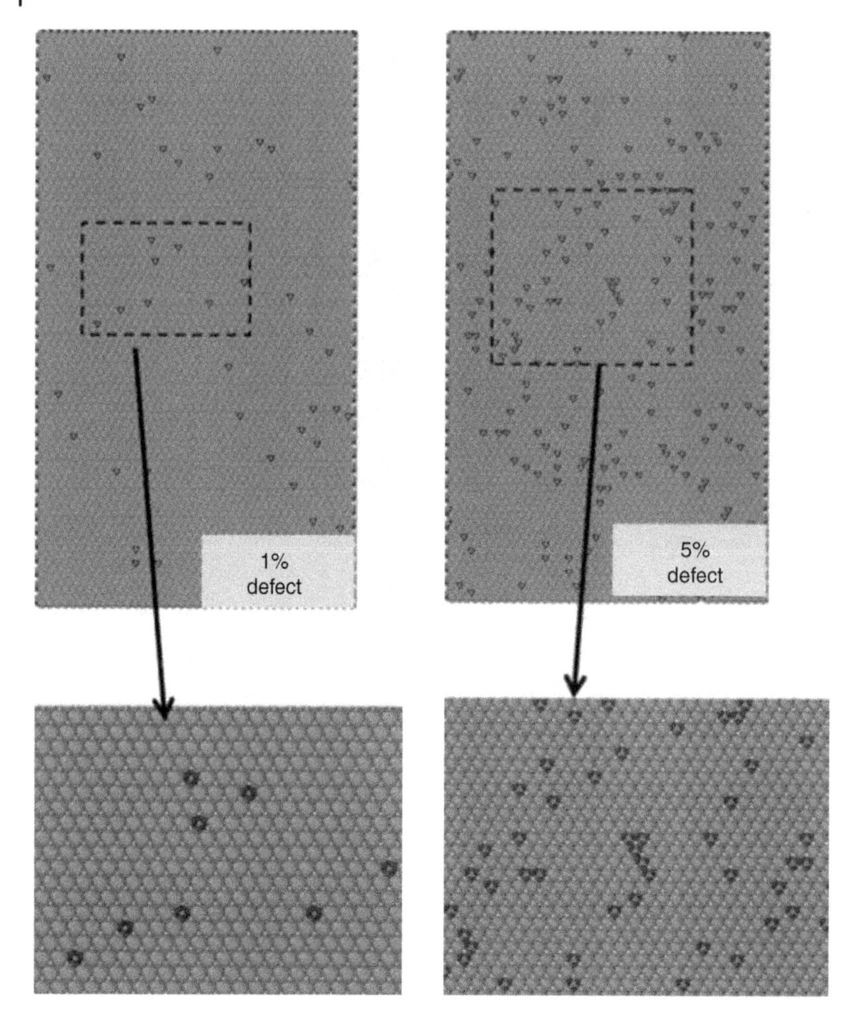

Figure 4.22 $SLMoS_2$ lattices with different concentrations of vacancies.

4.2 Microstructure

4.2.1 Generation of Microstructures

The concept of digital material representation (DMR) and its application in practical research at the microstructure level is dynamically evolving. The main objective of DMR is to create the digital representation of microstructure with its features (i.e. grains, grain orientations, inclusions, cracks, different phases, etc.) represented explicitly during the simulations. Generating material microstructure with specific geometrical features and properties is an essential algorithmic part of systems based on DMR. Such DMR is further used in numerical simulations of thermomechanical processing operations or simulations of microstructure behaviour under exploitation conditions. In the approach, the more accurate the digital representation is in the case of geometry and properties, the more accurate results can be obtained. That is why much research is put into developing methods responsible for creating the 2D and 3D representations of analyzed microstructures. A review of various approaches used to recreate microstructure morphology and acquire properties at the microscale level for the DMR approach is presented next.

4.2.1.1 Voronoi Tessellation

The Voronoi tessellation is a commonly used method dedicated to generating initial digital microstructures and interpreting some metallographic phenomena. The idea of this method is based on the mapping of the bounded area onto the group of specific polygons $P = \{p_1, ..., p_n\}$, generated around a set of n distinct initial points $S = \{s_1, ..., s_n\}$. The two following features characterize each polygon p_x:

- It is connected precisely to one point s_x, where $x = 1...n$.
- Each point inside the polygon is closer to s_x than to any other point from S.

For the present work, points from the S set represent grain nuclei, while the polygon areas around these points, called Voronoi cells, represent final grains in the microstructure.

The brute-force approach that identifies bisectrices between each pair of points and then eliminates the ones that do not satisfy Voronoi assumptions [14] is one of the most straightforward but also very time-consuming algorithmic solutions. A common alternative for 2D applications is the Fortune algorithm based on the sweep line. A more general approach is the Delaunay triangulation, often used for both 2D and 3D applications. In this case, for example, the Bowyer–Watson algorithm can be used. It starts by forming the super rectangle or triangle, enclosing all the points from set V that have to be considered. Then incrementally, subsequent points X from V are investigated and added to the triangulation set. After such an insertion step, a search is done to find the triangles whose circumcircles enclose X ($d < r$), and eliminate them from the set. As a result, an insertion polygon containing X is created. Edges between the vertices of the polygon and X are inserted and form the new triangulation. The algorithm is performed iteratively until all the points are evaluated, as seen in Figure 4.23.

Finally, when all the centres of the circumcircles of the triangles are connected, the Voronoi diagram is obtained. A similar procedure is performed in 2D and 3D computational domains, as presented in Figures 4.24 and 4.25, respectively.

The Voronoi tessellation can also provide digital microstructure models with periodic character, as presented in Figure 4.26.

4.2.1.2 Cellular Automata Grain Growth Algorithm

The basic definitions of the cellular automata method that are necessary to create a robust numerical model are presented in Chapter 3. This method can be easily adapted to develop a basic grain growth model to create a digital representation of the microstructure morphology of both single-phase and multiphase materials. However, more advanced algorithms based on the energy conditions or inverse analysis can also be found in the scientific literature [38, 62].

In the CA method, the computational domain in 2D space most of the time is composed of a regular grid of squares that represent CA cells. In 3D spaces, cubes are used as CA cells. However, other discretizations of computational domain based on, e.g. triangles or hexagons can also be used during the analysis, as seen in Figure 4.27.

The simple unconstrained grain growth model assumes that each CA cell can be in the two states Υ. The first is represented by '0' and is used to indicate cells that are empty, while the second represented by '1' is used to identify cells that already belong to a grain. Additionally, each CA cell can be described by a vector of internal variables ς that is used to store additional

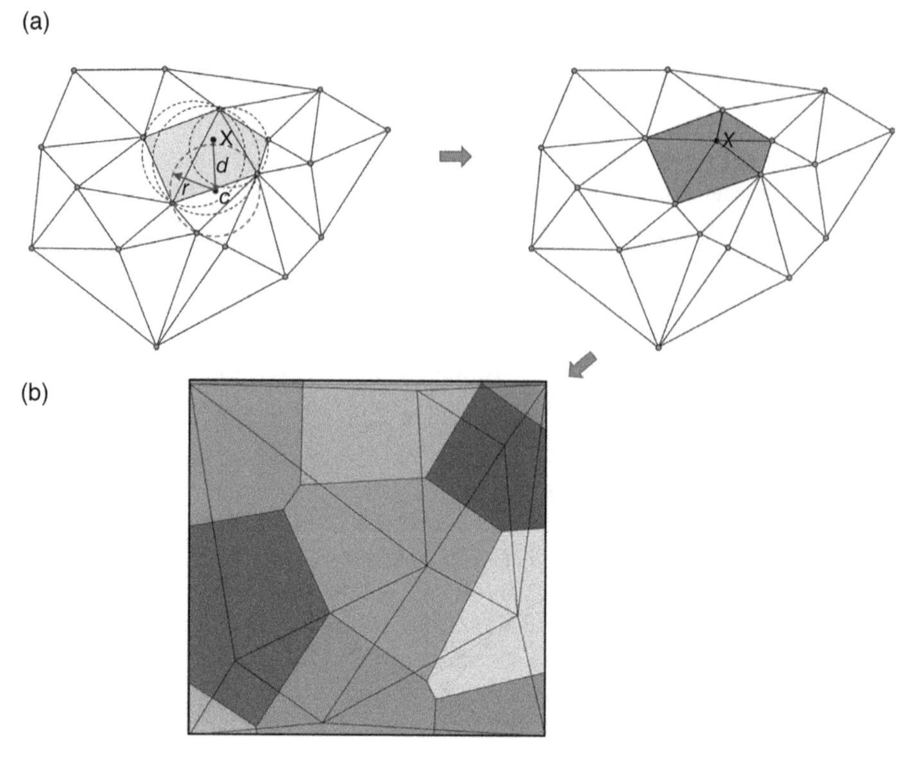

Figure 4.23 Illustration of (a) concept of the Delaunay triangulation steps and (b) Voronoi diagram.

information. In the simplest version of the algorithm, one internal variable is used that describes to which grain the cell belongs. In the more advanced versions of the grain growth algorithm, other internal variables are also defined, e.g. describing the crystallographic orientation of each CA cell by three Euler angles.

The basic simulation starts with all the CA cells in the state '0'. Then some of them are randomly selected, and their state is changed to '1' as they represent grain nuclei. At the same time, each of the grain nuclei is represented by a different value of the internal variable associated with grain assignment. Then, CA cell states progressively evolve towards the state '1' during the grain growth stage. The transition rule for this stage assumes that when a neighbour of a particular CA cell in the state '0' in the previous step was in the state '1', this particular cell can also change its state and adopt the value of the internal variable. In the unconstrained grain growth algorithms, grains grow with no restrictions until the impingement with other grains. After that, the growth is constrained only toward the regions with cells in the state '0', as seen in Figure 4.28. The algorithm is performed iteratively until all the CA cells change the state to '1'.

As mentioned in Chapter 3, various types of CA neighbourhood definitions can be used in the CA method. In the basic unconstrained grain growth model, the final microstructure morphology can be controlled by a proper selection of the neighbourhood, as presented in Figure 4.29.

Due to the stochastic character of the nucleation stage, final results obtained with the random neighbourhoods in a natural way replicate Gaussian grain size distribution. Therefore these neighbourhoods can be used to generate microstructure morphologies that are typical

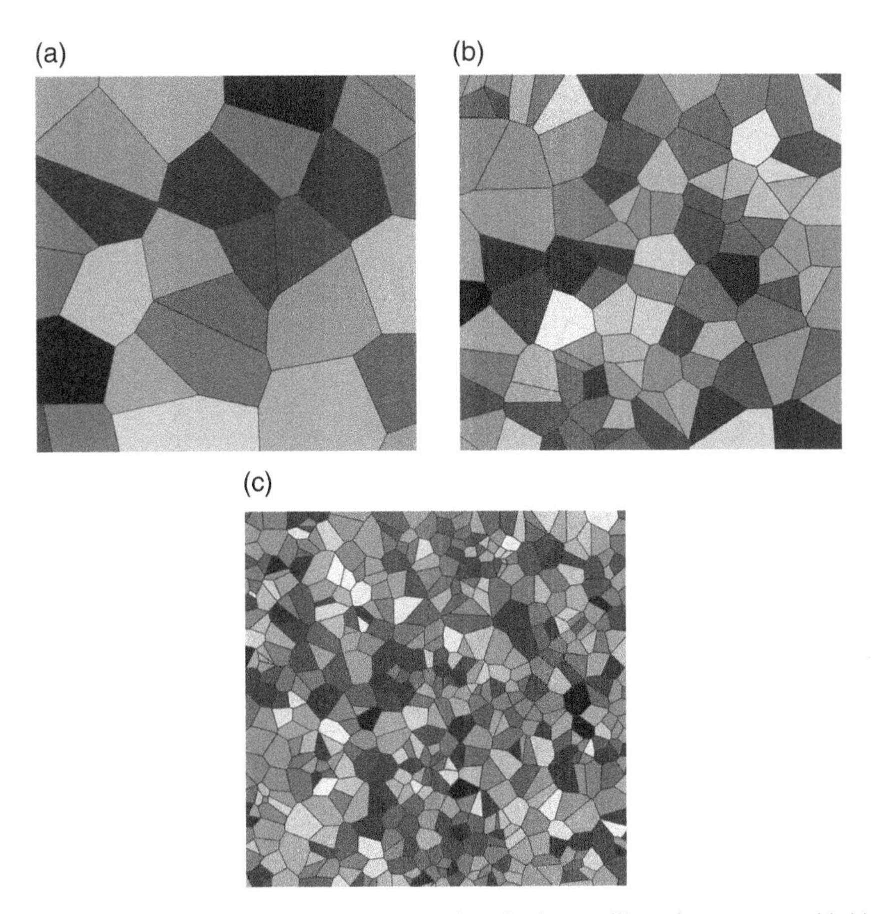

Figure 4.24 2D digital material representation of polycrystalline microstructure with (a) 30, (b) 100, and (c) 500 grains obtained by Voronoi tessellation.

for a material subjected to, e.g. annealing or static/metadynamic recrystallization. However, other neighbourhoods can also be defined to recreate microstructure morphologies after deformation, e.g. rolling (Figure 4.30).

The same applies to modifications in the nucleation stage that can be introduced to deliver gradient-type microstructures (Figure 4.31) or uniform grain distribution (Figure 4.32). In the former case, the initial nuclei are being distributed with a specific density, which is a function of distance from a selected sample edge. In the latter case, nuclei are equally spaced across the computational domain or generated with a specific radius to maintain the desired distance between them.

Similar to the Voronoi tessellation algorithm, the periodic boundary conditions can be introduced during the simulation, as presented in Figure 4.33.

The CA grain growth algorithm can also be easily modified to generate the substructure within the grains. In this case, first, the grain growth algorithm is executed, and then the algorithm is reinitiated for each grain separately. Therefore, the growth is constrained by the initial grain boundaries, as presented in Figure 4.34.

The CA algorithm also allows the identification of CA cells located between the grains that represent grain boundaries. Additionally, a distinction of grain boundaries between a pair of

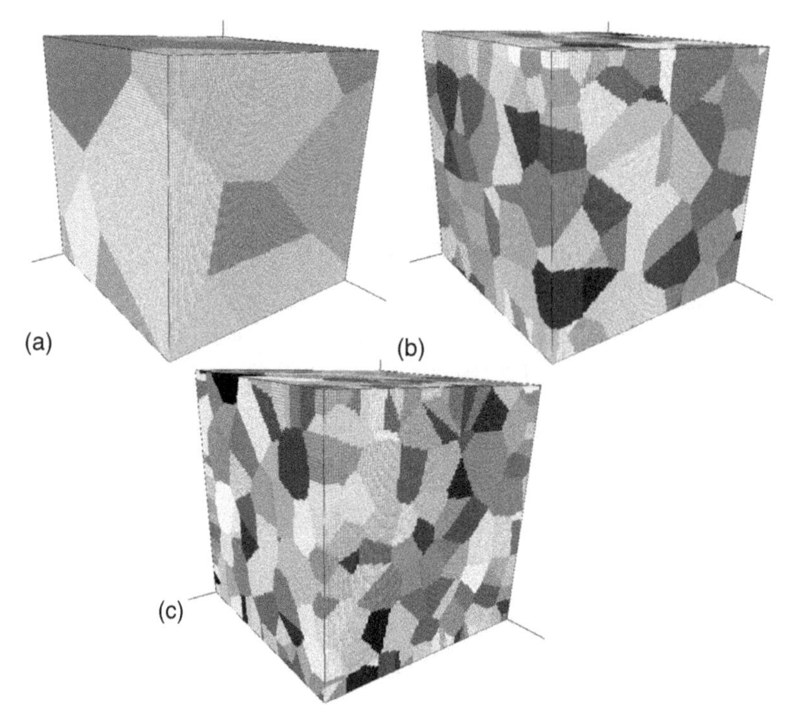

Figure 4.25 3D digital material representation of polycrystalline microstructure with (a) 20, (b) 250, and (c) 450 grains obtained by Voronoi tessellation.

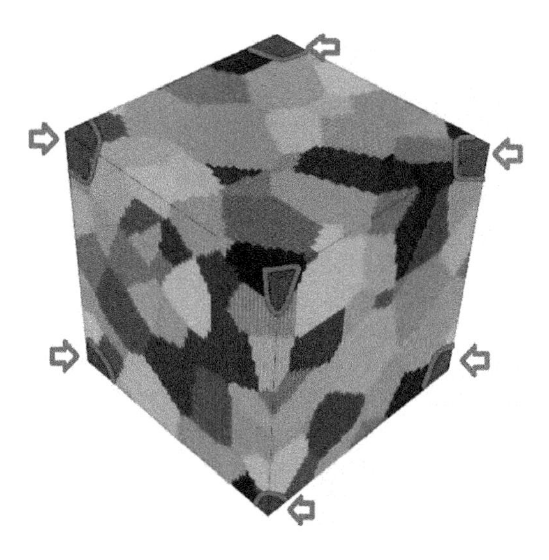

Figure 4.26 3D digital material representation of periodic polycrystalline microstructure obtained by Voronoi tessellation.

neighbouring grains allows assigning different material properties to these regions during further finite element calculations (Figure 4.35). The grain boundary thickness can be precisely controlled according to the requirements.

Similar approaches to digital microstructure generation were implemented in 3D space. The equivalent of the hexagonal or pentagonal random neighbourhoods is especially interesting in the 3D space because there are six possibilities in neighbour selection. Examples of

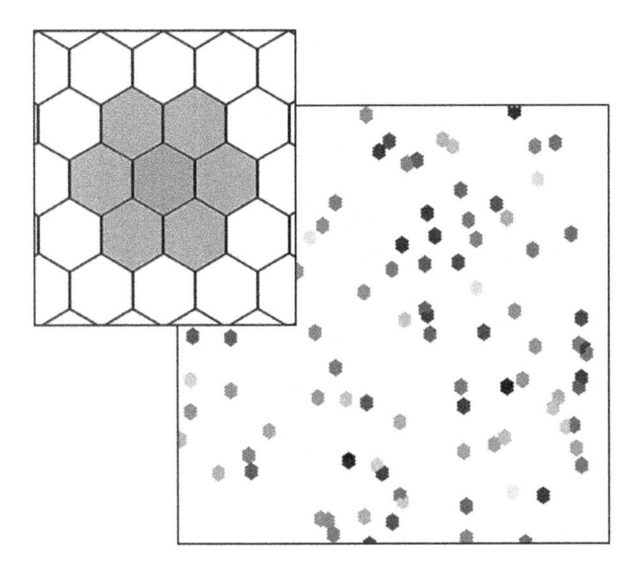

Figure 4.27 Examples of CA grain growth within the hexagon-based discretization of the computational domain.

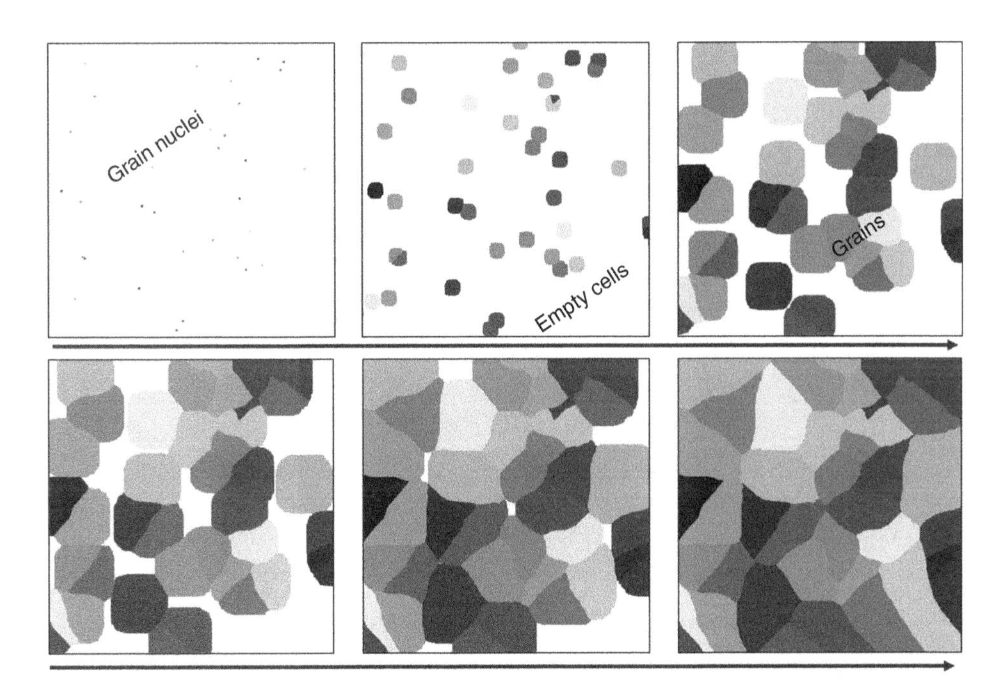

Figure 4.28 Subsequent stages of CA grain growth algorithm.

obtained microstructures on the basis of three different types of neighbourhoods are shown in Figure 4.36.

Finally, it has to be emphasized that the CA grain growth algorithm can be initiated in any shape of the computational domain and is not restricted to square or cubic domains. Examples of digital microstructure models generated within various geometries of the

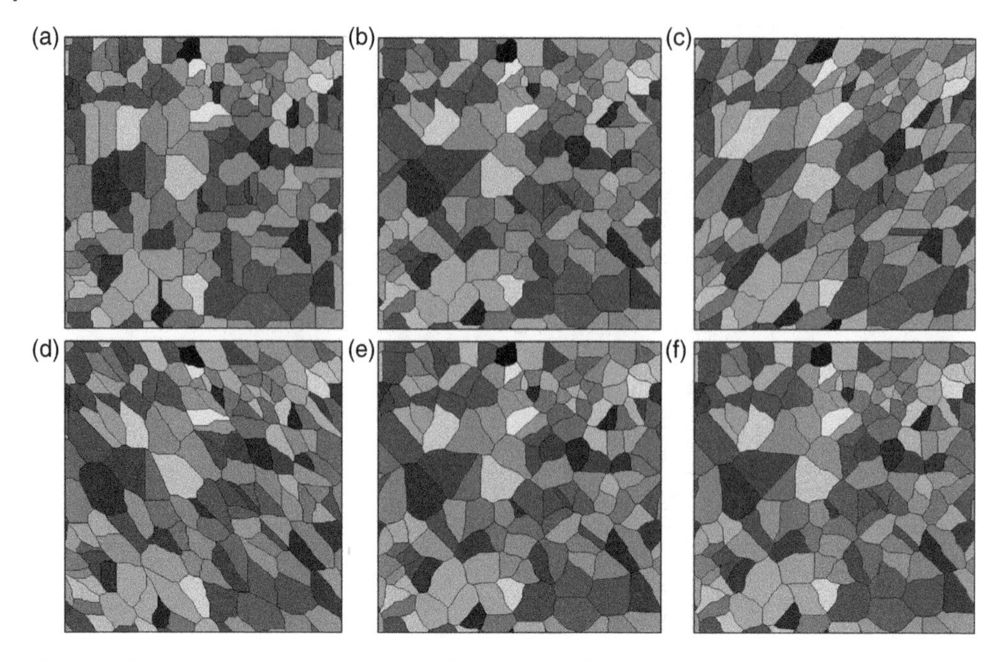

Figure 4.29 CA grain growth results with different types of neighbourhoods: (a) von Neumann, (b) Moore, (c) hexagonal left, (d) hexagonal right, (e) hexagonal random, and (f) random.

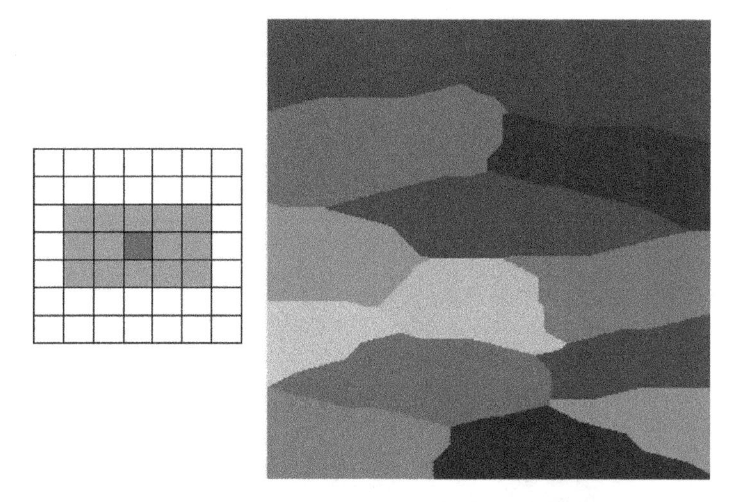

Figure 4.30 Digital microstructure with elongated grains obtained with the modified Moore-type neighbourhood.

computational domains prepared in the CAD software are presented in Figure 4.37. Such DMR models can be applied to simulate the behaviour of microcomponents during finite element calculation of processing or exploitation conditions.

As presented, the basic CA grain growth model with the appropriate neighbourhood can replicate the geometry of a microstructure after various processing routes. Such digital microstructure can then be considered as a representative volume element during numerical simulations of, e.g. loading conditions. It should also be mentioned that the CA algorithm can also

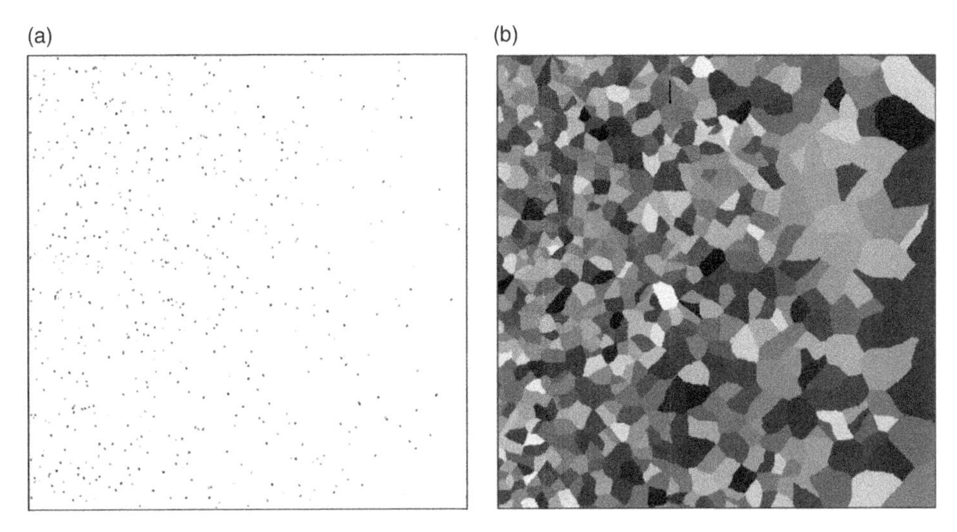

Figure 4.31 Digital microstructure with gradient grain size obtained with the modified nucleation algorithm: (a) nuclei and (b) final microstructure.

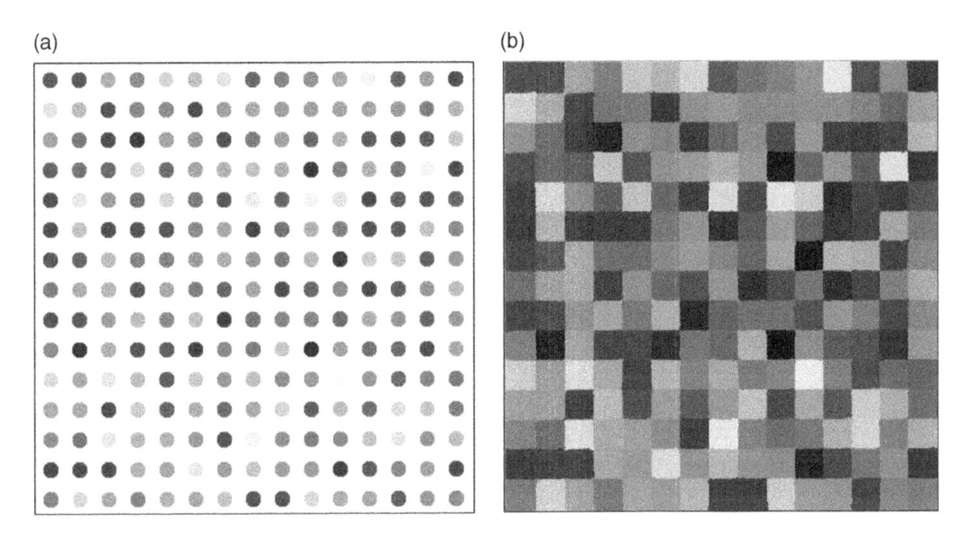

Figure 4.32 Digital microstructure with uniform grain distribution obtained with modified nucleation algorithm: (a) nuclei and (b) final microstructure.

be the basis of other more sophisticated solutions to generate digital microstructures like the close-packed sphere growth algorithm.

4.2.1.3 Close-Packed Sphere Growth CA-Based Grain Growth Algorithm

The sphere/ellipsoid packing algorithm combined with the CA growth algorithm is often used to create statistically representative microstructures [21, 62, 70]. The sphere packing algorithms can be divided into two major solutions: dynamic and constructive. In the former case, spheres update the position or the radius during the packing process controlled by a shrinking algorithm [29, 77], compression forces algorithm [3], or gravitational algorithm [24]. In the latter

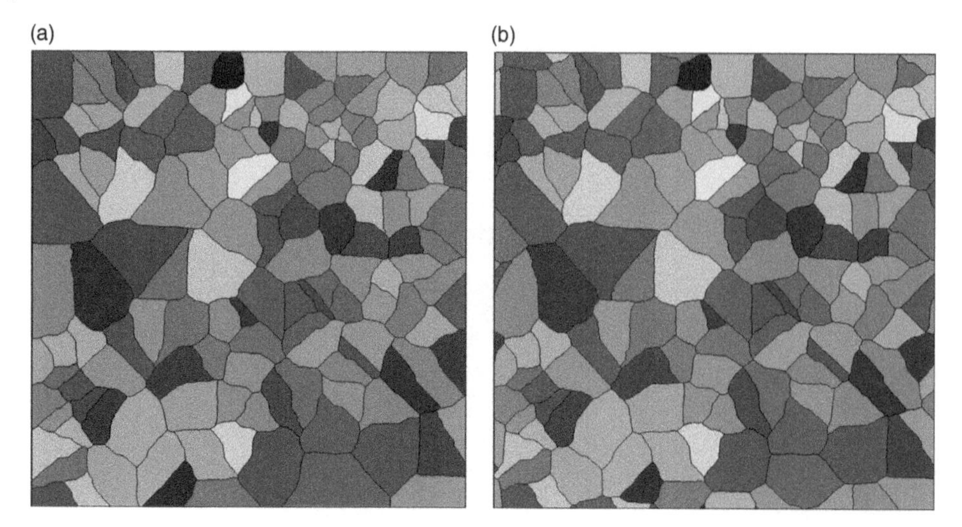

Figure 4.33 Microstructure with 10 grains generated (a) without and (b) with periodic boundary conditions.

Figure 4.34 Results of microstructure with the substructure within the grains, (a) initial and (b) final growth stages.

group of approaches, sphere radius or position are preserved during the packing process, and as a result, they are less computationally expensive, but the level of high-density packing is moderate [11]. The concept of gravitational packing was selected in [45] for the generation of digital microstructure and is based on four main steps:

- definition of a 2D or 3D computational domain,
- creation of circles/spheres with a radius R generated randomly according to the Gauss distribution function [6]. The initial position along the x-axis is also generated randomly,
- packing the computational domain with circles/spheres. The packing process is based on the iterative approach, where each feature is moving down exactly one pixel at a time, as seen in Figure 4.38. It is assumed that if the distance between the two centre points of the circles/spheres is smaller than the sum of their radii, the collision event is initiated. In this case, the feature is rotated under the gravity force to the right or left and continues the drop,

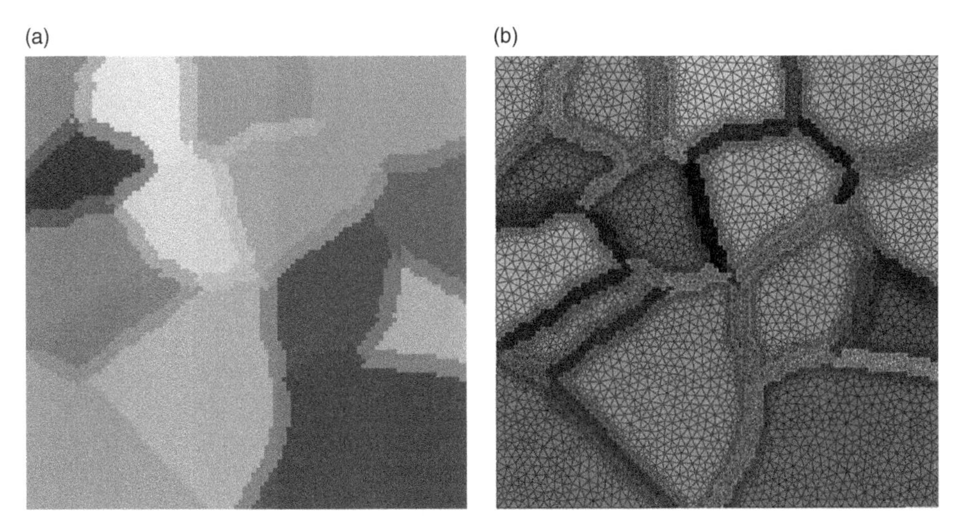

Figure 4.35 Results of microstructure with the explicitly represented grain boundaries between subsequent grains, (a) without and (b) with finite element mesh.

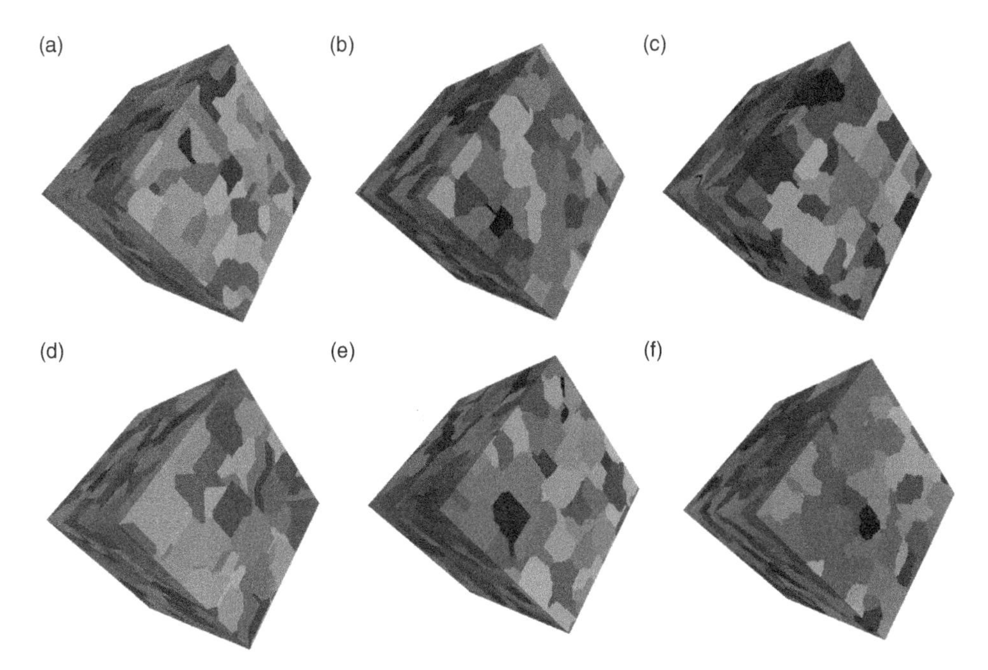

Figure 4.36 3D CA grain growth results with different types of neighbourhoods: (a) von Neumann, (b) Moore, (c) hexagonal left, (d) hexagonal right, (e) hexagonal random, and (f) random.

- CA grain growth initiated on the closed packed circles/spheres. As in the classical CA grain growth model, the growth continues until the computational domain is filled with grains (Figure 4.39).

As seen in Figure 4.40, combined sphere packing and CA grain growth algorithm ensure the creation of statistically representative digital microstructures, where the average grain

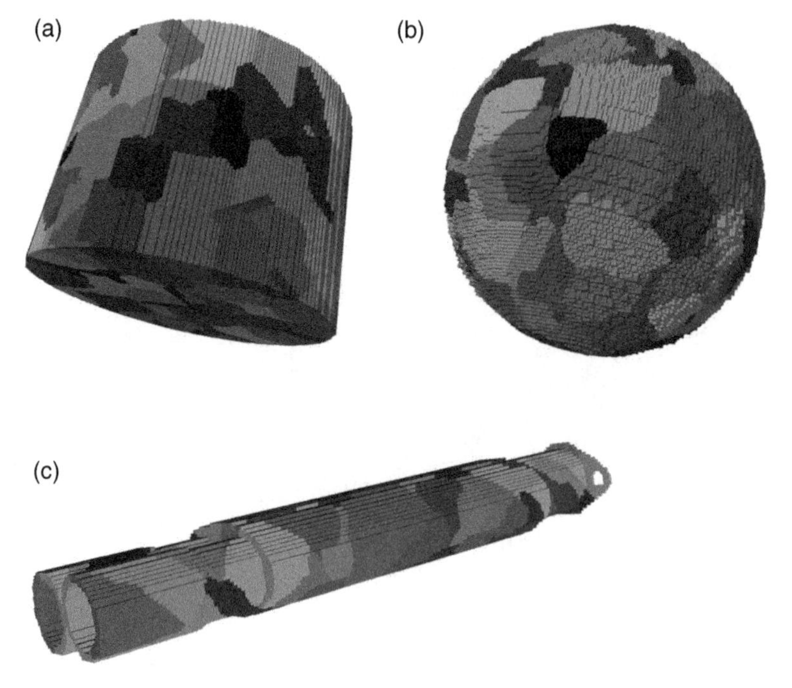

Figure 4.37 Digital microstructure models generated within various geometries of the computational domain, (a) cylindrical, (b) spherical, and (c) complex.

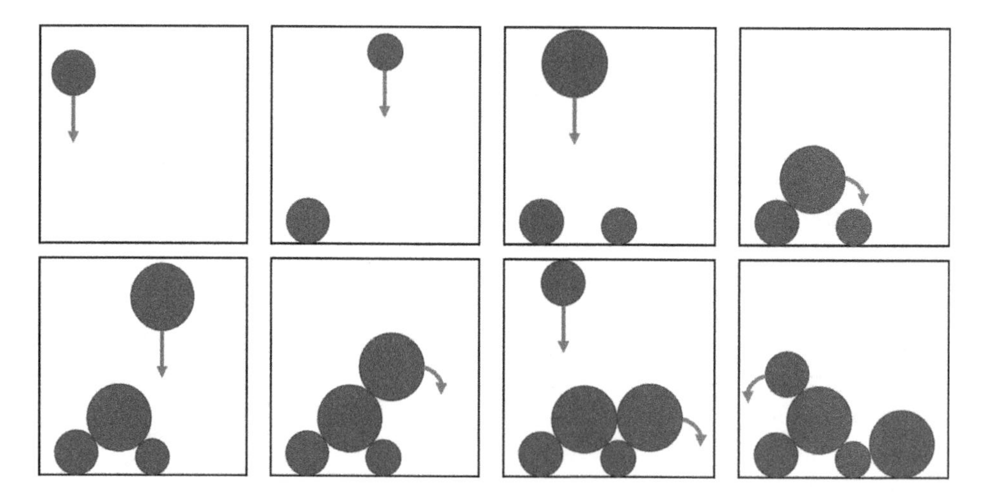

Figure 4.38 Concept of the modified sphere growth algorithm.

size distribution follows, e.g. the Gaussian curve. However, as presented in [21], other grain size distributions, including multi-modal ones, can be easily replicated.

The packing density of this approach can also be increased by the application of optimization algorithms based on, e.g. genetic algorithms as presented in [45]. As a result, the packing density and eventually the control level of the final grain size distribution increase but at the expense of increased computational time. If a digital microstructure that follows a desired

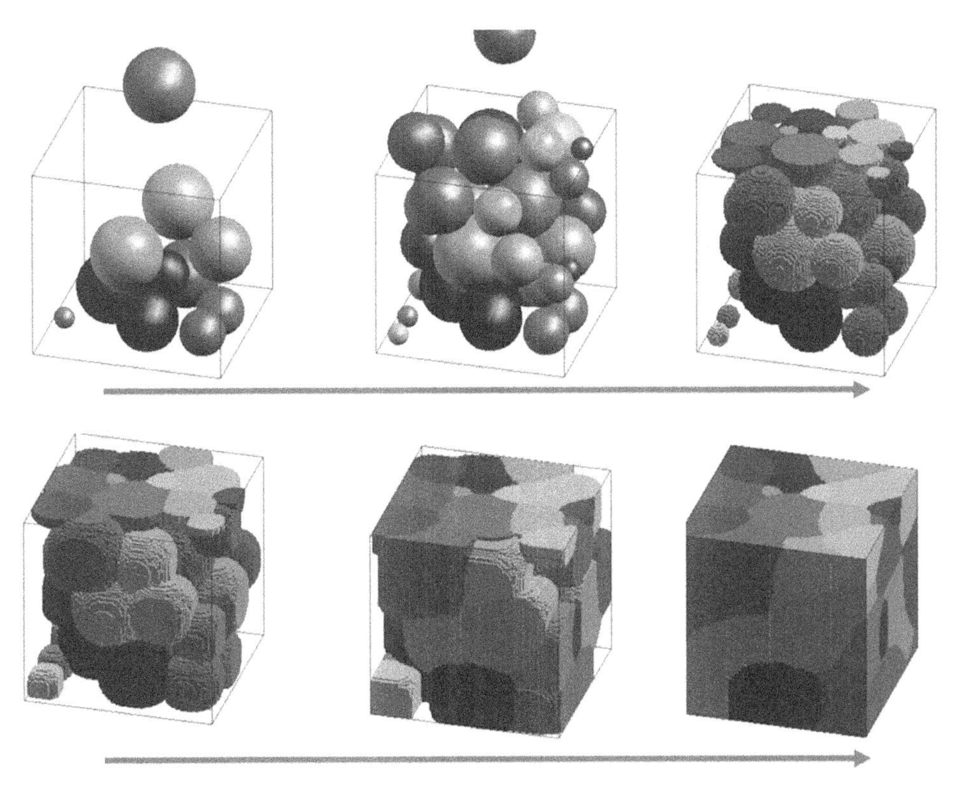

Figure 4.39 Results obtained in 3D space after subsequent steps of the algorithm.

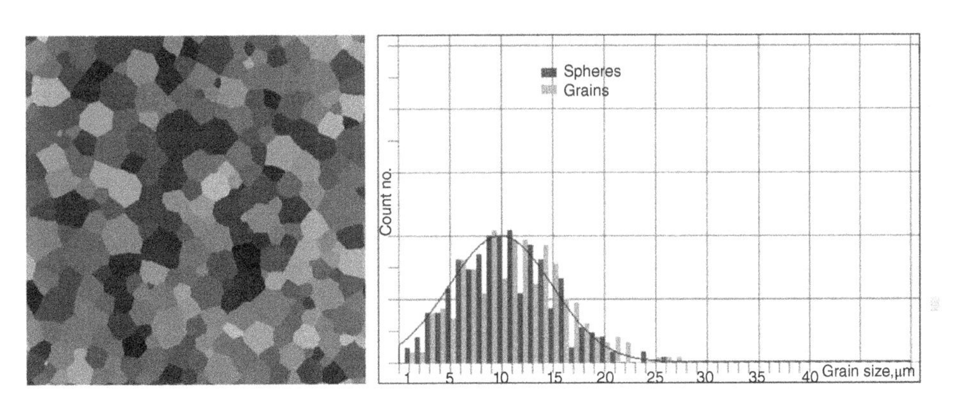

Figure 4.40 Result of modified sphere grown algorithm. The microstructure contains 204 grains.

grain sizes distribution is required, another method based on an inverse analysis can also be applied; see e.g. [62]. The approach is also quite time consuming as it requires a series of recalculations to minimize the defined goal function.

Similar results can also be obtained using another method that was mentioned earlier: the Monte Carlo (MC) approach.

4.2.1.4 Monte Carlo Grain Growth Algorithm

As mentioned in Chapter 3, the MC is a general name for a group of algorithms based on a completely random sampling of a solutions space Ω for application in mathematical and physical simulations [4, 18, 39, 83]. Therefore, the algorithm can be easily adapted to develop a grain growth model.

There can be Q states in the lattice in the MC grain growth model, where Q is the algorithm's coefficient that represents a number of subsequent grains in the microstructure. First, a computational domain that is similar to the CA method is generated. Then the algorithm is again executed in an iterative manner according to the following steps:

- random selection of a lattice site from the computational domain,
- calculation of the site energy based on the Hamiltonian from Chapter 3:

$$E_i = -J \sum_{j=1}^{Z} \left(\delta_{S_i S_j} - 1\right). \tag{4.5}$$

where: J – energy between two MC lattice sites that belong to different grains (grain boundary energy), Z – number of neighbours adjacent to the site, δ – Kronecker's delta with the value of 1 for the lattice sites that belong to the same grain ($S_i = S_j$), and the value of 0 for the lattice sites that belong to different grains ($S_i \neq S_j$),

- random reorientation of the investigated MC lattice site. The new state is randomly selected from all the available Q states,
- recalculation of the MC site energy based on Equation (4.5),
- determination of a difference in the energy ΔG_i between the initial and reoriented states. The new site orientation is accepted with a probability P_i when the new value of energy is lower or equal to the initial one (Figure 4.41):

$$P_i = \begin{cases} \exp\left(\frac{-\Delta G_i}{kT}\right) & \text{for} \quad \Delta G_i > 0 \\ 1 & \text{for} \quad \Delta G_i \leq 0 \end{cases}, \tag{4.6}$$

where: k – Boltzmann's constant, T – MC model temperature.

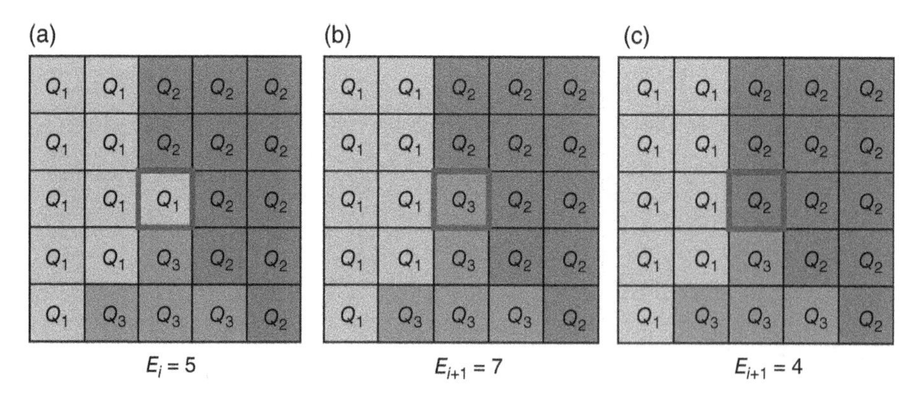

Figure 4.41 Illustration of the energy calculation process for (a) initial state and (b) and (c) two possibilities of the final state.

With successive MC steps, the energy value of the entire system is reduced, and the digital microstructure representation with clearly visible grains is obtained.

Each iteration of the MC algorithm is performed within the MC step (MCS) concept. A single MCS is considered as the n reorientation attempts, where n is the total number of all lattice sites. Therefore, in the classical algorithm, in a particular MCS, each cell has been analyzed at least once. Examples of microstructures in 2D and 3D computational domains obtained from the MC algorithm are shown in Figures 4.42 and 4.43.

Microstructures presented in Figures 4.42 and 4.43 were obtained using Moore's neighbourhood (8 and 26 neighbours in 2D and 3D space, respectively). However, other kinds of neighbourhoods can also be introduced, e.g. pentagonal, hexagonal, etc. [70]. It should also

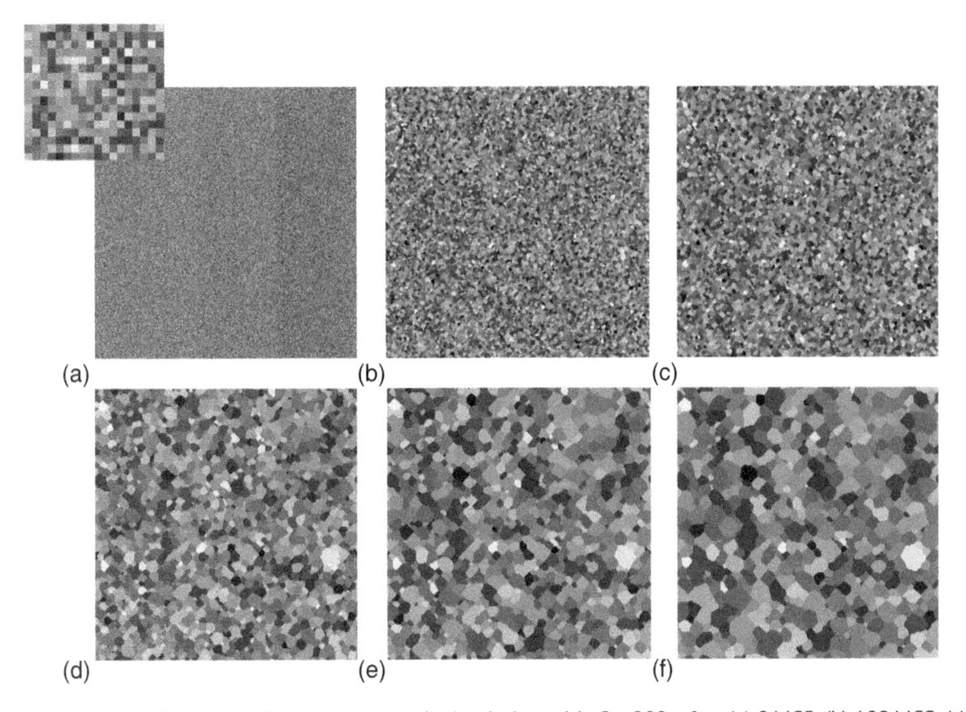

Figure 4.42 The Monte Carlo grain growth simulation with $Q = 200$, after (a) 0 MCS, (b) 100 MCS, (c) 200 MCS (d) 300 MCS, (e) 400 MCS, and (f) 500 MCS.

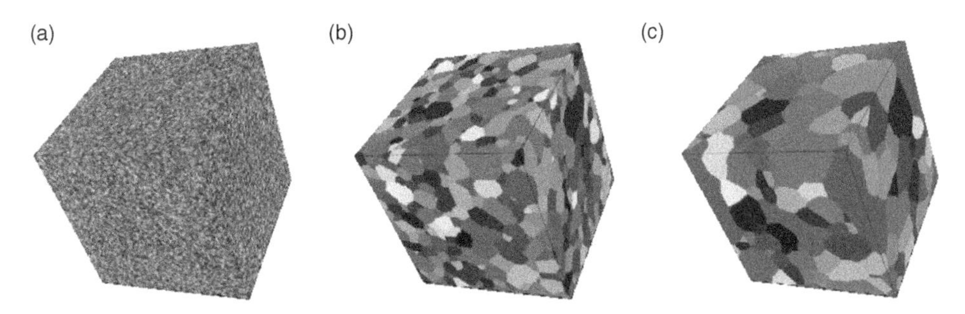

Figure 4.43 The process of grain growth in the 3D using the Monte Carlo algorithm; $Q = 50$: (a) 0 MCS, (b) 100 MCS, and (c) 400 MCS.

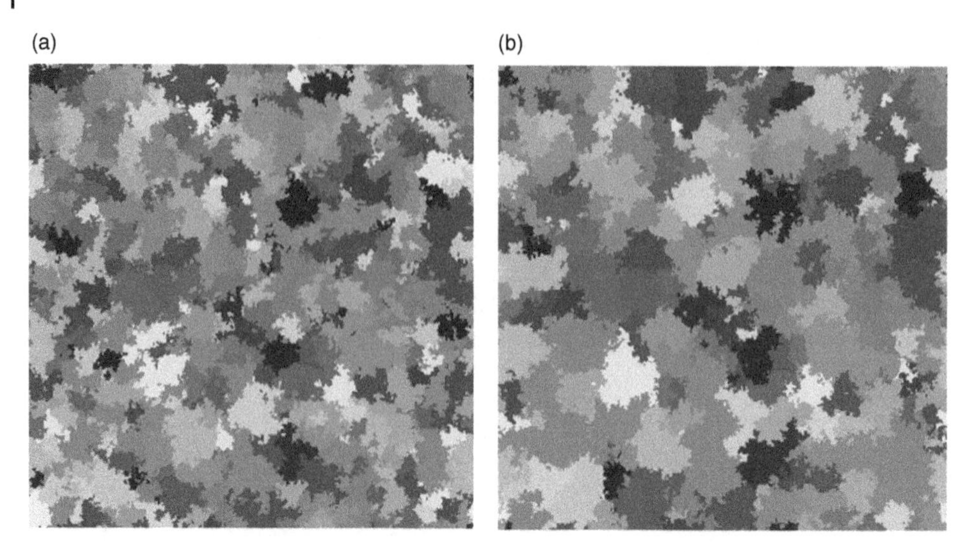

Figure 4.44 The Monte Carlo grain growth simulation for $kT = 6$, $Q = 50$; von Neumann's neighbourhood: (a) MCS = 200, (b) MCS = 400.

be pointed out that the kT parameter is not connected with the real temperature and is rather a model parameter that has to be set up in order to avoid desirable results of the microstructure morphology. The kT directly influences the probability of new cell state acceptance, and when too large, it can provide jagged grain boundaries, as seen in Figure 4.44. The typical value of the kT ratio that provides acceptable geometry of grain boundaries for the metallic polycrystalline microstructures is within the range (0.3–0.6) [59].

Due to the stochastic nature of the MC method, the major limitation is the significant computational time. In order to accelerate the algorithm, several modifications to this standard approach can be implemented. The first concerns the sampling of the new state of the cell. Contrary to the mentioned approach, where sampling of a new state was from the entire $Q-1$ available states, in the modified version, sampling is realized only from states among those directly adjacent to the analyzed cell. Another modification of the algorithm is based on dividing the cells into two groups lying inside the grain and those at the grain boundary. As a result, only cells lying along the grain boundary can change the state and minimize the energy of the system. Change of the cell lying inside the grain always leads to an increase in energy. Finally, it is assumed that during a particular MC step, each cell can be selected exactly once.

The CA and MC methods are often combined to obtain a required DMR of, e.g. multiphase, polycrystalline microstructures. In that case, the MC algorithm with a limited number of Q states, representing different phases in the material, can be initiated first, and then within each phase, the CA grain growth model is performed within each phase separately, as presented in Figure 4.45.

To facilitate the generation of such DMR models, a dedicated computational library was developed and implemented. Details of the approach and examples of its capabilities can be found in the paper by Bogun et al. [5].

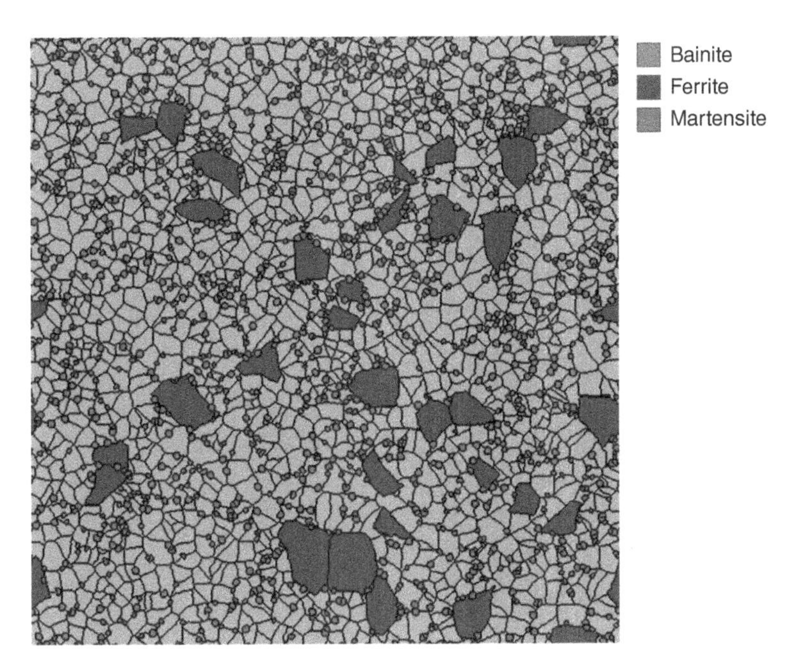

Figure 4.45 A digital material representation model for the polycrystalline complex phase microstructure obtained by combining MC and CA grain growth models.

4.2.1.5 DigiCore Library

The library is based on a set of developed classes and methods, which can be considered base elements to be inherited and extended in other libraries. Each additional dynamic-link library (DLL) implements particular algorithms for generating specific microstructure morphologies with various features in the digital model. The computational domain is assumed to be the cellular automata space. The developed functional concept of the DigiCore system is presented as the use case diagram (Figure 4.46).

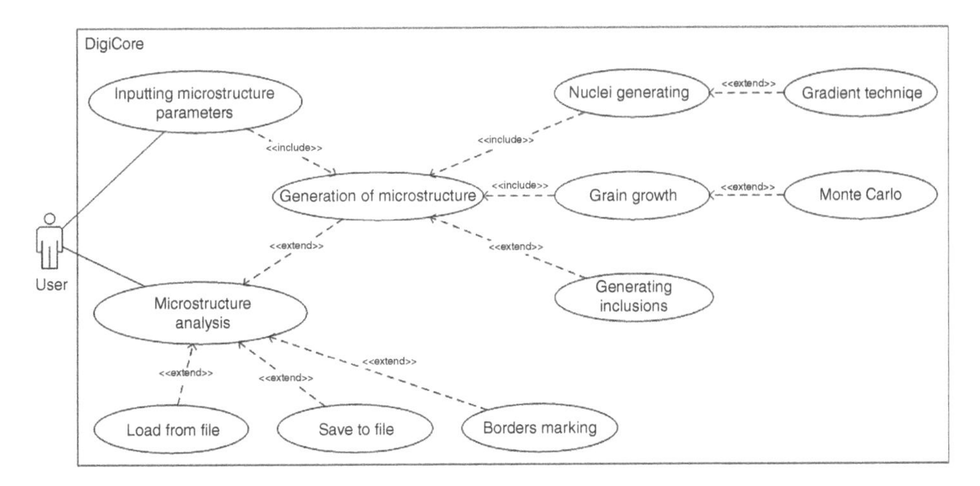

Figure 4.46 The use case diagram with basic functional requirements.

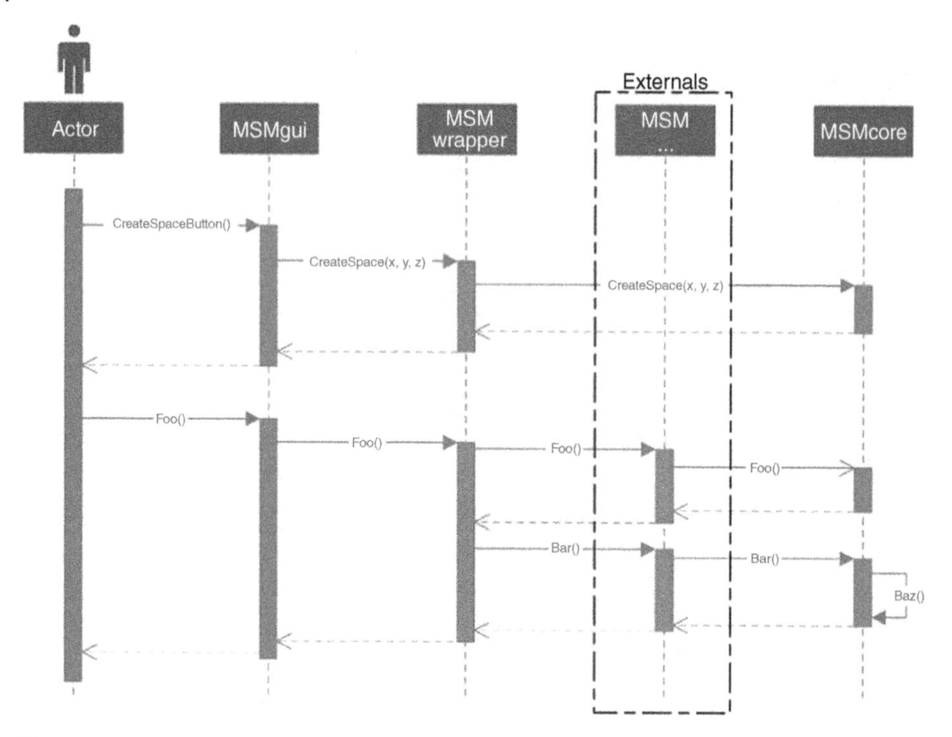

Figure 4.47 Sequence diagram presenting the role of the wrapper in the application.

The software is implemented within the object-oriented C++ language in the Microsoft Visual Studio environment. The Graphical User Interface (GUI) is implemented in C#. With that, there is no need to install external libraries, and that facilitates its use. However, to use native libraries written in C++ in the GUI, it is necessary to create a wrapper. This is a packaging class whose only task is to call methods from the other libraries. Figure 4.47 shows a universal outline of interaction between developed libraries.

All the elements listed in the diagram are integral parts of the program. Elements that are necessary for running the DigiCore with its basic functionalities are:

- *MSMcoreLib* – the core of the application,
- *MSMwrapperLib* – the connector between the GUI and modules, and
- *MSMgui*.

All crucial elements and extensions of the core functionality are contained in *MSMcoreLib*. This is the main project library, and like all others, has a form of a DLL. DLLs are loaded into the computer's operating memory only after their first call, which directly fits the requirements of this software, as unused application modules will not use the available memory. All the typical definitions for the CA method are gathered in the *MSMcoreLib* project, e.g. microstructure features, simulation type, boundary conditions, and neighbourhood types. These definitions are split according to the CA space under consideration, namely 2D or 3D. The relationships between developed neighbourhood classes are presented in Figure 4.48.

All user-developed additional modules for microstructure generations are located between the wrappers and the core of the system because they cannot act as separate independent programs but only within the *MSMcoreLib*. This structure of application ensures the required

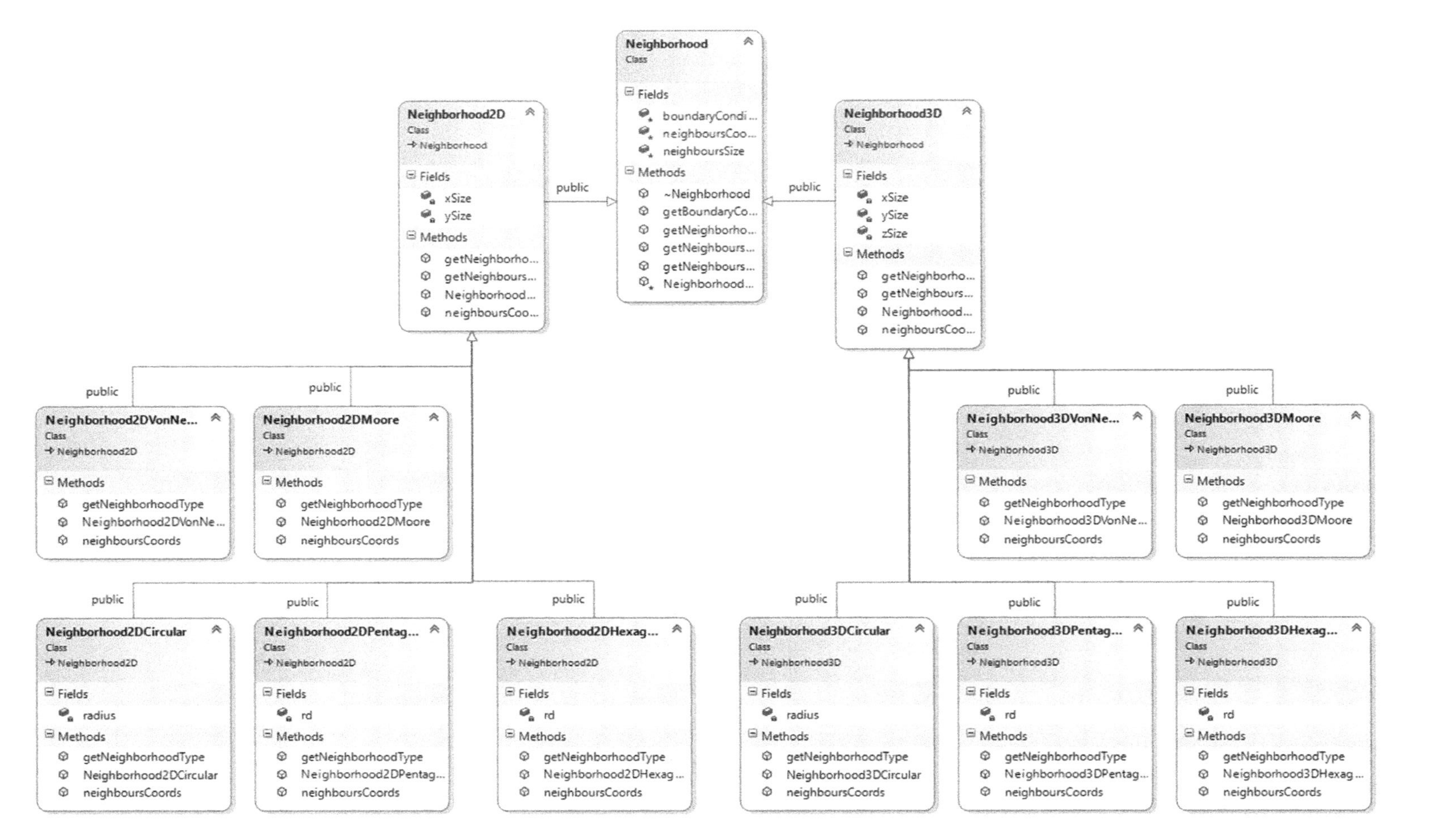

Figure 4.48 Neighbourhood classes and their relations.

modularity of the system, i.e. it is based on a single base library that can be extended by other system components independent of each other but working within the single application.

Within the DigiCore library, algorithms based on cellular automata or the MC approach can be used in various configurations to create a digital representation of polycrystalline materials. Examples of its capabilities in generating various complex heterogeneous microstructure morphologies are presented in [5].

However, when an exact representation of a specific microstructure is required, an image processing method can be used to obtain the digital format.

4.2.1.6 Image Processing

The image processing algorithm applied in this work was developed by [66] and adopted for the processing of light and electron microscopy images of microstructure [66–69]. A brief description of the main assumptions of this method is summarized next. The image preprocessing stage consists of the two supplementary approaches: filtering and reconstruction. Both are often required, especially in light microscopy and scanning electron microscopy with the electron backscattered diffraction (SEM/EBSD) images. The dynamic particles (DP) algorithm was designed and implemented [66] for the filtering operation. The main advantage of this solution is its universality, which allows the application of the DP method for images and signals and sophisticated multidimensional data. The basis of this method is in the definition of the particle, which can be treated as an N-dimensional object, which in the case of images is a 3D vector [X-axis, Y-axis, greyscale value] related to pixels inside analyzed images. The main idea of the DP algorithm consists of the appropriate movement of each particle according to the following set of differential equations:

$$\begin{cases} m_i \dfrac{d\vec{v}_i}{dt} = -\nabla V_{ij} - f_c \vec{v}_i \\ d\vec{r} = \vec{v}_i dt \end{cases}, \tag{4.7}$$

where: m_i – mass, v_i – a velocity of the i-th particle, V_{ij} – a potential between particles i and j, depending on the distance between them, f_c – a friction coefficient responsible mainly for the convergence of calculations.

The magnitude of the force that causes the movement of the considered particle is reduced by the friction coefficient $f_c < 1$. It was found that according to the Newtonian laws of motion, if all of the pre-conditions would be fixed properly, the entire system would remain stable and convergent to the expected results. Additionally, the friction coefficient f_c can be modified during the algorithm's calculations, which influences the final smoothness of the results and the stop criterion of the DP algorithm. A critical issue in this method is to normalize data before calculations to equal the impact of each dimension on the finally obtained results. Otherwise, the process of filtering will fail, giving an inappropriate shape of the denoised image. The normalization of the input data does not influence the data's density on each axis, and the data is re-scaled after calculations to its primary range. However, the problem of the boundary points of the analyzed image has not been solved yet. They remain fixed during calculations, which may cause some problems, while the boundary points are superimposed by the noise of a higher ratio than other particles.

The example of denoising results, shown in Figure 4.49, is further analyzed to distinctly separated areas inside images to gather information about boundaries of grains and subgrains and visible micro-shear bands or slip surfaces.

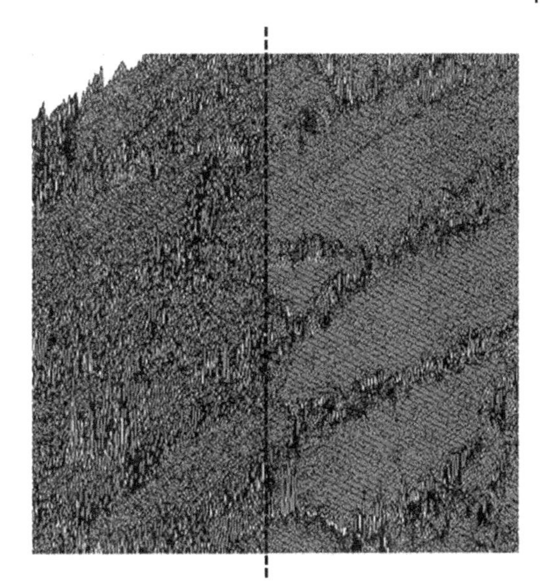

Figure 4.49 Original image of the microstructure from an optical microscope and after filtering in the form of 3D plots.

However, in the case of SEM/EBSD images, it is hard to decide which pixel is superimposed with noise and which one is a result of non-indexed measurements. Therefore, instead of DP algorithm, which cannot be applied automatically for such images, an image reconstruction algorithm has to be used [40]. Such a method is used to remove all pixels related to points in material, which were not indexed during the measurements using SEM, e.g. scratches, grain and subgrain boundaries, or dislocations. This algorithm is required to perform the next steps aiming at grains and subgrains detection. It prepares the output image without distortions, where one grain is marked with one main colour and some inherited sub-colours palette, which facilitates distinction between different image areas. The distortions from the original pictures are removed using RGB colours comparison in the 3D space, while the empty spaces are replaced with the average colour of neighbouring pixels (the von Neumann neighbourhood is used for this purpose). Examples of original and reconstructed results are presented in Figure 4.50.

Processing of single-phase microstructures after filtering and reconstruction processes resolves to edge detection based on the modified Canny Detector algorithm. The proposed approach consists of the following steps:

- Edge detection and enhancement by using undirected Laplace convolution filter – this algorithm enhances the borders between different areas of the image by calculating the colour gradients of the neighbouring pixel. The main disadvantage of this technique is the effect of twin borders, which is obtained through the convolution of the two matrices, giving the left and right borders. The suppression of the twin-border effect is performed in the subsequent step of the proposed approach.
- Detection of edges directions – this step is based on previously detected borders. Four main directions are taken into consideration, i.e. horizontal, vertical, and two diagonals marked by four different grey colours. This assumption can be stated due to the regular deployment of pixels inside an image.
- Tracing of edges – detected edges and their directions are used in the process of edge tracing. This is the most important step of calculations, giving efficient detection of edges,

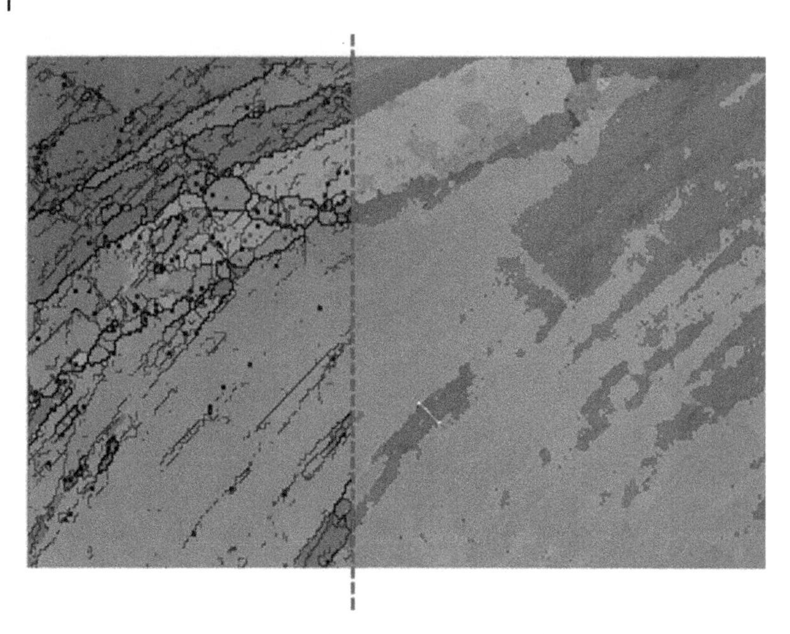

Figure 4.50 An original image from EBSD analysis and a reconstructed image of deformed copper sample.

which are poorly visible. Two thresholds are used in this process, namely higher (TH) and lower (TL). TH is used to determine initial edges characterized by the brightest pixels. Then TL is applied to detect darker pixels, which are placed nearby already detected borders. Therefore, the reliability of such a solution is higher than in the case of traditional approaches.

- Elimination of small groups of pixels (blobs) – the last step of calculations is focused on the thinning of thick borders and deletion of small groups of pixels. The thinned borders detected between grains are, in most cases, long and narrow, while the remaining distortions are still visible in the form of compact groups of pixels. The size of such groups does not usually exceed several pixels.

The resultant structure of the data after edge analysis is in the form of an array composed of 0 values (non-border pixels) and 1 values (border pixels). This structure is used to map an image into a set of different areas related to particular grains and subgrains in the material microstructure presented in Figure 4.51.

Image processing algorithms are especially useful to create a digital representation of two-phase or complex-phase microstructures [41, 42]. Obtaining such a representation using CA or MC algorithms is reported in the literature [22, 62, 73] but requires the development of a physically based phase transformation model. Therefore, in the case of image processing, the algorithms for the generation of digital representation of microstructure can be considered as straightforward.

Many algorithms dedicated to the analysis of multiphase microstructures were already developed at the end of the last century and then modified later to improve the quality of results [46, 68]. One of the modifications is based on the automated analysis of image histograms is proposed. Several images of dual-phase (DP) steel were analyzed as a case study. It was observed that there is a possibility to automatically find the greyscale threshold, which

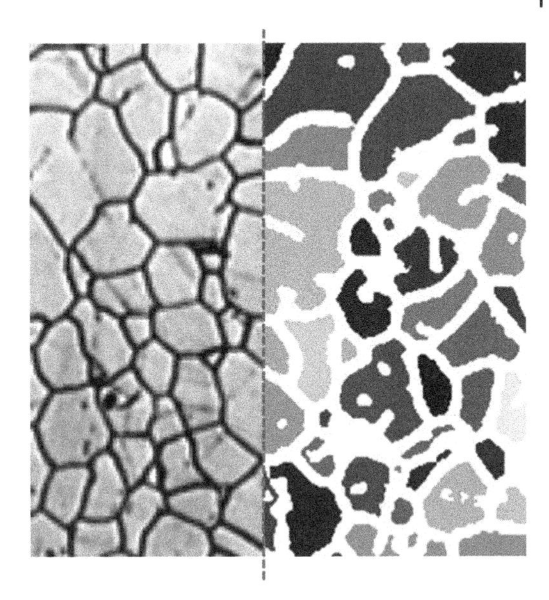

Figure 4.51 An original image from optical microscopy and equivalent processed image picture of a single-phase material.

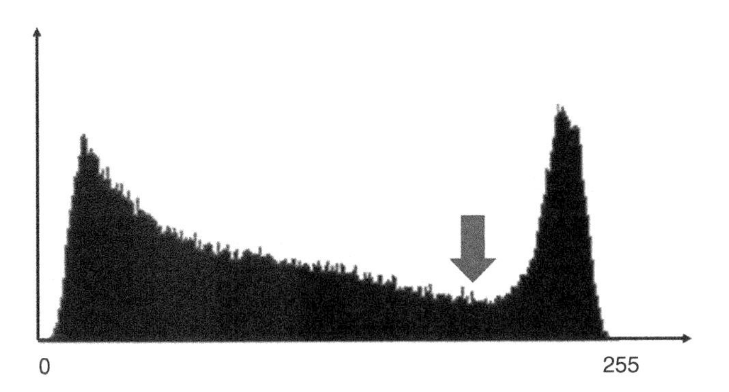

Figure 4.52 The histograms of various images of microstructures of the DP steel with clearly visible minimum and inflection points.

separates the light and dark phases on the image. The value of such threshold usually covers the first minimum or the first inflection point located on the left side of the peak, which is related to the ferrite phase (Figure 4.52). Detailed grain boundary detection algorithms can further process separated areas of the analyzed image to identify particular grains within each phase.

The ferrite phase visible in Figure 4.53 is the result of phase distinction performed for the material, in which the martensite dominates a major part of the microstructure. Due to this, the detection of areas occupied by particular ferrite grains can be obtained by the application of the simple tracing algorithm. Otherwise, the algorithm of edges detection has to be used. In most cases, it is enough to implement the modified Canny Detector algorithm, which proved its efficiency in the detection of grain borders for one-phase materials. However, these calculations often have to be preceded by the preprocessing of the image, focused on normalization and equalization of greyscale levels among all the grain areas.

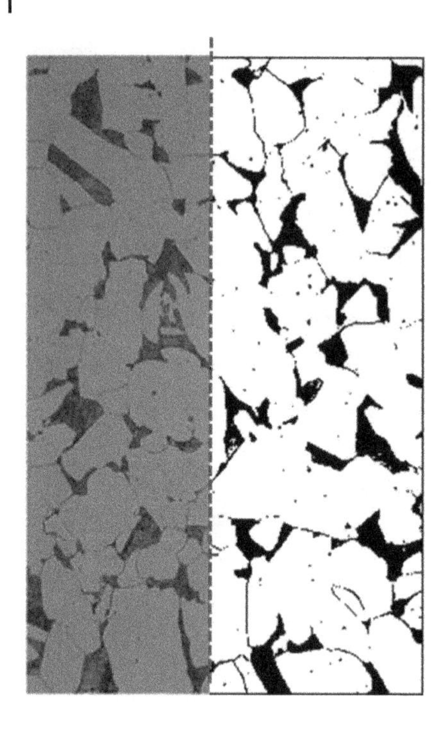

Figure 4.53 An original image of a two-phase material with the result of extraction of a single phase.

Obtaining digital microstructures of two-phase or multiphase microstructures from an image is very accurate; however, it is also very costly. If multiphase digital microstructures after various cooling conditions are required as input for further simulations, then a series of experimental work combined with metallographic analysis is obligatory. This leads to increased time and costs and is especially laborious for 3D investigations based on, e.g. serial sectioning [72] or more frequently used computer tomography [46]. The solution is to develop mentioned earlier numerical models capable of generating various initial digital microstructure morphologies based on simple grain growth or physics-based grain growth models.

As presented, the digital microstructure can be obtained using various techniques. As a result, different morphologies of single-phase or multiphase microstructures can be obtained. However, to create a functional DMR that is used during subsequent numerical simulations, the obtained microstructure morphology with all its features has to be filled with specific material properties.

4.2.2 Properties of the Microstructure Features

When the initial microstructure is created in morphology, then it is analyzed and filled with certain properties (e.g. chemical composition, crystallographic lattice, crystallographic orientation, material density, dislocation density, stacking fault energy, stress state, thermomechanical variables, yield stress, tensile strength, material moduli, anisotropy parameter, phase assignment, grain boundary energy, residual stress fields, etc.)

This possibility to assign material properties to particular grains is the main key in the DMR models. These properties can be represented as a scalar value or a vector containing several scalar elements. There is also a possibility of assigning other data structures, such as tensors (e.g. representing stress state), to particular grains. When a scalar value is considered, the

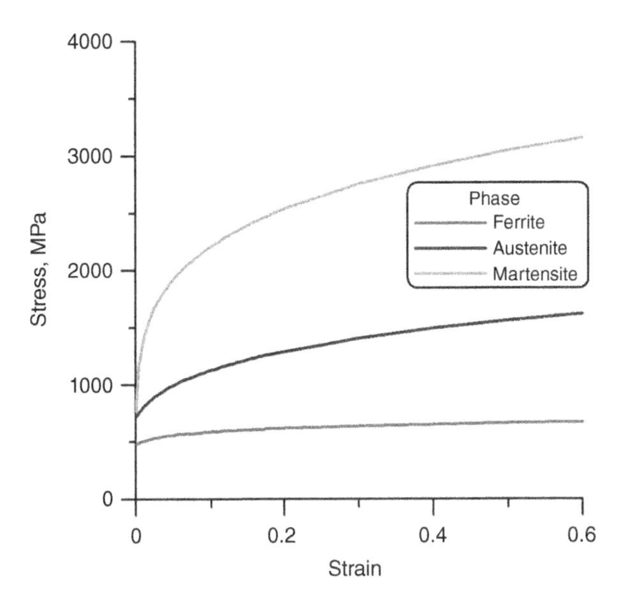

Figure 4.54 Flow stress models for different phases in steel [2, 15]. *Source:* Based on Delannay et al. [15], Asgari et al. [2].

assigned information can represent one parameter in the simple flow stress model. When a vector $\boldsymbol{r}$ of four scalar elements is assigned, it can represent subsequent parameters in the flow stress model, e.g. $\boldsymbol{r} = [K, n, m, Q]$:

$$\sigma_p = \sigma_0 + K\varepsilon_i^n \dot{\varepsilon}_i^m \exp\left(\frac{Q}{RT}\right), \tag{4.8}$$

where: σ_0 – stress at the beginning of plastic deformation, ε – strain, $\dot{\varepsilon}$– strain rate, T – temperature, n, m – coefficients of sensitivity of flow stress to strain and strain rate, respectively, K – hardening coefficients, Q – activation energy, which represents sensitivity to temperature, R – gas constant.

That way, particular grains can be described by different flow stress models (Figure 4.54). Parameters of the flow curves that describe particular material and that are assigned to particular grains are reported in the literature (e.g. steel [15], aluminium [78, 80]) or are identified experimentally during simple monotonic deformation of a sample in a particular phase or a particular crystallographic orientation [76, 78, 80]. When multiphase polycrystalline materials are used during deformation, other experimental techniques (e.g. nanoindentation tests [15, 64] or micropillar compression [44]) can be used to extract the behaviour of particular phases. When a crystal plasticity model is used, the properties vector can represent three Euler angles that determine grain crystallographic orientation [76].

However, if a digital material is input data for a specific numerical model, such as phase transformation, dynamic recrystallization, or fracture models, other properties (e.g. different elements concentration, accumulated energy, dislocation density, etc.) can be assigned to particular grains [47, 63, 73].

It can be concluded that the main advantage of the DMR model is that various material properties (e.g. chemical composition, crystallographic lattice, crystallographic orientation, stress state, thermomechanical variables, anisotropy parameter, phase assignment, etc.)

can be explicitly assigned to the subsequent microstructure features and then used during numerical modelling of material behaviour under thermomechanical conditions within various other computational tools. For more details on the DMR concept, refer to [43].

As a result, a multi-scale model that has the capability to model material both at the microscale and macroscale level can be created. As mentioned earlier, various methods can be used for the simulation of material behaviour (e.g. cellular automata or finite element). Some of the examples of multi-scale models available in the scientific literature are discussed in the following part of this book.

References

1 Akhter, M.J., Kuś, W., Mrozek, A., and Burczyński, T. (2020). Mechanical properties of monolayer MoS2 with randomly distributed defects. *Materials* 13 (6): 1307.

2 Asgari, A., Hodgson, P., Lemiale, V. et al. (2008). Multi-scale particle-in-cell modelling for advanced high strength steels. *Advanced Materials Research*, Trans Tech Publications Ltd 32: 285–288.

3 Baranau, V. and Tallarek, U. (2014). Random-close packing limits for monodisperse and polydisperse hard spheres. *Soft Matter* 10 (21): 3826–3841.

4 Blikstein, P. and Tschiptschin, A.P. (1999). Monte Carlo simulation of grain growth. *Materials Research* 2: 133–137.

5 Bogun, K., Sitko, M., Mojzeszko, M., and Madej, L. (2021). Cellular Automata-based computational library for development of digital material representation models of heterogenous microstructures. *Archives of Civil and Mechanical Engineering* 21: 1–15.

6 Box, G.E.P. and Muller, M.E. (1958). A note on the generation of random normal deviates. *The Annals of Mathematical Statistics* 29: 610–611.

7 Buehler, M.J. (2006). Mesoscale modeling of mechanics of carbon nanotubes: self-assembly, self-folding, and fracture. *Journal of Materials Research* 21 (11): 2855–2869.

8 Chatterjee, S.K. (2008). *Crystallography and the World of Symmetry, Springer Series in Material Sciences*, vol. 113. Springer.

9 Chu, L., Shi, J., and Ben, S. (2018). Buckling analysis of vacancy-defected graphene sheets by the stochastic finite element method. *Materials* 11: 1545.

10 Cranford, S. and Buehler, M.J. (2011). Twisted and coiled ultralong multilayer graphene ribbons. *Modelling and Simulation in Materials Science and Engineering* 12: 054003.

11 Cui, L. and O'Sullivan, C. (2003). Analysis of a triangulation based approach for specimen generation for discrete element simulations. *Granular Matter* 5: 135–145.

12 Daw, M.S. and Baskes, M.I. (1984). Embedded-atom method: derivation and application to impurities, surfaces, and other defects in metals. *Physical Review B* 29: 6443–6453.

13 Daw, M.S., Foiles, S.M., and Baskes, M.I. (1993). The embedded-atom method: a review of theory and applications. *Materials Science Reports* 9 (7–8): 251–310.

14 De Berg, M., Van Kreveld, M., Overmars, M., and Schwarzkopf, O. (2000). *Computational Geometry Algorithms and Applications*. Springer-Verlag.

15 Delannay, L., Doghri, I., and Pierard, O. (2007). Prediction of tension-compression cycles in multiphase steel using a modified incremental mean-field model. *International Journal of Solid and Structures* 44: 7291–7306.

16 Elizondo, A. (2007). Horizontal coupling in continuum atomistics. PhD thesis.

17 Enyashin, A.N. and Ivanovskii, A.L. (2011). Graphene allotropes. *Physica Status Solidi* 248 (8): 1879–1883.

18 Francis, A., Roberts, C.G., Cao, Y. et al. (2007). Monte Carlo simulations and experimental observations of templated grain growth in thin platinum films. *Acta Materialia* 55: 6159–6169.

19 Foiles, S.M., Baskes, M.I., and Daw, M.S. (1986). Embedded atom method functions for fcc metals Cu, Ag, Au, Ni, Pd, Pt and their alloys. *Physical Review B* 33: 7983–7991.

20 Girifalco, L.A. and Weizer, V.G. (1959). Application of the Morse potential function to cubic metals. *Physical Review* 114: 687–690.

21 Hajder, L. and Madej, L. (2020). Sphere packing algorithm for the generation of digital models of polycrystalline microstructures with heterogeneous grain sizes. *Computer Methods in Materials Science* 20: 22–30.

22 Halder, C., Bachniak, D., Madej, L. et al. (2015). Sensitivity analysis of the Finite Difference 2-D Cellular Automata model for phase transformation during heating. *ISIJ* 55: 285–292.

23 Henck, H., Pierucci, D., Chaste, J. et al. (2016). Electrolytic phototransistor based on graphene-MoS2 van der Waals p-n heterojunction with tunable photoresponse. *Applied Physics Letters* 109: 113103.

24 Hitti, K. and Bernacki, M. (2013). Optimized Dropping and Rolling (ODR) method for packing of poly-disperse spheres. *Applied Mathematical Modelling* 37 (8): 5715–5722.

25 Holian, B.L., Voter, A.F., Wagner, N.J. et al. (1991). Effects of pairwise and many-body forces on high-stress plastic deformation. *Physical Review A* 43 (6): 2655–2661.

26 Jacobsen, K.W., Norskov, J.K., and Puska, M.J. (1987). Interatomic interactions in the effective-medium theory. *Physical Review B* 35 (14): 7423–7442.

27 Jiang, J.-W. (2015). Graphene Versus MoS2: a mini review. *Frontiers of Physics* 10: 106801.

28 Jiang, J.-W., Park, H.S., and Rabczuk, T. (2013). Molecular dynamics simulations of single-layer molybdenum disulphide (MoS2): Stillinger-Weber parametrization, mechanical properties, and thermal conductivity. *Journal of Applied Physics* 114: 064307.

29 Jodrey, W.S. and Tory, E.M. (1986). Computer simulation of close random packing of equal spheres. *Physical Review A* 32 (4): 2347–2351.

30 Kandemir, A., Yapicioglu, H., Kinaci, A. et al. (2016). Thermal transport properties of MoS2 and MoSe2 monolayers. *Nanotechnology* 27: 055703.

31 Kang, M.A., Kim, S.J., Song, W. et al. (2017). Fabrication of flexible optoelectronic devices based on MoS2/graphene hybrid patterns by a soft lithographic patterning method. *Carbon* 116: 167–173.

32 Krivtsov, A.M. and Wiercigroch, M. (2001). Molecular dynamics simulation of mechanical properties for polycrystalline materials. *Materials Physics and Mechanics* 3: 45–51.

33 Krivtsov, A.M. (2003). Molecular dynamics simulation of impact fracture in polycrystalline materials. *Meccanica* 38: 61–70.

34 Large-scale Atomic/Molecular Massively Parallel Simulator (2021). www.lammps.org ().

35 Li, M., Wan, Y., Tu, L. et al. (2016). The effect of VMoS3 point defect on the elastic properties of monolayer MoS2 with REBO potentials. *Nanoscale Research Letters* 11: 155.

36 Liang, T., Phillpot, S.R., and Sinnot, S.R. (2009). Parametrization of a reactive many-body potential for Mo-S systems. *Physical Review B* 79: 245110.

37 Liang T., Phillpot S. R., Sinnot S. R., Erratum: Parametrization of a reactive many-body potential for Mo-S systems, *Physical Review B*, 85, 2012, 199903(E).

38 Liu, Y., Baudin, T., and Penelle, R. (1996). Simulation of normal grain growth by cellular automata. *Acta Metallurgica* 34: 1679–1683.

39 Liu, J.S. (2008). *Monte Carlo Strategies in Scientific Computing*. Springer Series in Statistics.

40 Madej, L., Yazdipour, N., Rauch, L., and Hodgson, P.D. (2008). Numerical simulation of the static recrystallization on the micro shear bands.Proceedings MS&T 2008, Pittsburgh, 832–843.

41 Madej, L., Malinowski, L., Perzynski, K. et al. (2019). Considering influence of microstructure morphology of epoxy/glass composite on its behavior under deformation conditions – digital material representation case study. *Archives of Civil and Mechanical Engineering* 19: 1–12.

42 Madej, L., Chang, Y., Szeliga, D. et al. (2021). Criterion for microcrack resistance of multiphase steels based on property gradient maps, *CIRP Annals Manufacturing Technology*, 70, 243–246.

43 Madej, L. (2017). Digital/virtual microstructures in application to metals engineering – A review. *Archives of Civil and Mechanical Engineering* 17: 839–854.

44 Madej, L., Wang, J., Perzynski, K., and Hodgson, P.D. (2014). Numerical modelling of dual phase microstructure behavior under deformation conditions on the basis of digital material representation. *Computational Material Science* 95: 651–662.

45 Madej, L., Pasternak, K., Szyndler, J., and Wajda, W. (2014). Development of the modified cellular automata sphere growth model for creation of the digital material representations. *Key Engineering Materials* 611–612: 489–496.

46 Madej, L., Legwand, A., Mojzeszko, M. et al. (2018). Experimental and numerical two- and three-dimensional investigation of porosity morphology of the sintered metallic material. *Archives of Civil and Mechanical Engineering* 18: 1520–1534.

47 Madej, L., Sitko, M., and Pietrzyk, M. (2016). Perceptive comparison of mean and full field dynamic recrystallization models. *Archives of Civil and Mechanical Engineering* 16: 569–589.

48 Mortazavi, B., Ostadhossein, A., Rabczuk, T., and van Duin, A.C.T. (2016). Mechanical response of all-MoS2 single-layer heterostructures: a ReaxFF investigation. *Physical Chemistry Chemical Physics* 18 (34): 23695–23701.

49 Mrozek, A. (2019). Basic mechanical properties of 2H and 1T single-layer molybdenum disulfide polymorphs: a short comparison of various atomic potentials. *International Journal for Multiscale Computational Engineering* 17 (3): 339–359.

50 Mrozek, A. and Burczynski T. (2012). Computational Models of Nanocrystalline Materials, Applied Sciences and Engineering (ECCOMAS 2012) (eds. J. Eberhardsteiner et al.).

51 Mrozek, A. and Burczyński, T. (2013). Examination of mechanical properties of graphene allotropes by means of computer simulation. *Computer Assisted Methods in Engineering and Science* 20 (4): 309–323.

52 Mrozek, A. and Burczyński, T. (2015). Computational models of polycrystalline materials. *International Journal for Multiscale Computational Engineering* 13 (2): 145–161.

53 Mrozek, A., Kuś, W., and Burczyński, T. (2015). Nano level optimization of graphene allotropes by means of a hybrid parallel evolutionary algorithm. *Computational Materials Science* 106: 161–169.

54 Mrozek, A., Kuś, W., and Burczyński, T. (2017). Method for determining structures of new carbon-based 2D materials with predefined mechanical properties. *International Journal for Multiscale Computational Engineering* 15 (5): 379–394.

55 Narita, N., Nagai, S., Suzuki, S., and Nakao, K. (2000). Electronic structure of three-dimensional graphyne. *Physical Review B* 62 (16): 11146–11151.

56 Nasiri, S. and Zaiser, M. (2016). Rupture of graphene sheets with randomly distributed defects. *AIMS Materials Science* 3: 1340–1349.

57 Naylor, C.H., Kybert, N.J., Schneier, C. et al. (2016). Scalable production of molybdenum disulfide based biosensors. *ACS Nano* 10 (6): 6173–6179.

58 Nguyen, T.H., Ahmad, R.T.N., and Saito, R. (2018). Two-dimensional MoS2 electromechanical actuators. *Journal of Physics D: Applied Physics* 51: 7.

59 Okuda, K. and Rollett, A.D. (2005). Monte Carlo simulation of elongated recrystallized grains in steels. *Computational Materials Science* 34: 264–273.

60 Ostadhossein, A., Rahnamoun, A., Wang, Y. et al. (2017). ReaxFF reactive force-field study of Molybdenum disulfide (MoS2). *The Journal of Physical Chemistry Letters* 8 (3): 631–640.

61 Park, H., Fellinger, M.R., Lenosky, T.J. et al. (2012). *Ab initio* based empirical potential used to study the mechanical properties of molybdenum. *Physical Review B* 85: 214121.

62 Pietrzyk, M., Madej, L., Rauch, L., and Szeliga, D. (2015). *Computational Materials Engineering: Achieving High Accuracy and Efficiency in Metals Processing Simulations*. Butterworth Heinemann Elsevier.

63 Pietrzyk, M. and Madej, L. (2017). Perceptive review of ferrous micro/macro material models for thermo-mechanical processing applications. *Steel Research International* 88: 1700193.

64 Perzynski, K., Cios, G., Szwachta, G. et al. (2019). Numerical modelling of a compression test based on the 3D Digital Material Representation of pulsed laser deposited TiN thin films. *Thin Solid Films* 673: 34–43.

65 Radisavljevic, B., Radenovic, A., Brivio, J. et al. (2011). Single-layer MoS2 transistors. *Nature Nanotechnology* 6: 147–150.

66 Rauch, L. and Kusiak, J. (2007). Edge detection and filtering approach dedicated to microstructure images analysis. *Computer Methods in Materials Science* 7: 305–310.

67 Rauch, L., Madej, L., and Wajda, W. (2008). Tool for modelling of deformation mechanisms based on digital representation of microstructure. Proceedings WCCM'08, Venice, CD.

68 Rauch, L. and Madej, L. (2008). Deformation of the dual phase material on the basis of digital representation of microstructure. *Steel Research International* 79: 579–586.

69 Rauch, L. and Madej, L. (2010). Application of the automatic image processing in modeling of the deformation mechanisms based on the digital representation of microstructure. *International Journal on Multi Scale Modeling* 8: 343–356.

70 Rollet, A.D., Saylor, D., and Frid, J. et al. (2004). Modelling polycrystalline microstructures in 3D. Proceedings Numiform (eds. Ghosh, S., Castro, J.C., Lee, J.K.), Columbus, 71–77.

71 Shengping, S. and Atluri, S.N. (2004). Atomic-level stress calculation and continuum-molecular system equivalence. *Computer Modelling in Engineering and Sciences* 6 (1): 91–104.

72 Sitko, M., Mojzeszko, M., Rychlowski, L. et al. (2020). Numerical procedure of three-dimensional reconstruction of ferrite-pearlite microstructure data from SEM/EBSD serial sectioning. *Procedia Manufacturing* 47: 1217–1222.

73 Sitko, M., Chao, Q., Wang, J. et al. (2020). A parallel version of the cellular automata static recrystallization model dedicated for high performance computing platforms – development and verification. *Computational Materials Science* 172: 109283.

74 Spirko, J.A., Neiman, M.L., Oelker, A.M., and Klier, K. (2004). Electronic structure and reactivity of defect MoS2 II. Bonding and activation of hydrogen on surface defect sites and clusters. *Surface Science* 572: 191–205.

75 Sunyk, R. and Steinmann, P. (2002). On higher gradients in continuum-atomistic modeling. *International Journal of Solids and Structures* 40: 6877–6896.

76 Szyndler, J., Grosman, F., Tkocz, M. et al. (2021). Through scale material flow investigation in novel incremental bulk forming process. *Journal of Materials Processing Technology* 287: 116487.

77 Torquato, S. and Jiao, Y. (2010). Robust algorithm to generate a diverse class of dense disordered and ordered sphere packings via linear programming. *Physical Review E* 82 (6).

78 Trębacz, L., Madej, L.,Wajda, W., and Paul, H. (2008). Simulation of plastic behaviour of FCC metals accounting for latice orientation. Proceedings WCCM'08, Venice, Italy, 1–2, CD.

79 Voiry, D., Mohiteb, A., and Chhowalla, M. (2015). Phase engineering of transition metal dichalcogenides. *Chemical Society Reviews* 44: 2702.

80 Wajda, W. and Paul, H. (2009). Modeling of microstructure and texture evolution of channel-die deformed aluminum bicrystals with {100}<001>/{110}<011> grains orientation. *Computer Methods in Materials Science* 9: 277–282.

81 Wang, W., Yang, C., Bai, L. et al. (2018). First-principles study on the structural and electronic properties of monolayer MoS2 with S-vacancy under uniaxial tensile strain. *Nanomaterials* 8: 74.

82 Xiong, S. and Cao, G. (2015). Molecular dynamics simulations of mechanical properties of monolayer MoS2. *Nanotechnology* 10: 185705.

83 Yu, Q. and Esche, S.K. (2003). A Monte Carlo algorithm for single phase normal grain growth with improved accuracy and efficiency. *Computational Materials Science* 27: 259–270.

84 Zhao, W., Pan, J., Fang, Y. et al. (2018). Metastable MoS2: crystal structure, electronic band structure, synthetic approach and intriguing physical properties. *Chemistry* 24: 15942–15954.

85 Zhou, M. (2003). A new look at the atomic level virial stress: on continuum-molecular system equivalence. *Proceedings of the royal society of London, Series A: mathematical physical and engineering sciences* 459 (2037): 2347–2392.

86 Zhou, R., Wang, J.G., Liu, H.Z. et al. (2017). Coaxial MoS2@carbonhybrid fibers: a low-cost anode material for high-performance Li-ion batteries. *Materials* 10: 174.

5

Examples of Multiscale Simulations

The development of reliable material models for metal-forming simulations has been of interest to scientists for many years [83]. When the hardening model is considered, several deterministic approaches, based either on closed-form equations describing the flow stress dependence on external variables [26] or on differential equations describing the evolution of internal variables [18, 68], have been published in the scientific literature. These models are commonly used in simulations for most metal-forming processes and give reasonably accurate results [5, 7, 11, 19, 27]. However, these models fail to describe the material behaviour when some special deformation conditions occur, e.g. fast changes in the deformation conditions or a strong tendency to strain localization. The main problem is to realistically describe the phenomena occurring in materials at lower length scales under the complex deformation conditions and incorporate this into the continuum-based approaches. Beyond this, accounting for the influence of such phenomena as cracks, shear bands, Luders bands, and Portevin-le-Chatelier bands at the macroscale and such discontinuities as microshear bands, grain boundaries, phase boundaries at the microscale becomes critical when the model aims to predict the properties of the final products. Finally, several of these phenomena are stochastic, and their realistic description by deterministic models may be limited.

Thus, the search for models that account more accurately for microscale and even nanoscale phenomena has been the research objective during the last two or even three decades [24, 29, 52]. The improvements in experimental techniques were major factors stimulating this research. New experimental techniques have made it possible to visualize physical processes and to measure relevant parameters at fine scales. Examples connected with new experimental equipment allowing research on mechanics of deformation are atomic force microscopy (AFM), nanoindentation tests, micropillar compression, computed tomography, electron backscattered diffraction (EBSD), atom probe tomography, etc. [20, 85, 86]. These techniques are supported by established approaches such as scanning and transmission electron microscopy [3]. The knowledge obtained about deformation mechanisms from these experiments has been recently combined with the idea of multiscale modelling and incorporated into the integrated computational material engineering concept [72] to develop models with new predictive capabilities.

5.1 Classification of Multiscale Modelling Methods

The modelling of discontinuities is now playing a more critical role in analyzing material behaviour during plastic deformation. However, as mentioned previously, the conventional

Multiscale Modelling and Optimisation of Materials and Structures, First Edition. Tadeusz Burczyński, Maciej Pietrzyk, Wacław Kuś, Łukasz Madej, Adam Mrozek, and Łukasz Rauch.

material models are not amenable to capture discontinuities. Accordingly, after multiphysics phenomena and multiscale analysis, the capture of these discontinuities is the third major challenge in contemporary computational mechanics of materials [12]. Finally, a fourth challenge constitutes the further development of computational methods to assess the probability of failure. The methods to cope with these challenges are usually classified into upscaling methods and concurrent multiscale computing [79].

In the upscaling class of methods, constitutive models at higher scales are constructed from observations and models at lower, more elementary scales. The idea of the representative volume element (RVE) is employed here [69]. By a sophisticated interaction between experimental observations at different scales and numerical solutions of constitutive models at increasingly larger scales, physically based models and their parameters can be derived at the macroscale; see for example [9, 48]. The methods of computational homogenization, e.g. [50, 69], are considered to belong to this group of methods.

In concurrent multiscale computing, the problem is solved simultaneously at several scales by an a priori decomposition. Two-scale methods, whereby the decomposition is made into the coarse and fine scales, are most frequent [59].

Both groups of methods have been intuitively used for decades. The solution of microstructure evolution equations in each Gauss integration point in the finite element (FE) model, and returning information regarding the flow stress accounting for microstructure, is a typical example of upscaling [61]. In contrast, an approach presented in [30] is an example of concurrent computing. In addition, recent years have witnessed the rapid development of multiscale methods in various areas of science, and applications in deformation processes are widespread. Various numerical methods have been applied to describe material behaviour at the microscale and/or nanoscales, e.g. cellular automata (CA), molecular dynamics (MD), Monte Carlo (MC), Vertex, Phase Field (PF), or Level Set (LS) methods [49, 64, 70]. For example, analysis in [42, 88] and [67, 89, 90] can reveal differences in the application of CA and MC methods, respectively, to simulate recrystallization phenomena. As a result of these developments, the coupling between fine and coarse scales is now completed more systematically [63]. Figure 5.1 shows the general idea of the distinction between upscaling models and concurrent multiscale computing. And a perceptive general review of different multiscale approaches in material science applications is presented next. However, as mentioned, the multiscale modelling techniques are becoming an everyday tool; therefore, the number of publications in this area is enormous.

In concurrent multiscale computing, the method used to describe the fine scale is applied to a part of the whole domain of the solution. It can be the same method used in the coarse scale (e.g. FE method), or it can be one of the earlier mentioned discrete methods (CA, MC). In the former case, the extended finite element (XFE) and the multiscale extended finite element (MS-XFE) methods are good examples. As mentioned in Chapter 3, in the XFE solution, special elements capable of accommodating a discontinuity are introduced.

In the MS-XFE method, a very fine mesh is generated in the area of particular interest that is selected before simulation. An example is the area around the front of crack propagation in a fracture simulation, as shown in Figure 5.2 [2]. The methods of mesh adaptation, e.g. hp adaptation [14, 57], multigrid method [16], or variational multiscale (VMS) methods [84], are also considered to belong to the concurrent multiscale computing group.

An example of the upscaling method based only on the FE method is presented in [51]. The two-scale model based on the FE method combined with the RVE approach is proposed to simulate the influence of colonies composed of ferrite and cementite on the macroscopic

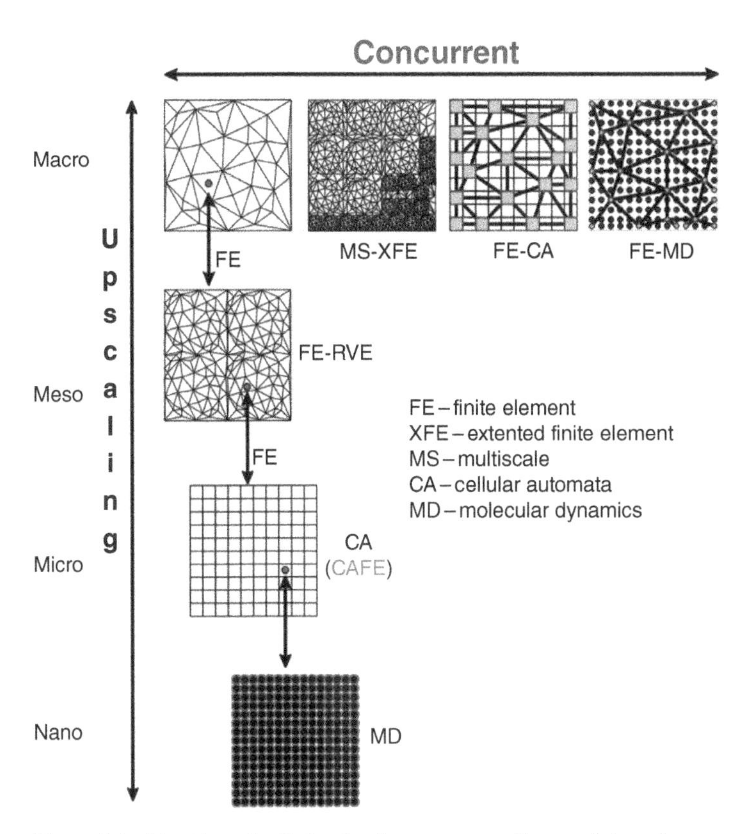

Figure 5.1 The idea of a distinction between upscaling models and concurrent multiscale computing.

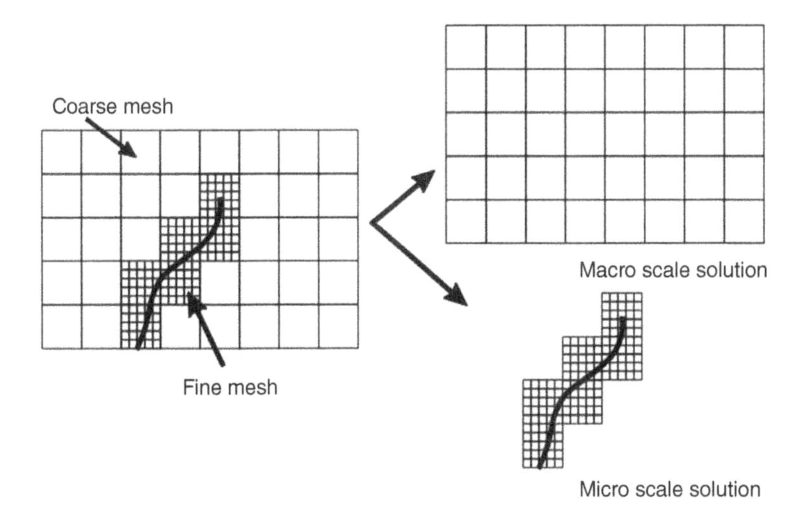

Figure 5.2 The idea of MS-XFE scale decomposition.

behaviour during wire drawing. The deformation of the pearlitic colony and changes in orientation of cementite lamellas during the multi-pass drawing are investigated. The macroscale model uses a two-dimensional rigid-plastic method that provides the boundary conditions for the microscale model. The penalty method was applied in this model to impose the boundary conditions on the grain boundaries. As a result of the microscale simulation, the friction coefficient is modified in the macroscale solution. This work can be classified as a conventional multiscale model based on the FE solution of the problem in two different scales within the upscaling group.

A large group of multiscale methods is represented by the multilevel approaches (i.e. [81]). Here, parameters for the macroscale model are identified by the multilevel model. Usually, multilevel models are composed of two or three levels (e.g. macro–meso scales or macro–meso–micro scales, respectively). The multiscale model can be implemented in the FE framework using the homogenization procedures to exchange information between scales all the way up to the macroscale. These approaches are often based on the crystal plasticity models (Lamel model, advanced Lamel model, grain interaction model, crystal plasticity model) that take into account interactions between particular grains and their rotations during deformation at the microscale [25]. In such cases, grains are usually represented by single FEs or in more advanced cases by many FEs, which increases the computational time [13, 40]. Therefore, the multilevel models can be classified as conventional multiscale models within the upscaling group.

However, not only FE method is used to create multiscale approaches. The boundary element method (BEM) is also widely used, e.g. [71]. In that work, a two-scale model based on the combination of the macro-BEM with the micro-BEM and macro-FE with micro-BEM is considered for simulation of intergranular microfracture in polycrystalline brittle materials. To investigate material behaviour at the lower scale, a set of RVEs is assigned to the points in the domain to the macroscale. The attached RVEs represent grains of the microstructure in the infinitesimal neighbourhood of the macroscale points. The macroscale solutions, related mainly to the stress and strain fields, are treated as boundary conditions for the microscale problem. The exchange of information is based on the averaging theorem [56]. On the other hand, the microscale solution is used to update the constitutive laws at the macroscale when microdamage is possible. This eventually results in strain-softening at the macroscale.

An interesting multiscale model based on the mesh-free methods has also been developed in [53]. A two-scale approach using a constrained natural element method (CNEM) is applied to simulate linear elasticity problems. The model has the ability to take into account the microstructure changes and their influence on macroscopic behaviour. The analyzed domain is divided into two regions, a microscopic one that represents microstructure evolution in the entire macroscopic domain and one called the complementary domain that describes the macroscale (Figure 5.3).

The constitutive relations are defined only at the microscale level and then extended by interpolating the complementary domain. *Source:* Based on Missoum-Benziane et al. [53].

Despite the large interest in conventional multiscale models based on, e.g. FE or BEM, there has been a recent interest in the development of alternative methods that represent concurrent as well as upscaling approaches. For example, in [67], authors used the MC method to describe the growth of layers in the chemical vapour deposition (CVD) process. This solution is coupled with FE, which predicts the temperatures and stresses in the resulting multi-layer coating. The interesting application in material science is also presented in [89]. In this case, a

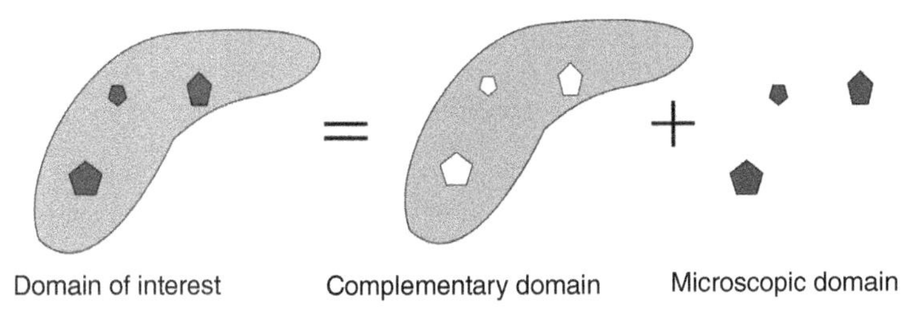

Figure 5.3 Division of domain applied by [53].

multiscale model based on the FE with the MC pots method was used to simulate static recrystallization and grain growth during annealing. The FE method is used to simulate the macroscopic plastic deformation. The field variables obtained in the macroscale are then input data for the MC model to simulate the microstructure evolution. To reduce the computational time, the microstructure model based on MC is only considered in a few representative elements of the FE mesh. Information about the microstructural changes is finally sent back to the FE code or compared with the experimental results to optimize the process parameters. Another alternative model developed in [54] uses the MD method coupled with the BEM to describe crack initiation and propagation. This approach will be described in detail in the following section. An exciting method that is based on the MD framework was developed in [1], for the rapid brittle fracture of a silicon slab under uniaxial tension. The main idea of the approach is to divide a domain into three regions. The first is located away from the crack tip and is simulated by the continuum FE method. The second is located near the crack tip and is modelled by MD, while the last region is located at the crack tip. This region is simulated using the tight-binding (TB) method, which is a quantum method containing electronic information and is commonly used in the description of matter.

When the interface between continuum and MD is considered, all FEs in this region are refined to match exactly with the atomic positions. Through these interfaces, information exchange between two methods is performed, making it an example of the concurrent approach.

Another similar approach by [78] uses the multiscale quasicontinuum (QC) method to simulate macroscale nonlinear deformation of crystalline material again using molecular mechanics. In the method, the material is divided into two main regions. The first is related to the surrounding material where no nonlinear behaviour is observed, and this region is simulated using only the FE approach. The second region is related to nonlinear behaviour, where the MD and FE approaches coexist. To reduce computational time in the regions of small deformation, atoms are segregated into two groups: non-representative and representative. In this region, the position of non-representative atoms is interpolated based on the position of representative atoms positioned at the FE mesh nodes (Figure 5.4). However, in the most interesting region, where we obtain larger deformation, the FE mesh is refined to match all atomic positions.

Another group of alternative methods is the CAFE approach based on the combination of the CA and FE methods. Four major directions of development of the CAFE approach are identified here. The first proposed by [21], can be classified as a concurrent multiscale method. In this work [21], the sample area is overlaid by the two separate meshes: FE and CA. The square CA

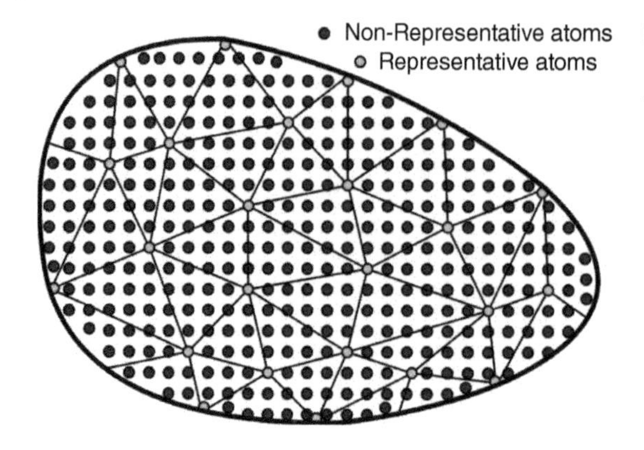

Figure 5.4 Material division to non-representative and representative atoms. *Source:* Based on Tadmor et al. [78].

space is defined to be denser than the FE triangular mesh. A CAFE method is applied to model the development of dendritic structures during solidification. In this approach, a differential equation describing the temperature field is solved using the FE model. However, a CA method was applied to model microstructure development and heat generation during solidification [22, 80]. A similar model for solidification simulations has also been applied by other authors [41, 65, 82]. Another interesting work based on the previous CAFE model [37] is also related to solidification, in particular dealing with the prediction of microsegregation and its influence upon microporosity during the casting of automotive components. The microscale approach combines stochastic and deterministic methods such as CA and the finite difference (FD) method and deals with multi-component diffusion in a three-phase system: liquid, solid, gas. During these microscale simulations, the nucleation and growth of grains and pores are calculated. The FE model provides information regarding heat transfer and fluid flow to the microscale model. This model is partly similar to the work by Gandin [21] and can also be classified as a concurrent method. A combination of the CA and FD method has also been applied to simulate the development of dendritic structures, but only at the microscale [87]. Application of the FD method to support CA calculations in the microscale is also commonly used during phase transformation [32, 33, 91]. FD is applied to solve the diffusion equation, while CA was usually used to track the state of particular cells in the domain to investigate the fraction transformed.

A second group represents the upscaling approaches, where separate CA are attached to each particular Gauss point in the FE mesh [9, 44, 73]. The main advantage is a backpropagation between methods in every time step, which increases the accuracy of computation. In [9], a multiscale CAFE method was applied to model fracture phenomena occurring during various forming processes. The model predicts the microstructure development and cracks propagation in the surface layer during rolling operations. Microstructural features such as grain boundaries, other phase particles, and grain interior are included during simulation in the CA spaces. The proposed algorithm incorporates three major steps: development of the primary microstructure, flow of the macroscopic parameters obtained from the FE simulation to the CA mesoscale, and finally, backpropagation of the parameters calculated in the CA spaces to the FE model.

This model [9] was the motivation for the CAFE model created by Shterenlikht [73]. This approach describes the propagation of the brittle and ductile fracture during a Charpy test. In this approach, two 3D CA spaces are introduced. The first, called the ductile space, describes phenomena connected with ductile fracture, while the second one, termed the brittle space, describes phenomena connected with brittle fracture during the Charpy test. Because the sizes of brittle and ductile fractures differ, different numbers and different dimensions of the CA cells are used in those CA spaces. The model then provides information about the cracks to the macroscale FE model. A further example of a CAFE model dealing with fracture can also be found in [31, 39]. A conceptually different approach for these issues based on the random CAFE model approach was recently developed in [60].

Another interesting application of the multiscale CAFE approach is applied to simulate dynamic recrystallization (DRX) [42]. Here three layers are representing the macroscale, mesoscale, and microscales. The first layer is based on FE and provides a macroscopic description of the material. The constitutive law is based on the Levy–Mises flow rule, in which flow stress σ_f is the only material parameter. The second layer connects the macroscopic and microscopic models and includes the calculation of σ_f, which is sent to the FE code. The local parameters, such as temperature and strain rate, are transferred in the opposite direction. The third layer, which is based on the CA calculations, simulates the evolution of the microstructure and dislocation density. A dedicated framework for the development of such CA models was presented in [66].

Application of the CAFE model to simulate another important phenomenon occurring in the material, the strain localization [44, 45], will be described as a case study in more detail in the following sections.

The third group of methods does not utilize backpropagation, and the flow of information is in one direction, e.g. from the FE to CA models [36]. Usually, the sample area is limited to a few grains to reduce calculation time because several FEs are applied to describe each particular grain. The amount of energy accumulated during deformation and the grain crystallographic orientation is obtained from the single crystal plasticity FE model and used as input data for the CA calculations of the austenite-ferrite phase transformation process after deformation. The detailed description of the CA model itself can be found in another work [35]. These methods may be classified as concurrent approaches. But the upscaling type solutions are also possible [47].

The final group of methods is related to various combinations of the CAFE method with other computational approaches such as Neuro Expert Cellular Automata Finite Element models (NESCAFE). NESCAFE modelling is a very interesting approach dealing with hybrid microstructural models to consider, explicitly, the development of microstructural features over various length scales. One example of such an approach is a combination of Neuro-Fuzzy (nF) and CA techniques attached to the FE code. The CA holds information concerning the material, i.e. grain boundaries and grain interiors, subgrains, misorientation, while the nF predicts the flow stress behaviour model based on this information [10]. Another model takes into account more microstructural details, e.g. microbands. The CA model describes grain interiors and grain boundaries and also the periodic function (φ). This function (φ) represents the spatial development of microbands, which depends on the initial orientation of CA cells. These hybrid methods are mainly classified as upscaling approaches.

Summarizing, it should be emphasized that at present, a significant variety of possible multiscale approaches are commonly used. These approaches are distinguished as upscaling, and concurrent multiscale computing but, in general, various continuum (e.g. FE, BE, natural element method, mesh-free etc.) and discrete (e.g. CA, MC) methods can be used in both approaches. The distinction between the two groups of approaches is based on how the domain of the solution is handled. In the upscaling group of methods, the coarse scale method covers the whole domain of the solution, and the lower scale methods are applied at selected points (the RVE idea), and they usually calculate the material parameters required in the constitutive law of the macroanalysis method. In the concurrent multiscale computing approach, various parts of the domain are covered by various methods. As shown previously, all possible associations of the continuum methods and discrete methods are possible.

However, a detailed case study is necessary for a full understanding of the differences between the concurrent and upscaling approaches. To provide this, several different multiscale models dealing with nano–micro, micro–macro, and nano–micro–macro data transfer will be discussed in the following sections, emphasizing the capabilities and limitations of these approaches.

5.2 Case Studies

5.2.1 Nano–Micro

The method of concurrent multiscale computing based on the continuum theory and discrete molecular model is presented in this section, along with numerical examples. Such methods usually combine the FE method (FEM) and MD or molecular statics (MS) [17, 34]. However, in the proposed algorithm, the BEM is utilized for modelling the continuum region. The equilibrium of the atomic lattice is obtained using the MS approach [54]. The BEM and MS have already been described in detail in Sections 3.1.2 and 3.2.1 of this book, respectively. The main advantage of the BEM is a simple mesh generation, as the continuum model in most cases has to be meshed only on its boundary. Additionally, the variant of BEM with various subregions and material properties can be used to model interactions between atoms and the continuum model at the transition area [4]. The MS was chosen due to a computationally effective algorithm, which does not require engaging of the time scale; however, in the concurrent approaches, the MS solver can be replaced with the MD code if necessary [17].

5.2.1.1 Multiscale Discrete-Continuum Model

The general schematic diagram of the proposed multiscale model is shown in Figure 5.5. The discrete atomic model occupies only a small area of the model, where the investigation of the behaviour at the molecular level is expected. The rest of the structure is modelled using BEM for the problems of linear elastostatics. The size of the continuum domain, in almost all practical cases, is much larger than the discrete molecular model, thus the total number of degrees of freedom of the whole model could be significantly reduced.

The interface between the discrete and continuum domains is created either by boundary-to-boundary coupling where the coordinates of outer atoms correspond to the appropriate nodes of boundary elements (Figure 5.5) or several BEM subregions divided by the internal boundaries. In the second case, the discrete atomic model is overlapped into the BEM continuum model at the interface area.

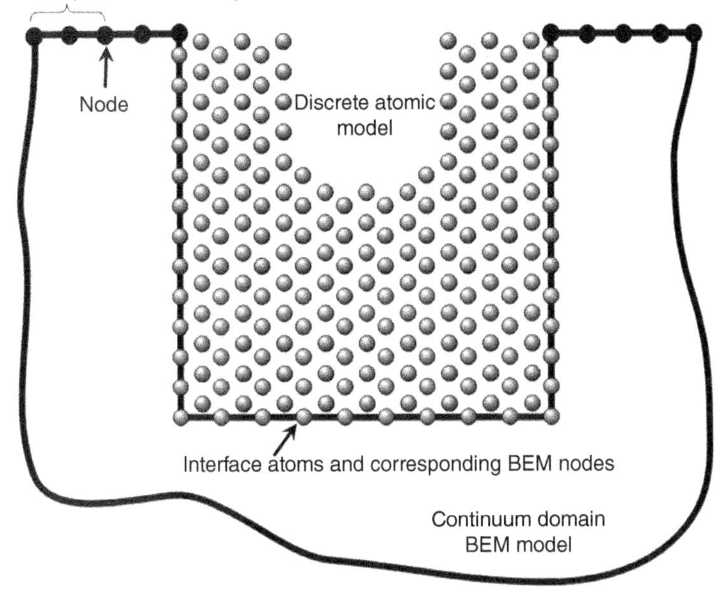

Figure 5.5 Structure of the nanoscale–microscale model.

The BEM with subregions was chosen due to the necessity of imposing the force boundary conditions in the interior of the continuum domain (i.e. on the nodes on the internal boundaries). Alternatively, the loads can be applied as the body forces; however, such an approach needs an additional discretization of the interior of the model, and thus eliminates the main advantage of the BEM. The structures of the models with one and two subregions are presented in Figure 5.6a and b, respectively. The coordinates of the interface atoms are equal to the corresponding nodes on the internal boundaries of the continuum model. The external loads or displacements are applied as the standard boundary conditions imposed to the BEM model.

The flowchart of solving the coupled discrete-continuum model, along with the main equations, is presented in Figure 5.7. The algorithm has an iterative nature and consists of several major steps:

1) In the first stage, the BEM model of a whole structure with a set of boundary conditions (displacements $\mathbf{u}^{bem0}$ and tractions $\mathbf{p}^{bem0}$ imposed respectively on the boundaries Γ_u^{bem} and Γ_p^{bem}) and internal points, corresponding to the appropriate atoms in the interface area, is solved (as presented in Figure 5.8).
2) Displacements of the internal points $\mathbf{u}_{int}^{bem}$ are obtained and introduced as initial deformations $\mathbf{u}^{at0}$ to the discrete molecular model. The atoms in the interface have to be fixed at their positions.
3) Equilibrium positions of the rest of the atoms are computed, using the MS method described in Chapter 3.
4) The forces $\mathbf{f}^{at}$, acting on fixed interface atoms and exerted by all atoms of the discrete model, are computed.

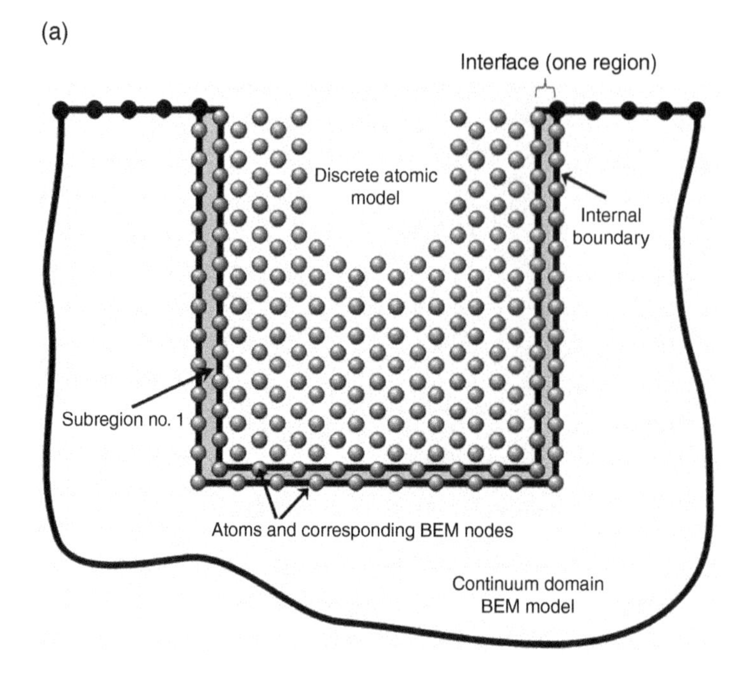

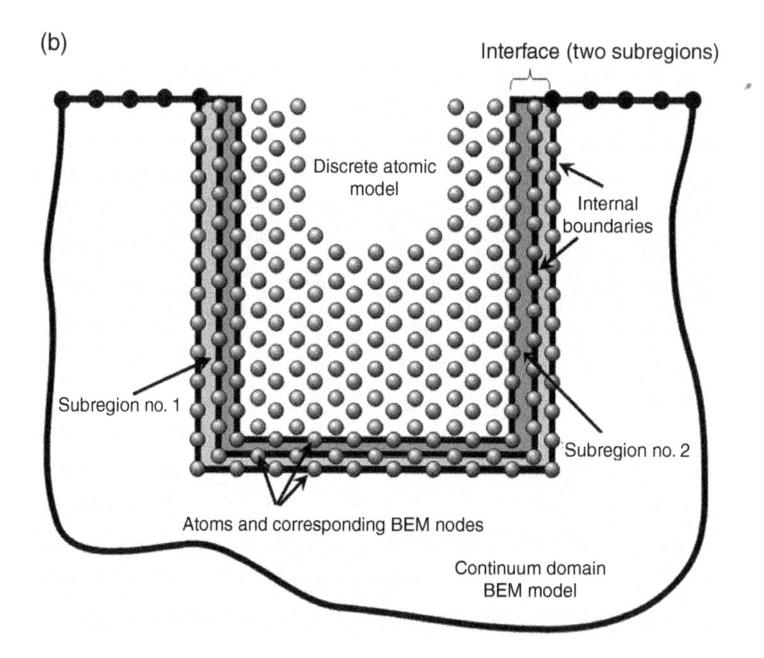

Figure 5.6 Interfaces with (a) one subregion and (b) two subregions.

5) Due to the formulation of the BEM, the computed interaction forces $\mathbf{f}^{at}$ have to be transformed into the nodal values of tractions $\mathbf{p}_{int}^{bem0}$ and applied to the continuum model. This conversion is explained later in this section and can be omitted when using FEM instead of BEM.

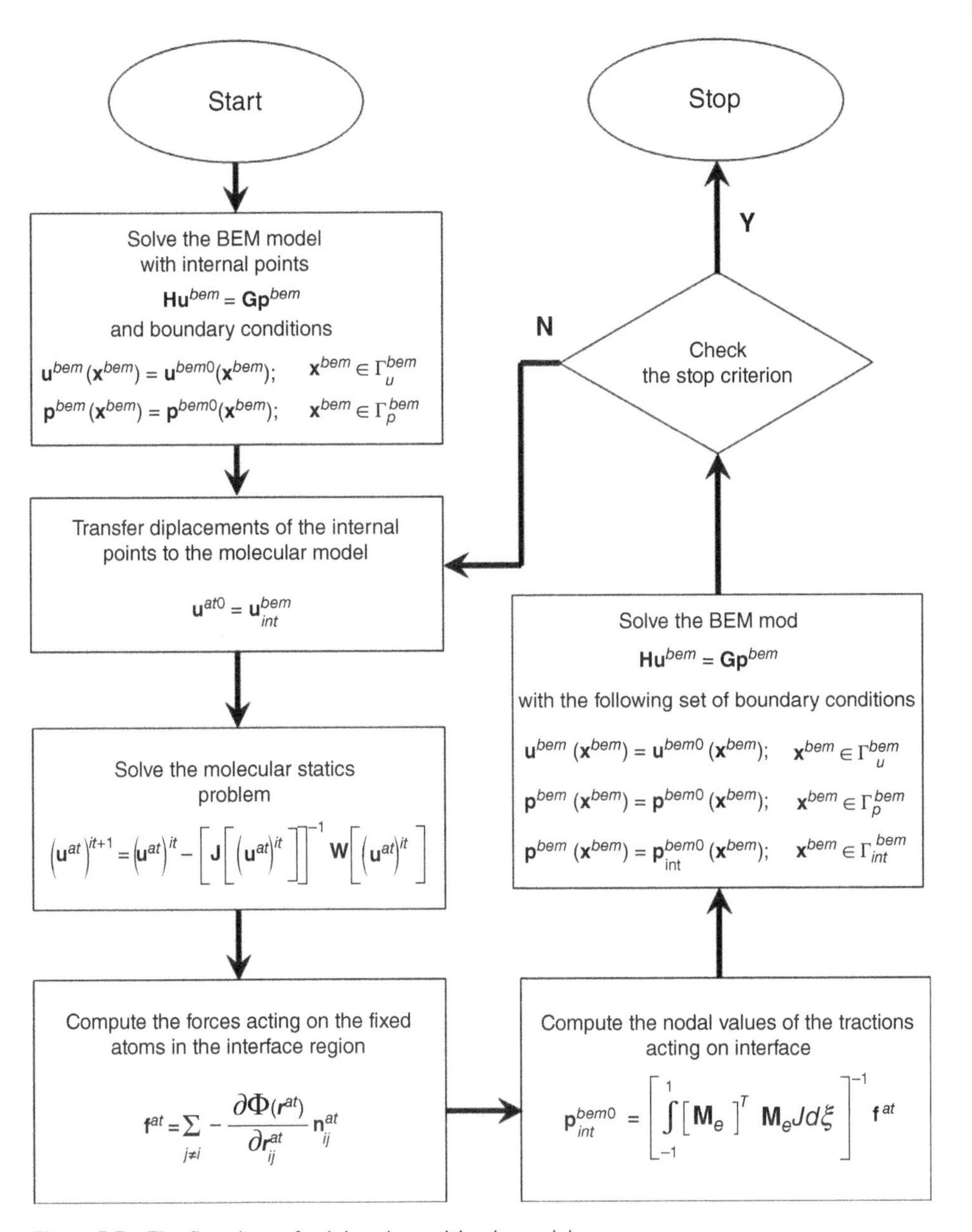

Figure 5.7 The flowchart of solving the multiscale model.

6) The BEM model with internal points is replaced with the BEM model with a vacancy or subregions (Figures 5.5 and 5.6, respectively), loaded with $\mathbf{p}_{int}^{bem0}$ and solved. All the external loads $\mathbf{p}^{bem0}$ and displacements $\mathbf{u}^{bem0}$ remain unchanged.

The computations are repeated iteratively until the stop condition is satisfied. The stop condition is executed when the differences between displacements of the atoms during two iterations or the Euclidean norm of the vector W(u) (see Eqs. 3.143 and 3.146 in Chapter 3) are less than an admissible value.

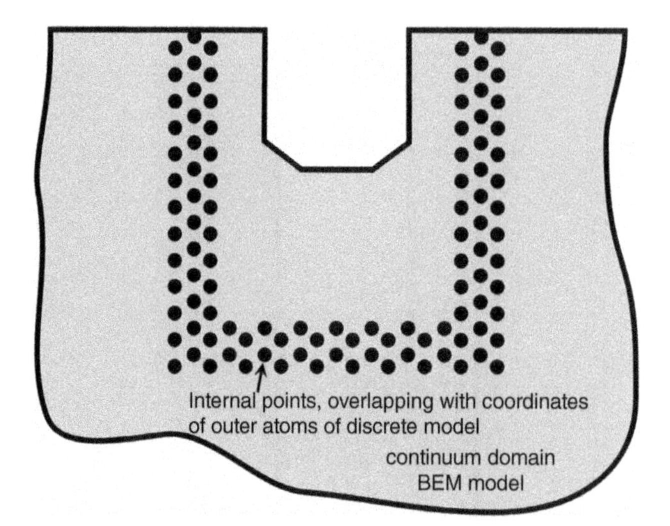

Figure 5.8 The BEM model with internal points.

5.2.1.2 Conversion of the Nodal Forces to Tractions

According to Betti's reciprocal theorem, the work of the nodal forces $\mathbf{f}_e^n$ on the displacements $\mathbf{u}_e^n$ of the boundary element's nodes is equal to the work of the tractions $\mathbf{p}$ on the displacements $\mathbf{u}$ of the boundary Γ_e of the considered boundary element:

$$L_f = L_p. \tag{5.1}$$

The works of the nodal forces and tractions equal respectively:

$$L_f = \{\mathbf{u}_e^n\}^T \mathbf{f}_e^n, \tag{5.2}$$

$$L_p = \int_{\Gamma_e} \{\mathbf{u}\}^T \mathbf{p} d\Gamma_e. \tag{5.3}$$

The displacements $\mathbf{u}$ and tractions $\mathbf{p}$ are approximated over the boundary element using their nodal values $\mathbf{u}_e^n$, $\mathbf{f}_e^n$ and the set of shape functions $\mathbf{M}_e$ (see Chapter 3 for details):

$$\mathbf{u} = \mathbf{M}_e \mathbf{u}_e^n, \tag{5.4}$$

$$\mathbf{p} = \mathbf{M}_e \mathbf{p}_e^n. \tag{5.5}$$

After substitution of the previous equations to Betti's theorem (5.1) and transformation to the boundary element's local system of coordinates $\xi \in [-1, ..., 1]$, the nodal values of the tractions can be computed:

$$\mathbf{p}_e^n = \left[\int_{-1}^{1} [\mathbf{M}_e]^T \mathbf{M}_e J d\xi \right]^{-1} \mathbf{f}_e^n. \tag{5.6}$$

In the case of the two-dimensional, three-node quadratic elements used in the presented multiscale model, the vector of the nodal displacements has the form:

$$\{\mathbf{u}_e^n\}^T = \left[u_x^1 \; u_y^1 \; u_x^2 \; u_y^2 \; u_x^3 \; u_y^3 \right], \tag{5.7}$$

and the matrix of the three parabolic shape functions is arranged as follows:

$$\mathbf{M}_e = \begin{bmatrix} M_e^1 & 0 & M_e^2 & 0 & M_e^3 & 0 \\ 0 & M_e^1 & 0 & M_e^2 & 0 & M_e^3 \end{bmatrix}$$
$$M_e^1 = \frac{1}{2}\xi(1-\xi) \qquad . \tag{5.8}$$
$$M_e^2 = \frac{1}{2}\xi(1+\xi)$$
$$M_e^3 = (1+\xi)(1-\xi)$$

Jacobian J is defined as:

$$J(\xi) = \sqrt{\left(\frac{\partial x_1}{\partial \xi}\right)^2 + \left(\frac{\partial x_2}{\partial \xi}\right)^2}, \tag{5.9}$$

where: variables x_1 and x_2 denote the global coordinates of the model.

5.2.1.3 Examples of the Nanoscale–Microscale Modelling

In the first example of the two-scale, continuum-discrete modelling, the shearing of the plate with a rectangular notch is considered. The continuum domain consists of 113 three-node quadratic elements, and the atomic model contains 884 atoms arranged in a regular hexagonal lattice. Both submodels are shown in Figure 5.9a and b, respectively. The whole left side of the plate was constrained, and the shear load was applied along the opposite side (Figure 5.10). The Lennard-Jones potential with the following parameters (taken from [77]): $\sigma = 0.2575$ nm, $\varepsilon = 0.1699$ nN nm was used. Dimensions of the plate are 40 nm × 27 nm, and the material properties of the BEM model are $E = 70$ GPa and $\nu = 0.35$.

The result of the simulation, the deformed plate, is presented in Figure 5.11a. The atoms moved to the new equilibrium positions. Opening of the cracks at the stress concentration areas, i.e. corners of the notch, can be observed (Figure 5.11b). The directions of the created cracks correspond to the closest packed directions of the HCP lattice.

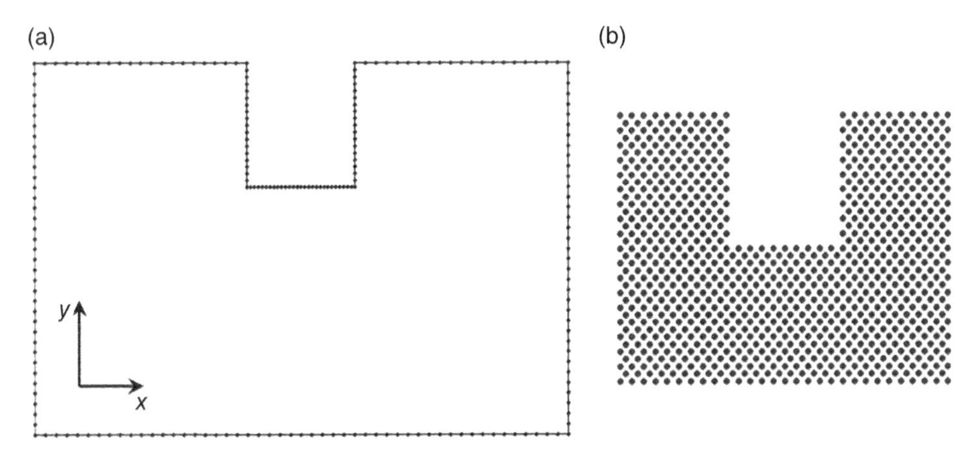

Figure 5.9 The plate with the rectangular notch: (a) continuum BEM model and (b) discrete molecular model.

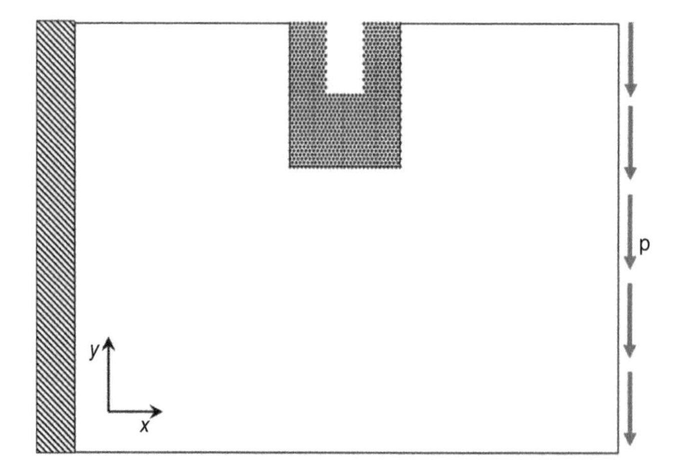

Figure 5.10 Continuum-discrete model of the plate with the rectangular notch.

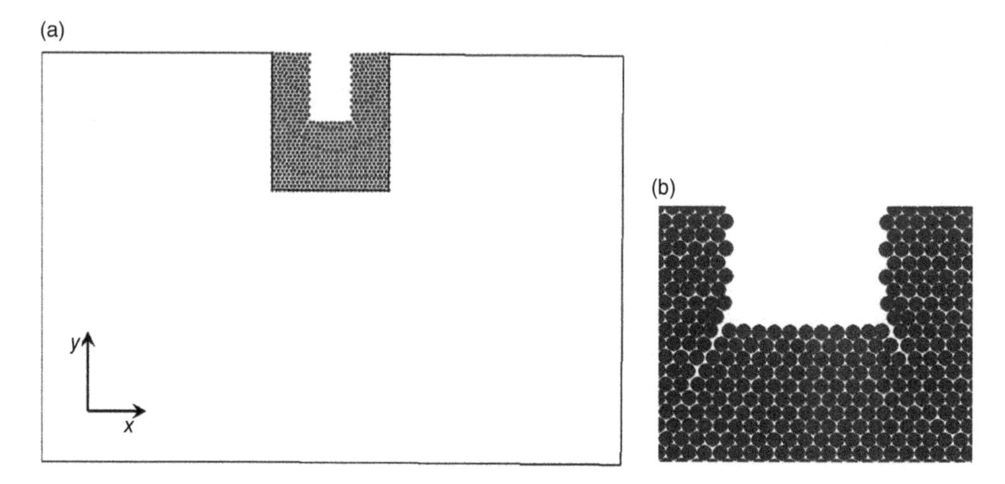

Figure 5.11 Results of the simulation: (a) deformed plate and (b) close-up of the interior of the notch.

The next example is shown in Figure 5.12. In this case, the plate is loaded and constrained similarly to the previous model, but the notch has a different V-shape. The dimensions of the plate are 90 nm × 45 nm. The continuum model contains 176 three-node quadratic elements, and the atomic model was built of 2033 atoms arranged into a regular hexagonal lattice. Both submodels and the whole assembled structure are presented in Figures 5.12 and 5.13, respectively. All the parameters of the atomic potential and material properties remain unchanged.

The result of the simulation can be seen in Figure 5.14a. Similarly, as in the previous case, the crack opening at the centre of the notch can be observed. The middle rows of the atoms moved in the opposite directions, and the new quasi-equilibrium state was achieved.

The previously presented example shows a drawback of the straight, boundary-to-boundary coupling of the two domains: a continuum and the discrete one. In interfaces of this kind, the displacements of the nodes of the BEM model are only imposed on one row of atoms. This fact, combined with the short-range character of the applied Lennard-Jones potential and few neighbours of the outer atoms, leads to errors in transferring the field of

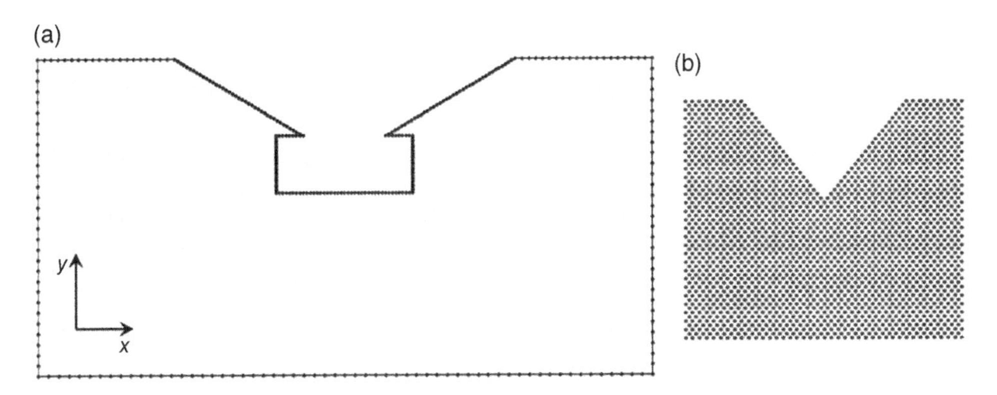

Figure 5.12 The plate with the V-notch: (a) continuum BEM model and (b) discrete molecular model.

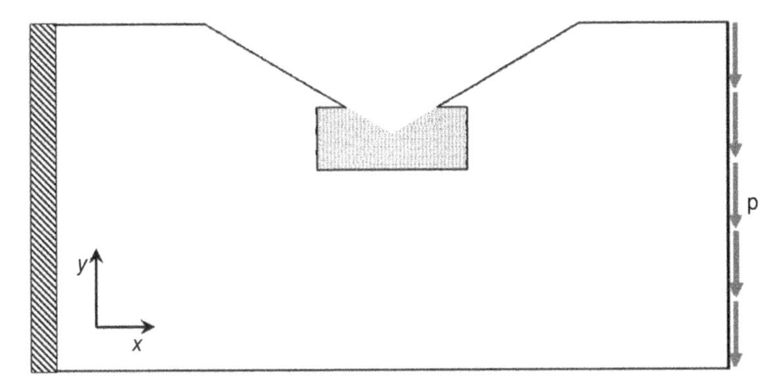

Figure 5.13 Continuum-discrete molecular model of the plate with the V-notch.

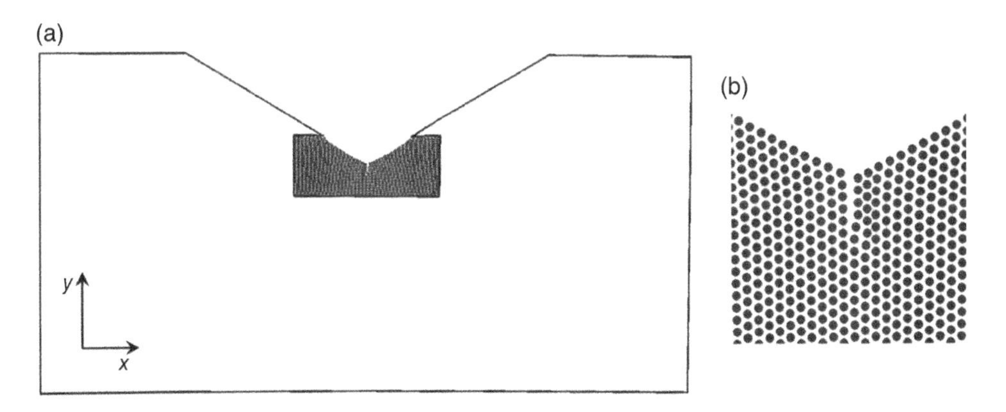

Figure 5.14 Results of the simulation: (a) deformed plate and (b) close-up of the interior of the notch.

displacement from the BEM model to the atomic lattice. This phenomenon, clearly visible as a separation of these two domains, is presented in Figure 5.15.

This drawback can be eliminated by different construction of the interface, based on the BEM with subregions. The application of such interfaces was illustrated in the following examples [55]. Again, shearing of the plate with the U-notch (Figure 5.16a) and the interface

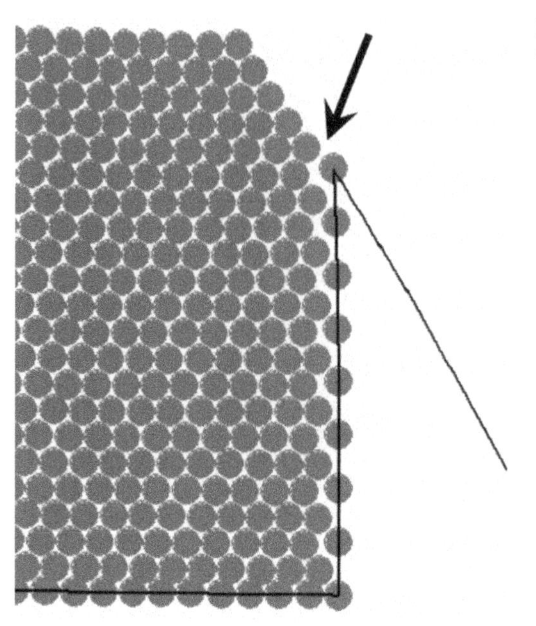

Figure 5.15 Separation of the atomic lattice from the BEM model.

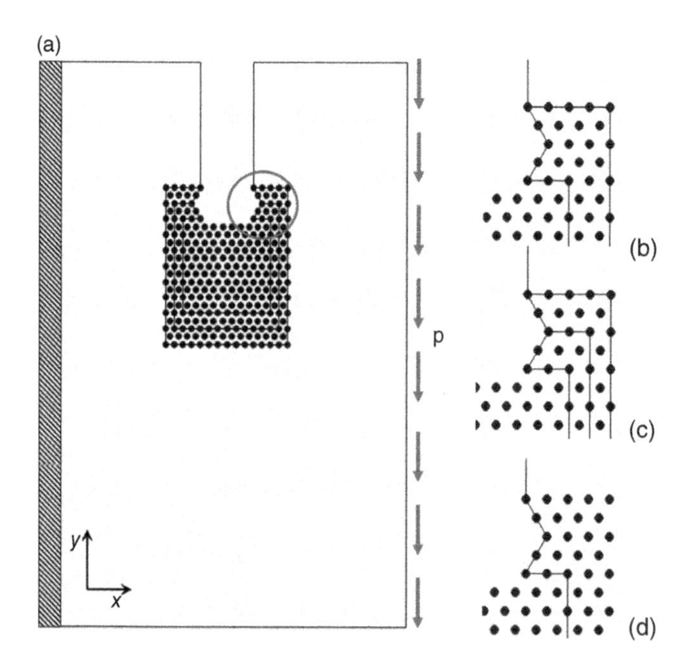

Figure 5.16 The plate with the U-notch (a), and different types of an interface: (b) one subregion, (c) two subregions, and (d) boundary-to-boundary coupling. *Source:* Based on Mrozek [55].

made of a various number of subregions (Figure 5.16b–d) was considered. The rectangular, 8 nm × 14 nm plate with the U-notch was constrained and loaded in an analogical way as previous two examples. The atomic model contains 276 atoms arranged in a regular hexagonal lattice, and the numbers of elements of each type of the BEM model are included in Table 5.1.

Table 5.1 A number of boundary elements of the continuum models.

Model/type of interface	Number of boundary elements
Initial all-BEM model	59 elements, 97 internal points
Boundary-to-boundary	69
One subregion	90
Two subregions	106

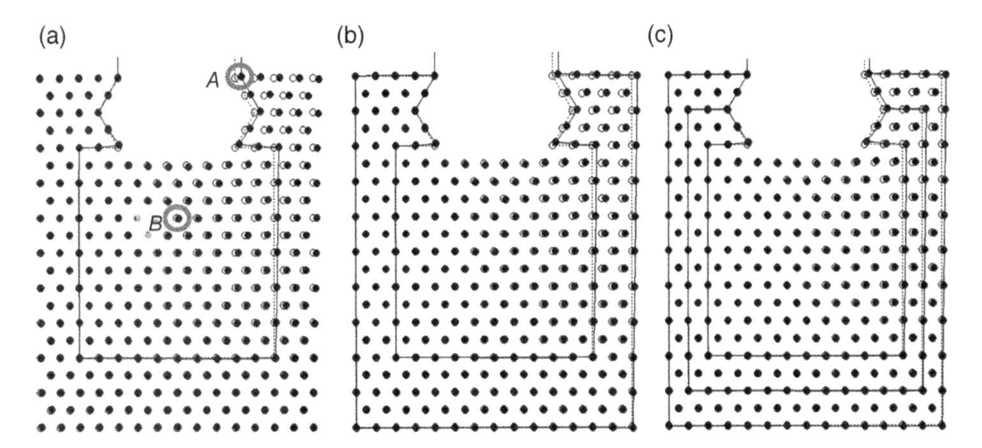

Figure 5.17 Results of the simulation – close-up of the deformed interior of the notch: (a) with boundary-to-boundary coupling, (b) one subregion, (c) two subregions. *Source:* Based on Mrozek [55].

The parameters of the Lennard-Jones potential were taken from [77], and the material properties of the BEM model were the same as in the previous two examples. The results of the multiscale simulation: the close-ups of the discrete and interface domains were illustrated in Figure 5.17a–c. The atomic lattice took a new equilibrium configuration under the loaded and deformed continuum model. The resultant values of displacements of the two selected atoms A and B, placed in the interface and in the centre of the lattice, are similar in both variants of the interface and are summarized in Table 5.2. The relative errors were computed with respect to the values obtained for the model with the two subregions. The convergence of the solution as the plot of the Euclidean norm of $\mathbf{W}(\mathbf{u})$ against the subsequent iterations is presented in Figure 5.18. The peak, corresponding to the second iteration, is caused by the transition from the whole-BEM model with internal points to the target, coupled continuum-discrete model.

The presented multiscale algorithm provides full coupling between the nanoscale and microscales and gives the possibility of analysis of all characteristic phenomena for the molecular level, e.g. slips, fractures, and crack propagation, combined with the simultaneous reduction of the total number of degrees of freedom since the size of the continuum domain could be much larger than the molecular one. Additionally, the application of boundary conditions such as external loads or displacements is more convenient at the microscale level. It is similar to the preparation of the conventional FEM or BEM macroscopic models.

This method requires that the material properties of the BEM model (i.e. the Young modulus and the Poisson ratio) have to correspond to the effective properties of the applied atomic

Table 5.2 Displacements of the selected atoms.

	Atom A		Atom B	
Interface	**Displacement (Å)**	**Relative error (%)**	**Displacement (Å)**	**Relative error (%)**
Boundary	0.6334	2.9	0.2231	1.7
One subregion	0.6215	1	0.2211	0.8
Two subregions	0.6153	—	0.2192	—

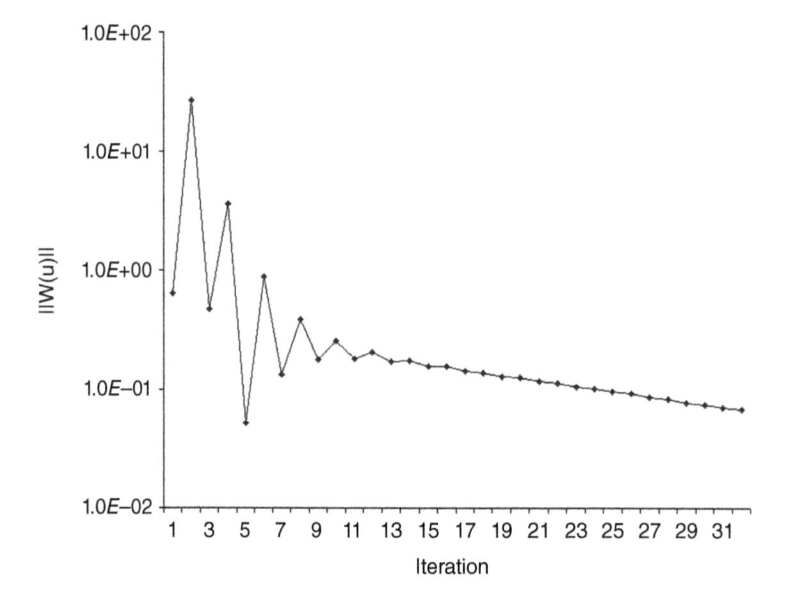

Figure 5.18 Convergence of the solution.

potential and molecular model; otherwise problems with the convergence of the solution may occur. The estimation of the effective mechanical macroscopic parameters of the molecular models is described in Chapter 4. Application of the interface with one or more subregions effectively eliminates the problems with transferring displacements from the continuum domain to the discrete one.

5.2.2 Microscale–Macroscale

A typical example of a microscale–macroscale model of an upscaling type can be the cellular automata finite element (CAFE) method. The models, which account for the influence of recrystallization, phase transformation, and microshear bands on the flow stress, are described briefly in this section as examples of phenomena occurring during the plastic deformation of metals under hot and cold deformation conditions. The CAFE technique is used in all these cases. The FE model deals with the phenomena occurring at the macroscale level, while the CA model describes phenomena at the microscale as schematically can be seen in

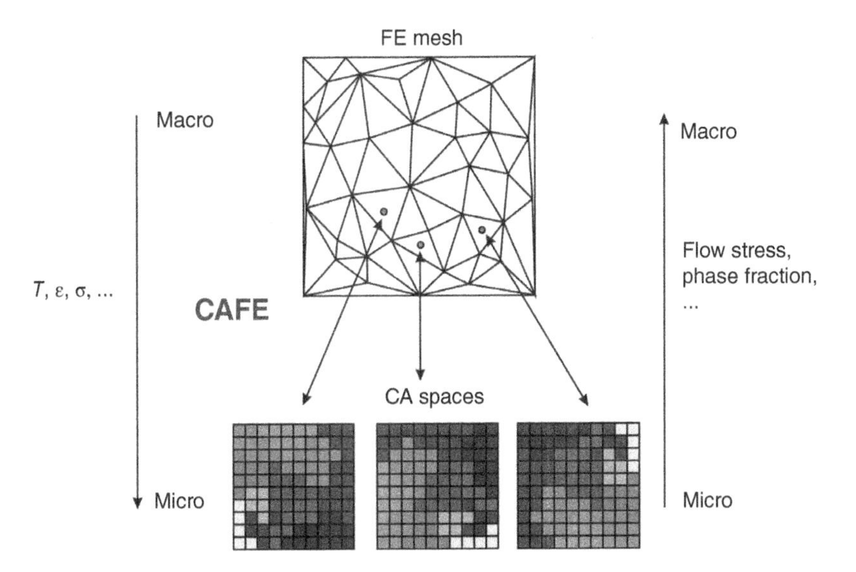

Figure 5.19 The idea of the CAFE approach with the information flow between scales.

Figure 5.19. Usually, a digital material representation described in Chapter 4 is input data for the microscale models [74].

The CA part of the CAFE model accounts for the evolution of microstructure and/or dislocation density and/or shear bands and microshear bands. Each cell has defined states and attributed variables in the CA model, which are specific to the considered process. Transition rules, which describe the new state of the cell on the basis of the state of this cell and the adjacent cells in the previous time step, are crucial for the accuracy of the method and its capability to recreate the considered process realistically. As mentioned in Chapter 3, the transition rules are, in general, given as logical functions. Transition rules are usually defined based on experimental observations and experts' knowledge about the considered process.

5.2.2.1 Dynamic Recrystallization

Prediction of microstructure evolution is crucial for realistic simulation of hot forming processes. The approach, which has been successfully used since the late 1980s, connects the FE method with closed-form equations describing recrystallization and grain growth [61]. Nonetheless, these kinds of approaches supply information only about average values of microstructural parameters, which is a drawback. Presently, it is expected that microstructure evolution models should supply wider information beyond the average values, and the distribution of the microstructural parameters should also be predicted.

One of the possible available solutions is a multiscale CAFE model. In the CAFE model of hot forming, the CA part accounts for the evolution of microstructure and dislocation density during deformation. The detailed description of the CAFE model developed by the authors is given in [42], and it is repeated briefly as an example. The 2D CA lattice with random hexagonal neighbourhood and with periodic boundary conditions is used. The digital material representation obtained by one of the methods described in Chapter 4 is used as input for the model. To use the information provided by the digital microstructure, the state of each CA cell is described by internal variables: (i) local dislocation density ρ, (ii) assignment to grain, and (iii) state: unrecrystallized or recrystallized. Eventually, each grain is described by its orientation, average dislocation density, and the number of CA cells belonging to this grain. The increment of average dislocation density ρ is calculated at grain level, separately for each grain, using the following differential equation:

$$\frac{d\rho(t)}{dt} = \frac{\dot{\varepsilon}}{bl} - k_2\rho(t), \tag{5.10}$$

where: l – average free path for dislocations and k_2 – the self-diffusion. Other parameters in the equations are calculated as:

$$l = A_0 Z^{-A_1} \quad k_2 = k_{20}\exp\left(\frac{Q_s}{RT}\right) \quad k_3 = k_{30}\exp\left(\frac{Q_m}{RT}\right), \tag{5.11}$$

where: Q_m – activation energy for grain boundary mobility, Q_s – activation energy for self-diffusion, A_0, A_1, k_{20}, k_{30} – coefficients, which have to be determined.

Nonuniform distribution of dislocation density inside the grain is imposed by nondeterministic algorithm controlling the incremental updates of ρ variable in the CA cells. All these interactions between the cells remain local, i.e. there is no global data required for the updating of the algorithm.

As mentioned in Chapter 3, the definition of the transition rules is crucial for the CA method. In the present model, these rules were created to replicate mechanisms leading to the initiation and propagation of the DRX phenomenon. As a result, these rules control the simulation of nucleation and subsequent grain growth. The rules themselves are logical functions, and the result of their application is not deterministic due to quasi-random neighbourhood definition and dislocation density distribution within the CA lattice. The rule describing the nucleation states that the nucleus appears if the cell is located at the grain boundary (GB) and dislocation density in the cell reaches a certain critical value ρ_c [42]. Due to the nonuniform distribution of dislocation density in grains, the value of ρ at the GB is also nonuniform. Thus, the model allows rather to introduce the nuclei due to actual process conditions than to introduce other premises on the nucleation rate.

Once the CA cell becomes a nucleus, the dislocation density in this cell is set to that in the annealed material, and the random orientation is assigned to the newly created grain. The second transition rule describes the growth of recrystallized grains. It is based on the GB velocity, which depends on the GB mobility M and the driving force for growth F; see [42] for details. As a result, the model predicts both the average grain size and the grain size distribution. The current state of the multi mesh CA model of DRX is presented in [42].

The CA model can then be incorporated into the FE code [38], e.g. rolling, as it is schematically shown in Figure 5.20. Due to symmetry, only the top half of the deformation zone is considered in this case study. Several CA spaces were created at the cross section of the sample, and their state was calculated using changes of external variables along the flow lines. Consequently, the local development of microstructure components is computed through models operating at different scales. The macroscale FE model calculates local values of temperature, strain, strain rate, and stress. Nucleation of new grains and GB motion, which are fundamental for the recrystallization model, is calculated at the CA model's microscale.

Such a model was validated by comparison of predictions with the results of measurements in the laboratory two-high mill. A detailed description of the experiment is given in [23]. Very good agreement with the experiment was obtained for the average grain size distribution along with the thickness of the strip. Beyond this, good agreement is observed for the

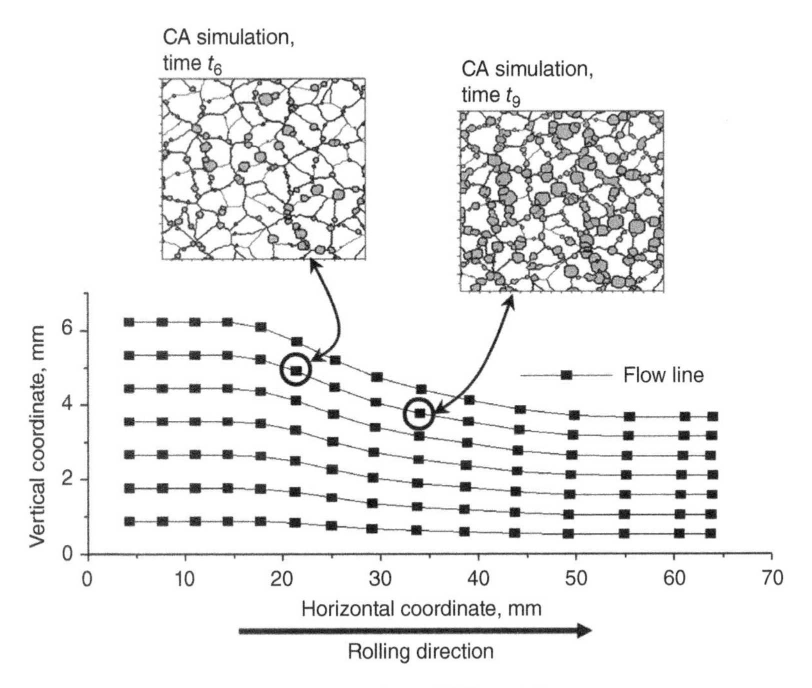

Figure 5.20 Schematic illustration of the CAFE model.

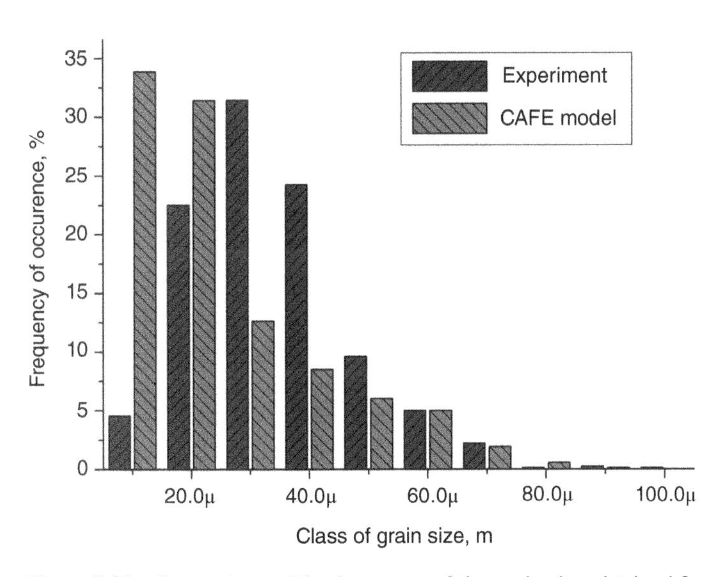

Figure 5.21 Comparison of the frequency of the grain size obtained from the experiment and calculated by the CAFE model at the central point in the sample.

frequency of the grain size; see Figure 5.21 for the centre of the strip. It is seen that the CAFE model predicts grain size distribution reasonably well.

The feedback between the CA model and FE code is added, and the flow stress in each Gauss point is calculated, accounting for the recrystallized volume fraction. This model was also tested in [43], and a very good agreement between measured material response to loading and calculated flow stress was obtained. A conceptually different application of the CAFE model for DRX that is based on an uncoupled approach is presented in [47].

In general, it can be concluded that the application of the CA approach as the rheological model in the FE code improves the accuracy of multiscale simulations.

5.2.2.2 Phase Transformation

Phase transformation is another factor which influences material properties during deformation at elevated temperatures. Similar to recrystallization, closed-form equations describing transformation kinetics can be implemented into the FE code, and mutual relation between macroscales and microscales can be created [46, 62]. However, the method has the same disadvantages, and only average microstructural parameters can be determined that way.

Again, that is why the micro–macro CAFE model needs to be developed [62]. Due to similarities to the previous model, only the microscale model is briefly described. The CA model simulates microstructure evolution during cooling in the 2D and 3D space. Each CA can be in one of the following states: ferrite (α), austenite (γ), or ferrite-austenite (α/γ). The last state describes cells located at the interface between austenite and ferrite grains. Each cell also contains information about internal variables, e.g. the volume fraction of ferrite, the carbon concentration, the growth length of the ferrite cell into the ferrite-austenite cell, or the growth velocity of an interface cell. These internal variables are used in the transition rules to replicate mechanisms involved in the phase transformation. Two major transition rules have been defined to describe the nucleation and growth of the ferrite phase. The nucleation mechanism, which is stochastic in nature, can be described in various ways [15]. In the presented model at the beginning of each time step, a number of nuclei is calculated in a probabilistic manner. The locations of the grain nuclei are generated randomly along the boundaries. When a cell is selected as a nucleus, the state of this cell changes from austenite (γ) to ferrite (α). At the same time, all the neighbouring cells of the ferrite (α) change their state to ferrite-austenite (α/γ).

The defined transition rules again describe the kinetics of ferrite growth to closely replicate experimental observations of these mechanisms [6]. The CA cell changes the state from austenite-ferrite into ferrite cell when the ferrite volume fraction in the cell exceeds the critical value. Otherwise, the cell remains in the austenite-ferrite state. When the cell changes its state to ferrite, all the neighbouring cells in the austenite state change their states into the austenite-ferrite state. When a change in the cell state occurs, the corresponding carbon concentration changes according to the FeC diagram. In these types of models, the diffusion problem is solved directly [28, 75] or indirectly, and then the carbon concentration in the austenite CA cells increases uniformly [62].

The model can be validated with the experimental CCT diagram obtained for low carbon steel, and typical results are presented in Figures 5.22 and 5.23. Grain structures shown in Figure 5.22 were obtained from the CAFE cooling simulation after hot rolling, and they are in qualitative agreement with commonly observed microstructures.

As seen, predicted phase transformation kinetics is also in good agreement with the experimental data (Figure 5.23). However, for the low cooling rates, the pearlite start temperature is not predicted accurately, and the CA model's quantitative accuracy is still under investigation.

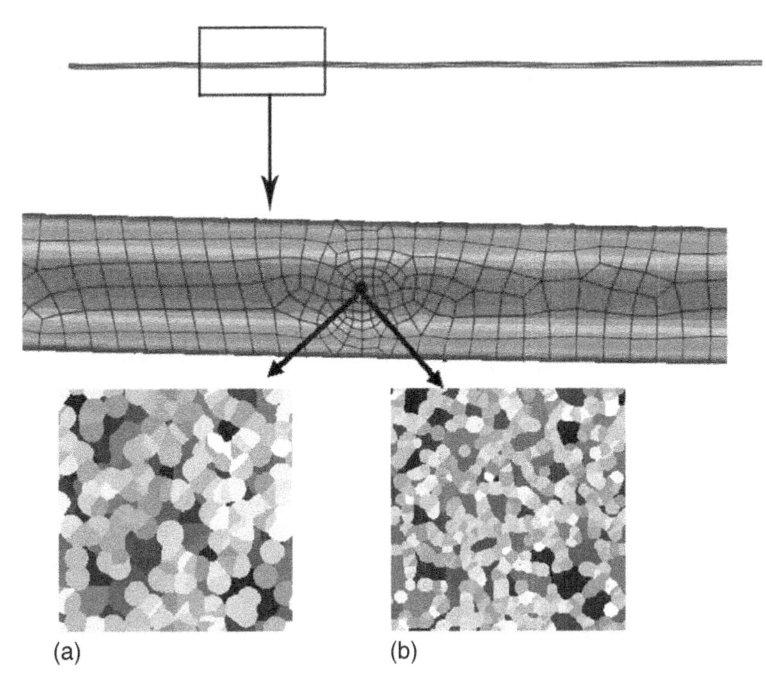

Figure 5.22 Geometrical representation of the obtained dual-phase microstructures for (a) 2 and (b) 10 °C/s.

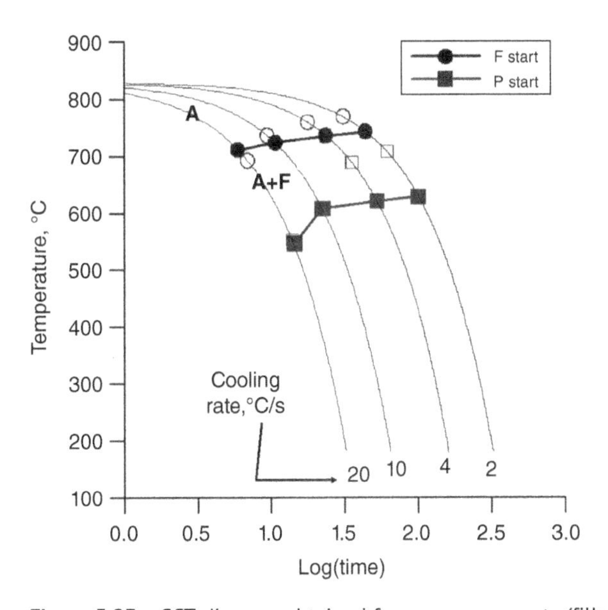

Figure 5.23 CCT diagram obtained from measurements (filled symbols) and calculated by the CA model (open symbols).

5.2.2.3 Microshear Bands, Shear Bands, and Strain Localization

Contrary to hot rolling, thermally activated phenomena are not important in cold rolling. Instead, another lower-scale phenomenon and its influence on the macroscopic behaviour have to be accounted for. Cold rolling is an industrial process where a plane state of strain exists, which fosters initiation of microshear and shear bands that became a dominant deformation mechanism along with a slip. The strain localization due to shear banding has been investigated by scientists for over 40 years, both experimentally and theoretically. Analyses were performed under various kinds of deformation, e.g. uniaxial tension, compression, and plane strain. Despite proofs of the discontinuous character of deformation, application of the continuum mechanics to describe the flow of polycrystals is common when industrial needs are taken into account. Models accounting for the strain localization, with the continuum assumption, have been proposed in the scientific literature; see for example [58]. Unfortunately, these models cannot be easily generalized to simulate complex industrial deformations. This is a significant disadvantage, as models can predict the influence of microshear bands only in the type of test that was analyzed. Thus, the continuum approach has limitations; they do not describe properly certain processes characterized by the strong influence of strain localization, which leads to discontinuities during real industrial deformation conditions. As a result, the problem of realistic description of strain localization in bulk processes is still not solved. An alternative approach based on the multiscale CAFE method proposed in [44, 45] is one of the possible solutions to mentioned limitations. In this model, two CA spaces dealing with microscale and mesoscales are attached to each Gauss integration point responsible for macroscale behaviour, as seen in Figure 5.24.

According to CA method requirements, cells in both CA spaces (microshear band [MSB] space and shear band [SB] space), are defined by state variables, as well as by a set of transition rules. The latter is defined based on experimental knowledge, e.g. [8], and they control changes of states in MSB and SB space in order to recreate physical mechanisms leading to the initiation and propagation of these bands. The details regarding cell states and transition rules are presented in [44, 45].

According to the defined mapping operations, information about the occurrence of microshear and shear bands is exchanged in both directions between the FE model and CA spaces

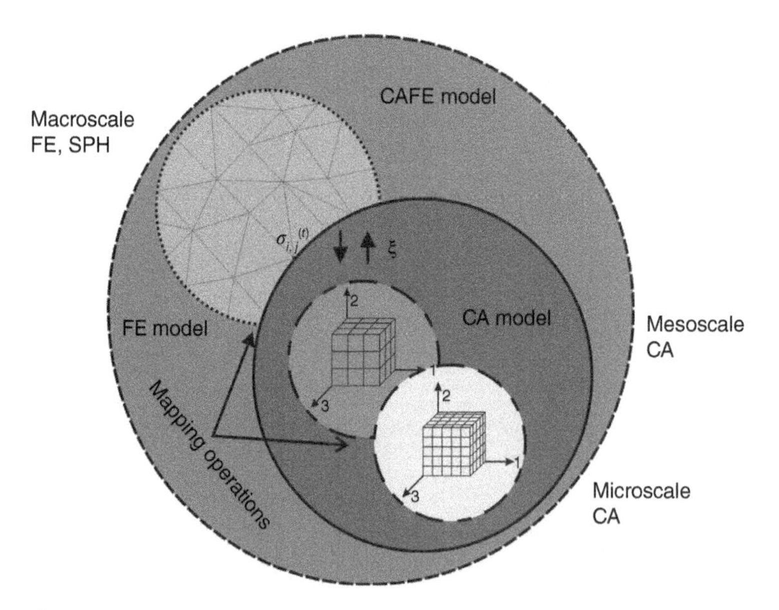

Figure 5.24 Illustration of the developed multiscale strain localization CAFE model.

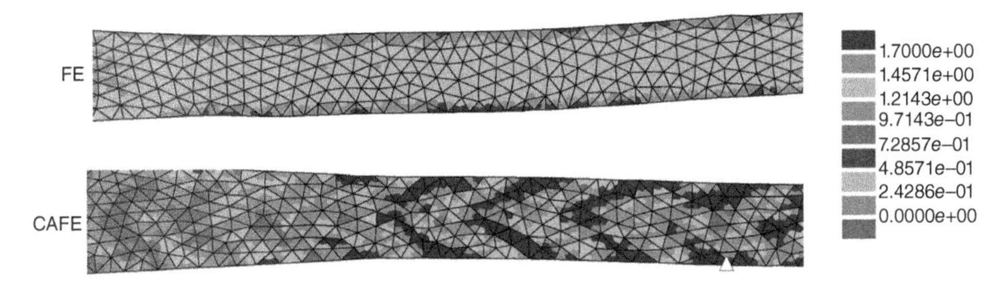

Figure 5.25 Strain distribution predicted by the FE and CAFE models after the second rolling pass.

during each time step. In each time increment, information about the stress tensor is sent from the FE solver to the MSB space, where the development of microshear bands is calculated according to the transition rules. After the exchange of information between CA spaces, transition rules for the SB space are introduced, and propagation of the shear bands is modelled. Based on the information supplied by the CA spaces, flow stress accounting for shear banding is calculated and is returned to the FE programme and used in the next step of calculations. As presented, the developed model is a typical example of a hierarchical multiscale CAFE approach.

Examples of differences obtained using only macroscale model and the multiscale CAFE model can be seen in Figure 5.25. Simulation of two-pass industrial cold rolling of copper with a total reduction in the height of 70% is selected as a case study.

As seen in Figure 5.25, the conventional model predicts uniform strain distribution along the sample. The CAFE model predicts localized bands aligned to the rolling direction at an angle of 40–50°. The strain localization is developing simultaneously in the two opposite families of bands that are crossing each other in the central part of the sample. Results obtained from the CAFE model are closer to the experimental results than those from the conventional model [45].

The presented model was an example of a three-scale model: micro–meso–macro CAFE model. The major disadvantage of this approach limiting its practical applications was computing time. However, recent development in the computer hardware and modelling capabilities within fog/cloud environments open new capabilities for such complex numerical models [76].

References

1 Abraham, F.F., Broughton, J.Q., Berstine, N., and Kaxiras, E. (1998). Spanning the continuum to quantum length scales in a dynamic simulations of brittle fracture. *Europhysics Letters* 44: 783–787.

2 Allix, O. (2006). Multiscale strategy for solving industrial problems. *Computational Methods in Applied Sciences* 6: 107–126.

3 Breitbarth, E., Zaefferer, S., Archie, F. et al. (2018). Evolution of dislocation patterns inside the plastic zone introduced by fatigue in an aged aluminium alloy AA2024-T3. *Materials Science and Engineering A*. 718: 345–349.

4 Burczyński, T., Mrozek, A., Górski, R., and Kuś, W. (2010). The molecular statics coupled with the subregion boundary element method in multiscale analysis. *International Journal for Multiscale Computational Engineering* 8 (3): 319–330.

5 Chamanfar, A., Valberg, H.S., Templin, B. et al. (2019). Development and validation of a finite-element model for isothermal forging of a nickel-base superalloy. *Materialia* 6: 100319.

6 Christian, J.W. (2002). *The Theory of Transformation in Metals and Alloys*. London: Pergamon.

7 Chen, K., Wang, Y., and Wang, J. (2020). An experimental and theoretical study on constitutive model of Al-2024T4 for sheet metal incremental forming. *Procedia Manufacturing* 50: 565–569.

8 Cizek, P. (2002). Haracteristics of shear bands formed in an austenitic stainless steel during hot deformation. *Material Science and Engineering A* 324: 214–218.

9 Das, S., Palmiere, E.J., and Howard, I.C. (2002). CAFE: a tool for modelling thermomechanical processes. In: *Proc. Thermomech. Processing: Mechanics, Microstructure & Control* (ed. E.J. Palmiere, M. Mahfouf and C. Pinna), 296–301. Sheffield.

10 Das, S., Abbod, M.F., Zhu, Q. et al. (2007). A combined neuro fuzzy-cellular automata based material model for finite element simulation of plane strain compression. *Computational Material Science* 40: 366–375.

11 Davey, K., Bylya, O., and Krishnamurthy, B. (2020). Exact and inexact scaled models for hot forging. *International Journal of Solids and Structures* 203: 110–130.

12 De Borst, R. (2008). Challenges in computational materials science, multiple scales, multi-physics and evolving discontinuities, Computational Material. *Science* 43: 1–15.

13 Delannay, L. (2018). Modelling of microscopic strain heterogeneity during wire drawing of pearlite. *Proceedia Manufacturing* 15: 1893–1899.

14 Demkowicz, L., Rachowicz, W., and Devloo, P. (2002). A fully automatic hp-adaptivity. *Journal of Scientific Computing* 17: 127–155.

15 Donnay, B., Herman, J.C., Leroy, V. et al. (1996). Microstructure evolution of C–Mn steels in the hot deformation process: the STRIPCAM model. In: *Proc. Modelling of Metal Rolling Processes* (ed. J.H. Beynon, P. Ingham, H. Teichert and K. Waterson), 23–35. London.

16 Dumett, M.A., Vassilevski, P., and Woodward, C.S. (2002). A multigrid method for nonlinear unstructured finite element elliptic equations. *Journal on Scientific Computing* 1–19.

17 Elizondo, A. (2007). Horizontal coupling in continuum atomistics. Ph.D. thesis.

18 Estrin, Y. (1996). Dislocation density related constitutive modelling. In: *Unified Constitutive Laws of Plastic Deformation* (ed. A.S. Krausz and K. Krausz). Academic Press.

19 Firat, M. (2007). Computer aided analysis and design of sheet metal forming processes: part II – Deformation response modelling. *Materials and Design* 28: 1304–1310.

20 Fujita, N., Ishikawa, N., Roters, F. et al. (2018). Experimental–numerical study on strain and stress partitioning in bainitic steels with martensite–austenite constituents. *International Journal of Plasticity* 104: 39–53.

21 Gandin, C.A. and Rappaz, M. (1994). A coupled finite element - cellular automaton model for the prediction of dendritic grain structures in solidification processes. *Acta Metallurgica* 42: 2233–2246.

22 Gandin, C.A. and Rappaz, M. (1996). 3D cellular automaton algorithm for the prediction of dendritic grain growth. *Acta Metallurgica* 45: 2187–2195.

23 Gawąd J., Madej W., Kuziak R., Pietrzyk M. (2008). Multiscale model of dynamic recrystallization in hot rolling. *Proc. ESAFORM 11*, Lyon, CD.

24 Gawad, J., Van Bael, A., Eyckens, P. et al. (2013). Hierarchical multiscale modelling of texture induced plastic anisotropy in sheet forming. *Computational Materials Science* 66: 65–83.

25 Gawad, J., Banabic, D., van Bael, A. et al. (2015). An evolving plane stress yield criterion based on crystal plasticity virtual experiments. *International Journal of Plasticity* 75: 141–169.

26 Grosman, F. (1997). Application of the flow stress function in programmes for computer simulation of plastic working processes. *Journal of Material Processing Technology* 64: 169–180.

27 Guo, Z., Lasne, P., Saundersc, N., and Schillé, J.P. (2018). Introduction of materials modelling into metal forming simulation. *Procedia Manufacturing* 15: 372–380.

28 Halder, C., Bachniak, D., Madej, L. et al. (2015). Sensitivity analysis of the finite difference 2-D cellular automata model for phase transformation during heating. *ISIJ* 55: 285–292.

29 He, W.J., Zhang, S.H., Prakash, A., and Helm, D. (2014). A hierarchical multiscale model for hexagonal materials taking into account texture evolution during forming simulation. *Computational Materials Science* 82: 464–475.

30 Hirt, G., Kopp, R., Hofmann, O. et al. (2007). Implementing a high accuracy multi-mesh method for incremental bulk metal forming. *Annals of the CIRP* 56: 313–316.

31 Khvastunkov, M.S. and Leggoe, J.W. (2004). Adapting cellular automata to model failure in spatially heterogeneous ductile alloys. *Scripta Materialia* 51: 309–314.

32 Kumar, M., Sasikumar, R., and Kesavan, N.P. (1998). Competition between nucleation and early growth of ferrite from austenite-studies using cellular automaton simulation. *Acta Metallurgica* 46: 6291–6303.

33 Kundu, S., Dutta, M., Ganguly, S., and Chandra, S. (2004). Prediction of phase transformation and microstructure in steel using cellular automation technique. *Scripta Materialia* 50: 891–895.

34 Kwon, Y.W. and Jung, S.H. (2003). Discrete atomic and smeared continuum modelling for static analysis. *Engineering Computations* 20 (8): 964–978.

35 Lan, Y.J., Li, D.Z., and Li, Y.Y. (2004). Modelling austenite decomposition into ferrite at different cooling rate in low-carbon steel with cellular automaton method. *Acta Materialia* 52: 1721–1729.

36 Lan, Y.J., Xiao, N.M., Li, D.Z., and Li, Y.Y. (2005). Mesoscale simulation of deformed austenite decomposition into ferrite by coupling a cellular automaton method with a crystal plasticity finite element model. *Acta Materialia* 53: 991–1003.

37 Lee, P.D., Chirazi, A., Atwood, R.C., and Wand, W. (2004). Multi scale modelling of solidification microstructures, including microsegregation and microporosity in an Al-Si-Cu alloy. *Material Science and Engineering A* A365: 57–65.

38 Lenard, J.G., Pietrzyk, M., and Cser, L. (1999). *Mathematical and Physical Simulation of the Properties of Hot Rolled Products*. Amsterdam: Elsevier.

39 Li, Y., Shterenlikht, A., Ren, X. et al. (2019). CAFE based multiscale modelling of ductile-to-brittle transition of steel with a temperature dependent effective surface energy. *Materials Science and Engineering: A* 755: 220–230.

40 Lin, F., Shi, Q., and Delannay, L. (2020). Microscopic heterogeneity of plastic strain and lattice rotation in partially recrystallized copper polycrystals. *International Journal of Solids Structures* 184: 167–177.

41 Ma, J., Wang, B., Zhao, S. et al. (2016). Incorporating an extended dendritic growth model into the CAFE model for rapidly solidified non-dilute alloys. *Journal of Alloys and Compounds* 668: 46–55.

42 Madej, L., Sitko, M., and Pietrzyk, M. (2016). Perceptive comparison of mean and full field dynamic recrystallization models. *Archives of Civil and Mechanical Engineering* 16: 569–589.

43 Madej, L., Gawad, J., and Pietrzyk, M. (2006). Numerical method of accounting for stochastic and discontinuous character of various phenomena occurring in materials subjected to thermo-mechanical processing. *Proc. MS&T 2006*, 351–362, Ohio, USA.

44 Madej, L., Hodgson, P.D., and Pietrzyk, M. (2007). The validation of a multiscale rheological model of discontinuous phenomena during metal rolling. *Computational Materials Science* 41: 236–214.

45 Madej, L., Hodgson, P.D., and Pietrzyk, M. (2009). Development of the multiscale analysis model to simulate strain localization occurring during material processing, Archive of Computational Methods. *Arhives of Computational Methods in Engineering* 16: 287–318.

46 Madej, L. (2010). *Development of the Modelling Strategy for the Strain Localization Simulation Based on the Digital Material Representation*. Krakow: AGH University Press.

47 Madej, L., Tokunaga, T., Matsuura, K. et al. (2015). Physical and numerical modelling of backward extrusion of Mg alloy with Al coating. *Annals of the CIRP* 64: 253–256.

48 Maire, L., Fausty, J., Bernacki, M. et al. (2018). A new topological approach for the mean field modelling of dynamic recrystallization. *Materials & Design* 146: 194–207.

49 Mellbin, Y., Hallberg, H., and Ristinmaa, M. (2015). A combined crystal plasticity and graph-based vertex model of dynamic recrystallization at large deformations. *Modelling and Simulation in Materials Science and Engineering* 23: 045011.

50 Mieche, C. (2003). Computational micro-to-macro transitions for discretized micro-structures of heterogeneous materials at finite strains based on the minimization of averaged incremental energy. *Computer Methods in Applied Mechanics and Engineering* 192: 559–591.

51 Milenin, A. and Muskalski, Z. (2007). The FEM simulation of cementite lamellas deformation in pearbibic colony during drawing of high carbon steels. In: *Proc. Numiform*, vol. 2007 (ed. J.M.A. Cesar de Sa and A.D. Santos), 1375–1380. Porto.

52 Militzer, M., Hoyt, J.J., Provatas, N. et al. (2014). Multiscale modelling of phase transformations in steels. *JOM* 66: 740–746.

53 Missoum-Benziane, D., Ryckelynck, D., and Chinesta, F. (2007). A new fully coupled two-scale modelling for mechanical problems involving microstructure: the95/5 technique. *Computer Methods in Applied Mechanics and Engineering* 196: 2325–2337.

54 Mrozek, A., Kuś, W., and Burczyński, T. (2007). Application of the coupled boundary element method with atomic model in the static analysis. *Computer Methods in Materials Science* 7: 284–288.

55 Mrozek, A. (2009). A continuum-atomistic model of material medium in the computer analysis of mechanical deformable structures. Ph.D. thesis.

56 Nemat-Nasser, S. (1999). Averaging theorems in finite deformation plasticity. *Mechanics of Materials* 31: 493–523.

57 Paszyński, M., Kurtz, J., and Demkowicz, L. (2006). Parallel fully automatic hp-adaptive 2D finite element package. *Computer Methods in Applied Mechanics and Engineering* 195: 711–741.

58 Pecherski, R.B. (1998). Continuum mechanics description of plastic flow produced by micro-shear bands. *Technische Mechanik* 18: 107–115.

59 Perzynski, K., Ososkov, Y., Jain, M. et al. (2017). Validation of the dual phase steel failure model at the micro scale. *International Journal for Multiscale Computational Engineering* 15 (5): 443–458.

60 Perzynski, K. and Madej, L. (2017). Complex hybrid numerical model in application to failure modelling in multiphase materials. *Archives of Computational Methods in Engineering* 24: 869–890.

61 Pietrzyk, M. (1990). Finite element based model of structure development in the hot rolling process. *Steel Research* 61: 603–607.

62 Pietrzyk, M., Kusiak, J., Kuziak, R. et al. (2014). Conventional and multiscale modelling of microstructure evolution during laminar cooling of DP steel strips. *Metallurgical and Materials Transactions A* 45A: 5835–5851.

63 Pietrzyk, M., Madej, L., Rauch, L., and Szeliga, D. (2015). *Computational Materials Engineering: Achieving High Accuracy and Efficiency in Metals Processing Simulations*. Butterworth Heinemann Elsevier.
64 Pietrzyk, M. and Madej, L. (2017). Perceptive review of ferrous micro/macro material models for thermo-mechanical processing applications. *Steel Research International* 88: 1700193.
65 Quested, T.E. and Greer, A.L. (2005). Grain refinement of Al alloys: micromechanisms determining as-cast grain size in directional solidification. *Acta Materialia* 53: 4643–4653.
66 Rauch, L., Madej, L., Spytkowski, P., and Golab, R. (2015). Development of the cellular automata framework dedicated for metallic materials microstructure evolution models. *Archives of Civil and Mechanical Engineering* 15: 48–61.
67 Rondanini, M., Barbato, A., and Cavalloti, C. (2006). A multi scale model of the Si CVD process. *Proc. MMM*, 98–101, Freiburg.
68 Roucoules, C., Pietrzyk, M., and Hodgson, P.D. (2003). Analysis of work hardening and recrystallization during the hot working of steel using a statistically based internal variable method. *Material Science and Engineering A* 339: 1–9.
69 Roters, F., Diehl, M., Shanthraj, P. et al. (2019). DAMASK – The Düsseldorf Advanced Material Simulation Kit for modelling multi-physics crystal plasticity, thermal, and damage phenomena from the single crystal up to the component scale. *Computational Materials Science* 158: 420–478.
70 Scholtes, B., Shakoor, M., Settefrati, A. et al. (2015). New finite element developments for the full field modelling of microstructural evolutions using the level-set method. *Computational Material Science* 109: 388–398.
71 Sfantos, G.K. and Alibadi, M.H. (2007). Multiscale boundary element modelling of material degradation and fracture. *Computer Methods in Applied Mechanics and Engineering* 196: 1310–1329.
72 Schmitz, G.J. and Prahl, U. (2016). *Handbook of Software Solutions for ICME*. Wiley-VCH.
73 Shterenlikht, A. and Howard, I.C. (2006). The CAFE model of fracture – application to a TMCR steel, Journal. *Fatigue and Fracture of Engineering Materials and Structures* 29: 770–787.
74 Sitko, M., Chao, Q., Wang, J. et al. (2020). A parallel version of the cellular automata static recrystallization model dedicated for high performance computing platforms – development and verification. *Computational Materials Science 172* 109283.
75 Sitko, M., Pietrzyk, M., Kuziak, R., et al., Development and identification of the cellular automata phase transformation model. Paper presented at ESAFORM. *24th International Conference on Material Forming*, Liège, Belgique. doi:10.25518/esaform21.2640.
76 Sourav, K., Arijit, S., and Ruhul, A. (2019). An overview of cloud-fog computing: architectures, applications with security challenges. *Security and Privacy* 2: 1–14.
77 Sunyk, R. and Steinmann, P. (2002). On higher gradients in continuum-atomistic modelling. *International Journal of Solids and Structures* 40: 6877–6896.
78 Tadmor, E.B., Oritz, M., and Philips, R. (1996). Quasicontinuum analysis in crystals. *Philosophical Magazine A* 73: 1529–1563.
79 Tekkaya, A.E., Allwood, J.M., Bariani, P.F. et al. (2015). Metal forming beyond shaping: predicting and setting product properties. *CIRP Annals Manufacturing Technology* 64: 629–653.
80 Thevoz, P. and Gandin, C.A. (2001). *Micro-Macroscopic Solidification Models: Encyclopaedia of Materials Science and Technology*, 5637–5642. Elsevier.
81 Van Houte, P., Kanjarla, A.K., Val Bael, A. et al. (2006). Multiscale modelling of the plastic anisotropy and deformation texture of polycrystaline materials. *European Journal of Mechanics – A/Solids* 25: 634–648.

82 Vandyoussefi, M. and Greer, A.L. (2002). Application of the cellular automata-finite element model to the grain refinement of directionally solidified 1l-4.15 wt% Mg alloys. *Acta Materialia* 50: 1693–1705.

83 Volk, W., Groche, P., Brosius, A. et al. (2019). Models and modelling for process limits in metal forming. *CIRP Annals Manufacturing Technology* 68: 775–798.

84 Volker, J., Songul, K., and Kaya, L. (2006). A two-level variational multiscale method for convection-dominated convection–diffusion equations. *Computer Methods in Applied Mechanics and Engineering* 195: 4594–4603.

85 Wang, J., Weyland, M., Bikmukhametov, I. et al. (2019). Transformation from cluster to nano-precipitate in microalloyed ferritic steel. *Scripta Materialia* 160: 53–57.

86 Wang, J., Ramajayam, M., Charrault, E., and Stanford, N. (2019). Quantification of precipitate hardening of twin nucleation and growth in Mg and Mg–5Zn using micro-pillar compression. *Acta Materialia* 163: 68–77.

87 Yang, X.L., Dong, H.B., Wang, W., and Lee, P.D. (2004). Microscale simulation of stray grain formation in investment cast turbine blades. *Material Science and Engineering A* A386: 129–139.

88 Yazdipour, N., Davis, C.H.J., and Hodgson, P.D. (2007). Simulation of dynamic recrystalization using random grid cellular automata. *Computer Methods in Materials Science* 7: 168–174.

89 Yu, Q. and Esche, S.K. (2005). A multiscale approach for microstructure prediction in thermo-mechanical processing of metals. *Journal of Material Processing Technology* 169: 493–502.

90 Yu, P., Wu, C.S., and Shi, L. (2021). Analysis and characterization of dynamic recrystallization and grain structure evolution in friction stir welding of aluminium plates. *Acta Materialia* 207: 116692.

91 Zhang, L., Zhang, C.B., Wang, Y.M. et al. (2003). A cellular automaton investigation of the transformation from austenite to ferrite during continuous cooling. *Acta Materialia* 51: 5519–5527.

6

Optimization and Identification in Multiscale Modelling

The present chapter is devoted to selected optimization and identification problems in the case of multiscale modelling. The multiscale modelling of materials and structures [21, 36, 68] is an important area of research that allows design of new materials and products with better quality, strength, and performance parameters. The multiscale approach allows creating reliable models taking into account product and material properties and topology in different length scales. As presented in Chapter 5, the multiscale models can be analyzed using many approaches. The bridging and homogenization methods are most popular [37]. The bridging method consists of connecting of scales on some boundaries; this method is especially important if some phenomenon occurs in a small part of the structure and should be taken into account. An example of the bridging method is a connection between an atomic model and a continuum model discretized by the finite element method for problems with crack initialization and propagation. The area near the crack is modelled using the discrete atomic model, and the rest of the structure is analyzed by means of finite elements [71, 72] or the boundary element method [4]. The two main approaches – mathematical and numerical homogenization are used for analyzing models with a locally periodical microstructure. The mathematical homogenization method is widely used in analyses of laminates. The micromodels have to be analyzed for each local microstructure when computational homogenization is used [22, 28]. Computational homogenization allows consideration of complicated microstructure with nonlinearities like the elasto-plastic material, the contact friction, wear, or phase changes in the micromodel. A short description of computational homogenization is presented in Chapter 3.

The optimization in multiscale modelling allows finding structures with better performance or strength in one scale with respect to design variables for another scale. The identification problem is formulated as the evaluation of some geometrical or material parameters of structures in one scale having measured information in another scale. This last problem can be formulated and considered as a special optimization problem.

The special and specific case of optimization problems, associated with multiscale approaches, is optimization of atomic clusters. This case is also considered in the present chapter. The problems discussed in the chapter tackle optimization and identification of microstructure parameters on the basis of objective functionals and measured data obtained for the macroscale level.

The analysis methods of multiscale models can be used in optimization and identification algorithms. The global optimization algorithms based on bioinspired algorithms described in Chapter 3 are widely used in optimization and identification problems in solid mechanics [9].

Multiscale Modelling and Optimisation of Materials and Structures, First Edition. Tadeusz Burczyński, Maciej Pietrzyk, Wacław Kuś, Łukasz Madej, Adam Mrozek, and Łukasz Rauch.

The most important advantages of bioinspired algorithms are their robustness, significant probability of finding the global optimum, and easy adaptation to new problems. The main disadvantage is the long computation time due to the need of solving hundreds or thousands direct problems during optimization. The computations can be shortened by using parallel multi-subpopulation approaches.

6.1 Multiscale Optimization

6.1.1 Optimization of Atomic Clusters

This section is devoted to the special case of optimization based only on one scale – the atomic scale. This special case has an important consequence, especially in the design of new 2D nanomaterials and nanostructures.

6.1.1.1 Introduction to Optimization of Atomic Clusters

The atomic clusters are small, isolated spatial molecular structures that contain up to several thousands of atoms. The ideal clusters have some unique mechanical, optical, and electronic properties due to their ideal, the most stable, spatial arrangement of atoms. Such a structure corresponds to the global minimum of the potential energy surface (PES) of the atomic cluster [35, 57].

Generally, searching for the global minimum on the PES is a non-trivial, NP-hard problem because the number of local minima increases almost exponentially with the cluster size [65]. Since the ab initio methods are still limited to the small atomic structures, many various heuristic and artificial intelligence–based approaches have been recently applied to deal with the problem of minimization of the potential energy of the cluster. Random searches and the Monte Carlo (MC) simulated annealing were adopted by [35] to study atomic clusters, while [64] applied the basin-hopping MC algorithm to the Morse and Lennard-Jones clusters. Genetic algorithm (GA) was applied, probably for the first time, to the Morse potential by [57]. Prediction of the aluminium atoms distribution using distributed evolutionary algorithm (DEA) was described in [47]. The authors of [60] used an adaptive immune algorithm for optimization of the Lennard-Jones clusters. Particle swarm optimization (PSO) of the nickel clusters was performed by [70].

In this section, the process of minimization of the potential energy of the small atomic clusters, based on the following methods of computational intelligence is presented: evolutionary algorithm (EA), artificial immune system (AIS), and PSO with pair-wise as well as many-body empirical potentials [44]. Listed methods of the non-classical optimization simulate biological processes of the natural environment and organisms such as the theory of evolution and the biological immune systems and have been already explained in Chapter 3. Note, these approaches generally do not need any information about the gradient of the fitness function and give a strong probability of finding the global optimum. The main drawback of non-classical methods is the longer time of computations compared to the conventional, e.g. gradient-based, algorithms.

The procedure of finding the global minima on the PES using EA, AIS, or PSO is performed in the same manner. Respectively, chromosomes, memory cells, and particles contain the real-valued Cartesian coordinates of each atom in the considered cluster. The total number of project variables equals 3 N, where 3 is the dimension of the problem and N denotes the number of atoms in the cluster. The initial population is generated randomly, and the atoms can move freely in the sphere of radius:

$$r_0\sqrt[3]{N}, \tag{6.1}$$

where r_0 is the equilibrium distance between two atoms. No other constraints are applied to the set of atoms. The formulation of the fitness function is independent of the chosen method of optimization and is expressed as the total potential energy of the considered atomic cluster. For the pair-wise potentials, the fitness function can be formulated as follows:

$$V_{tot} = \sum_{i,j>i} \Phi_2(r_{ij}), \tag{6.2}$$

and similarly for the many-body (here, three-body) interaction models:

$$V_{tot} = \sum_{i,j>i} \Phi_2(r_{ij}) + \sum_{i,j>i,k>j} \Phi_3(r_{ij}, r_{ik}, r_{jk}). \tag{6.3}$$

Additionally, the average binding energy (i.e. average energy per atom) E_B is estimated for each type of the cluster:

$$E_B = -\frac{V_{tot}}{N}, \tag{6.4}$$

and the regions of enhanced stability are determined by the second difference in the binding energy, which is related to the thermodynamic stability of the atomic cluster:

$$D_2(N) = 2E_B(N) - E_B(N+1) - E_B(N-1). \tag{6.5}$$

The process of minimization of the potential energy was performed using the Morse and Murrell-Mottram (MM) potentials (see Section 2.1 and works [48, 49] for details). Both potentials were fitted to the properties of the bulk aluminium. The following parameters of the Morse potential were taken from [23]: $D_e = 0.2703$ eV, $r_0 = 3.253$ Å, $\alpha = 1.1646$ Å^{-1}. The set of coefficients of the MM model are included in Table 6.1 [17].

The global minima for Morse ($N = 2...21$) and MM ($N = 3...15$) clusters were obtained using EA, AIS, and PSO. The lowest energy levels for each cluster are summarized in Table 6.2. The values for MM structures with nuclearity $N = 16...20$ (marked with $*$) are taken from [35]. The lowest average binding energies and their second differences, as a function of N, are presented in Figures 6.1–6.4, respectively, for Morse and MM clusters.

The peaks in the $E_B(N)$ and $D_2(N)$ functions at $N = 4$, 6, 13, and 19 correspond to the configurations of atoms with enhanced stability. This phenomenon was also predicted by [16]. Such clusters have an identical shape for both types of applied potential models and are shown in Figure 6.5.

The less stable structures are generally created by modifications of the super stable ones, i.e. additional atoms form extra vertexes attached to the main atomic structure, e.g. the 20 and the 21 atom clusters made on the base of $N = 19$ stable double icosahedron structure (Figure 6.6). The 29 atom Morse cluster is shown in Figure 6.7.

The small Morse and MM clusters ($N = 2...20$) in almost all cases have the same form. The exceptions are the structures with 9, 16, 18, and 20 atoms. The MM global minima, in these cases, correspond to the second most stable isomers (local minima) of the Morse clusters. It is due to differences in the shape of PES, described by pair-wise and many-body potentials: the PES of the pair-wise interactions has significantly more minima than the PES determined by

Table 6.1 Parameters of the Murrell-Mottram potential for Al.

Parameter	Value
a_2	7
a_3	8
D_e (eV)	0.9073
r_0 (Å)	2.7568
C_0	0.2525
C_1	−0.4671
C_2	4.4903
C_3	−1.1717
C_4	1.6498
C_5	−5.3579
C_6	1.6327

Table 6.2 Lowest energies of the Morse and Murrell-Mottram clusters.

	E_B (eV)	
N	Morse	Murrell-Mottram
2	0.13 515	—
3	0.2 703	0.831 7
4	0.40 545	1.137 83
5	0.49 642	1.324 7
6	0.59 286	1.503 9
7	0.65 393	1.604 6
8	0.70 494	1.692 3
9	0.75 344	1.764
10	0.80 528	1.828 2
11	0.85 051	1.880 5
12	0.91 256	1.948 2
13	0.98 672	2.031 3
14	0.99 274	2.037 5
15	1.01 499	2.071 9
16	1.03 278	2.104 5*
17	1.06	2.135 4*
18	1.0 933	2.161 8*
19	1.13 553	2.189*
20	1.15 096	2.207 18*
21	1.1 651	—

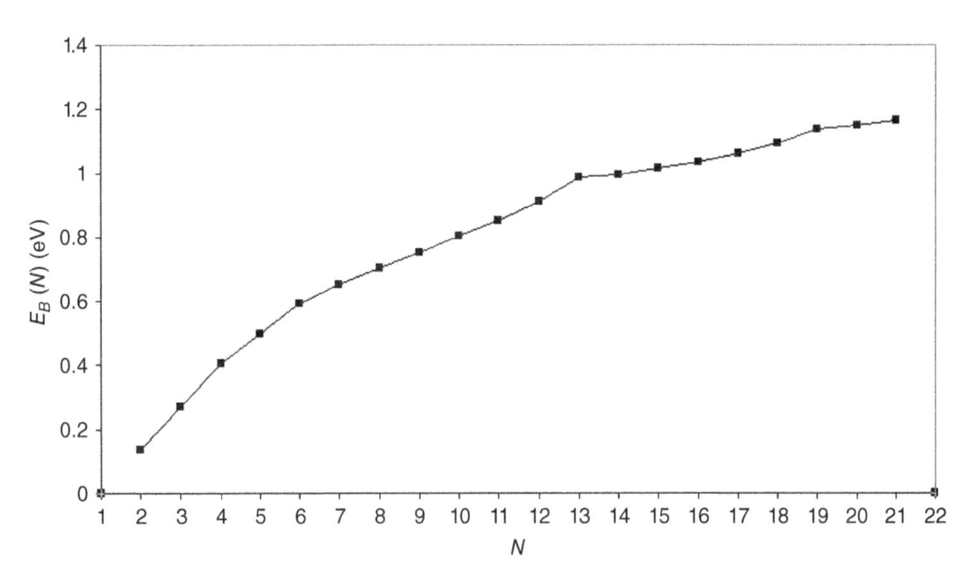

Figure 6.1 Average binding energies of the Morse clusters.

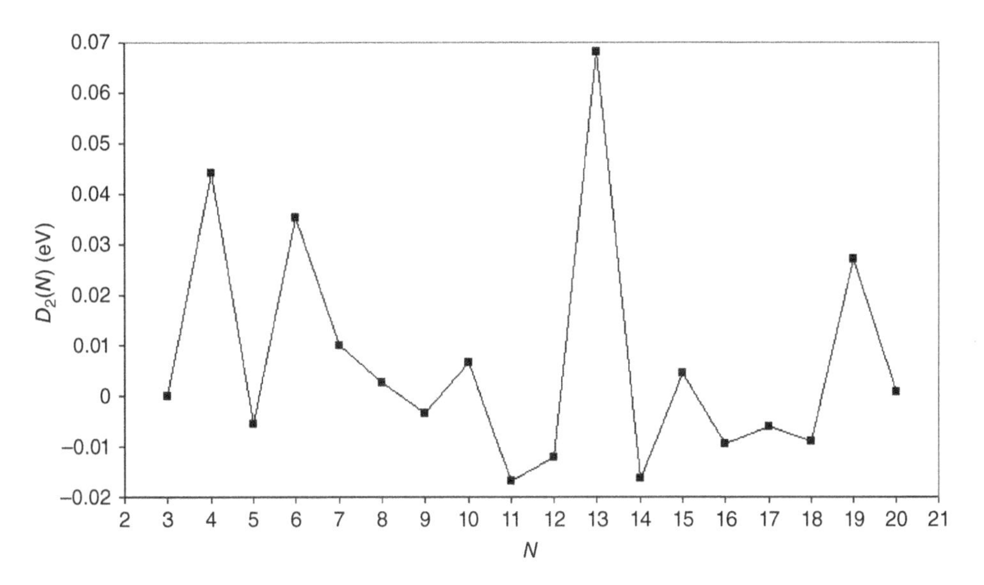

Figure 6.2 The second difference of the binding energy of the Morse clusters.

the many-body potential. Obtained results are comparable with ones presented in works [35, 60]. This phenomenon, for the case of $N = 9$, is shown in Figure 6.8.

The performance of the EA and the AIS in the minimization of the potential energy was similar, and these two algorithms found all the global minima within the considered range up to $N = 29$ (i.e. 87 project variables, see Figure 6.7). The energy minimization using PSO is generally faster than two others; however, the PSO had a tendency to get stuck at local minima. A more detailed analysis of the performance of applied methods is included in the paper [44].

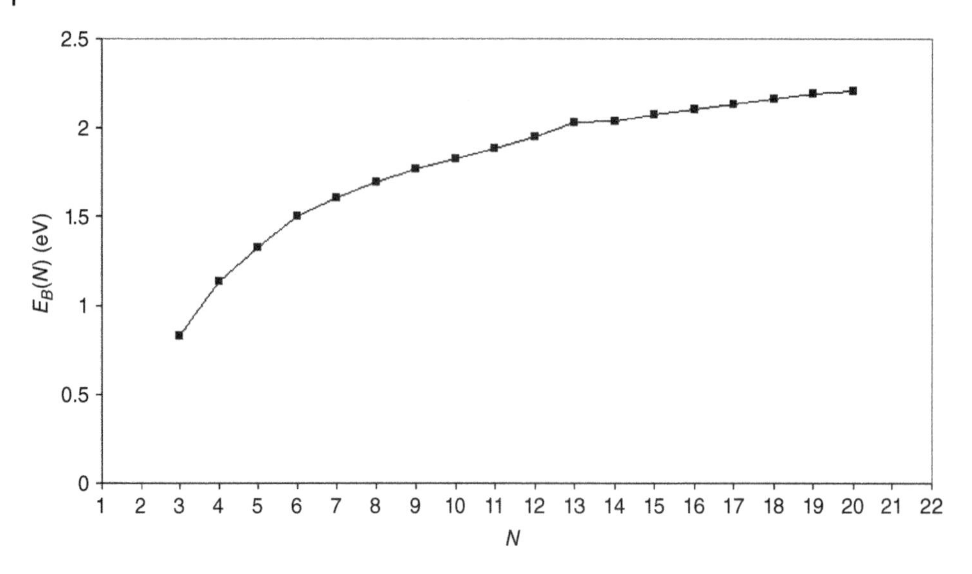

Figure 6.3 Average binding energies of the MM clusters.

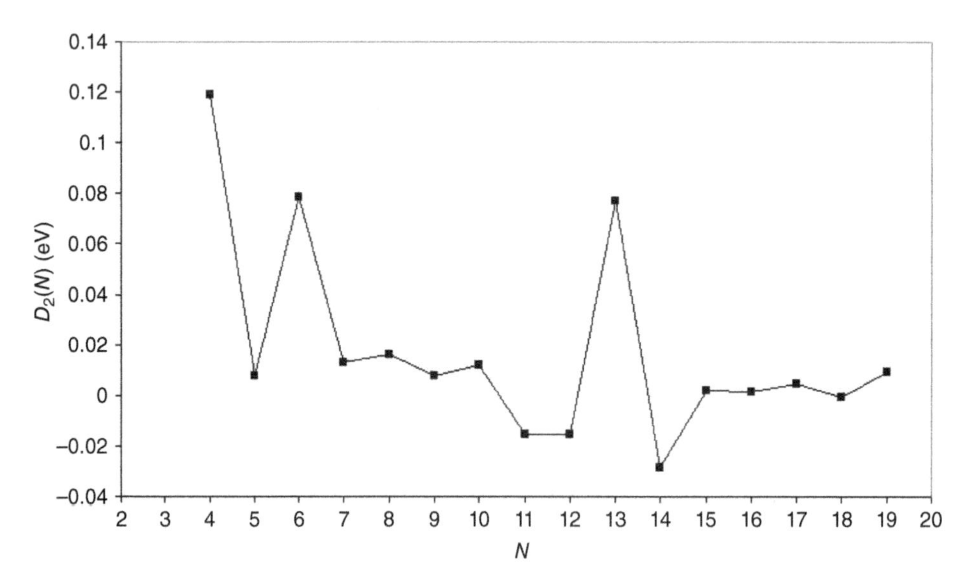

Figure 6.4 The second difference of the binding energy of the MM clusters.

6.1.1.2 Optimization of Carbon Atomic Clusters

Carbon atoms form various types of bondings and spatial configurations. This ability is determined by the atoms' hybridization states, which depend on their particular electronic configuration. This phenomenon is responsible for the existence of many different allotropes of carbon. Besides natural ones, like diamond, graphite, and amorphous carbon, numerous two-dimensional synthetic structures such as graphene and their derivatives have been the subject of interest of researchers in recent years [18, 20, 24, 42, 53, 59]. This is due to the unique electronic, thermal, and mechanical properties of such structures. Additionally, two-dimensional

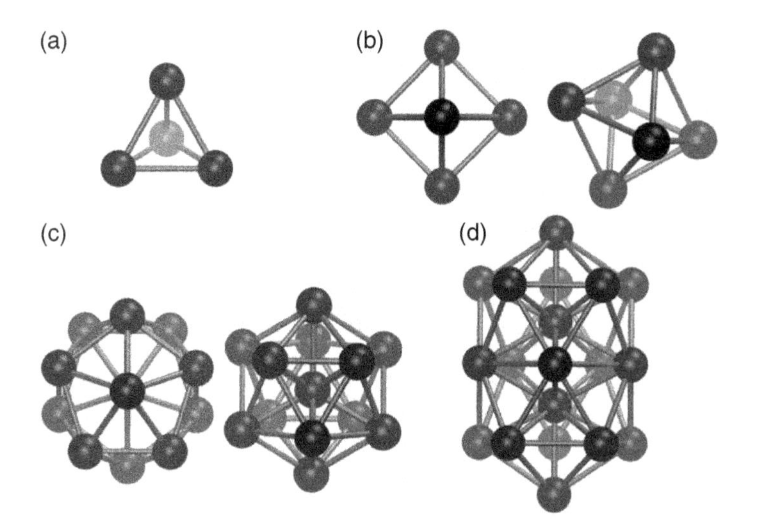

Figure 6.5 Super stable configurations of atoms: (a) tetrahedron $N = 4$, (b) octahedron $N = 6$, (c) icosahedron $N = 13$, and (d) double icosahedron $N = 19$.

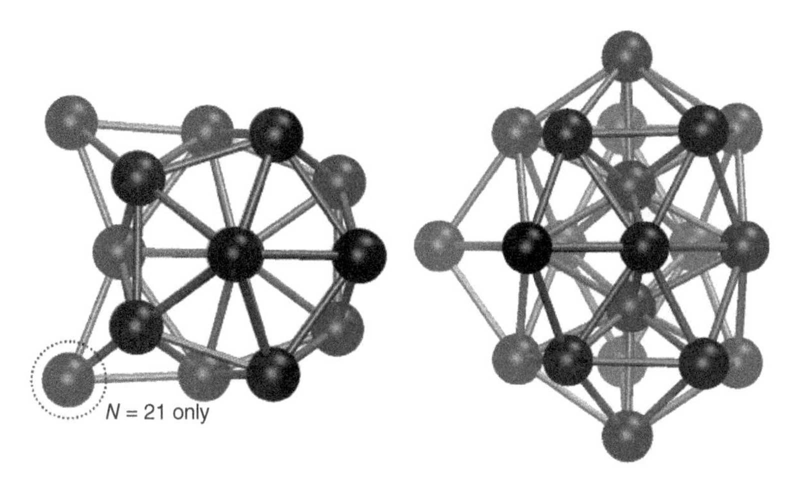

Figure 6.6 $N = 20$ and $N = 21$ clusters.

graphene-like materials can be used to create another, more complex class of nanostructures, such as nanotubes.

Two-dimensional graphene-like materials can be classified as periodic, flat atomic networks, made of stable configurations of carbon atoms in certain hybridization states. Looking from a higher level, it can be seen (Figure 6.9) that each flat carbon network can be decomposed into particular parts such as benzene rings (or other polygonal elements) and acetylenic linkages of different lengths (–C≡C–, –C≡C–C≡C–). Depending on the arrangement of the considered structure, a rectangular or triclinic unit cell of a given size and atomic density can be identified in each type of the flat periodic network. An overview of such structures (i.e. graphyne and supergraphene), along with a detailed description and investigation of their structural and electronic properties using the tight-binding method is presented in [20]. The stress–strain relations and mechanical properties were obtained by [42].

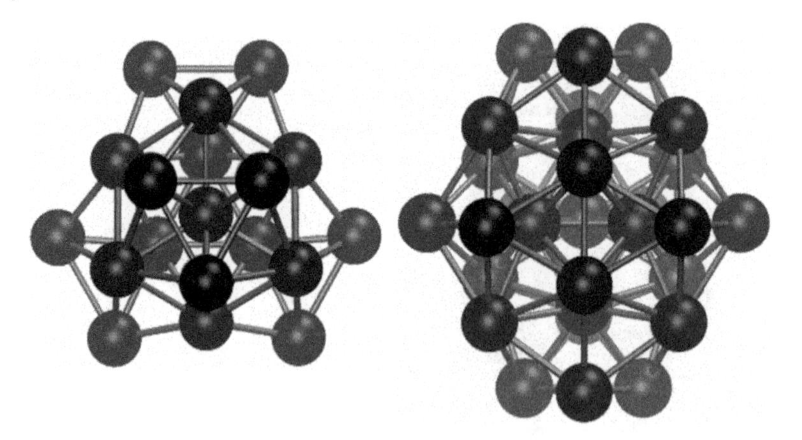

Figure 6.7 $N = 29$ Morse cluster.

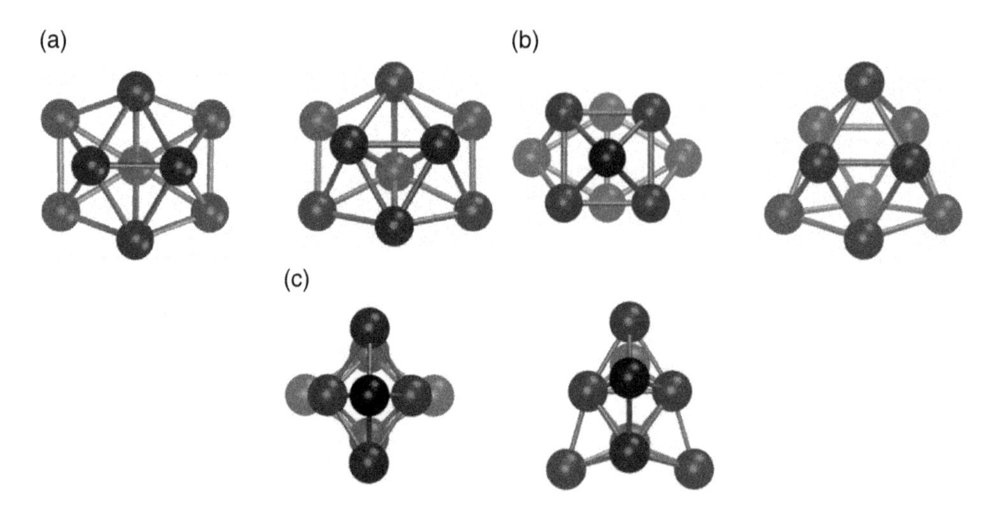

Figure 6.8 $N = 9$ structures: (a) global minimum of the Morse cluster ($E_B = 0.753\,44$ eV), (b) global minimum of MM ($E_B = 1.764$ eV), structurally corresponds to the second stable isomer of the Morse cluster ($E_B = 0.748\,16$ eV), and (c) another symmetric isomer of the Morse cluster ($E_B = 0.728\,31$ eV).

Examples of hypothetical and predicted flat carbon materials like graphyne [1], supergraphene [20], and squarographites [3] with their modular and periodic structures suggest that potentially another, similar stable arrangement of carbon atoms can be found. In combination with the unique properties of 2D carbon-based materials, these facts prompted [45] to develop a method for finding new types of stable 2D carbon networks.

Since stable configurations of atoms correspond to the global minima on PES, such a task can be considered as an optimization problem of an atomic carbon cluster. However, the number of local minima increases almost exponentially with the number of atoms in the considered structure, thus searching for the global minimum on a PES became a non-trivial, NP-hard problem.

Due to the high costs of computations, the full quantum ab initio methods are usually applied to relatively small atomic systems. On the other hand, classical optimization methods

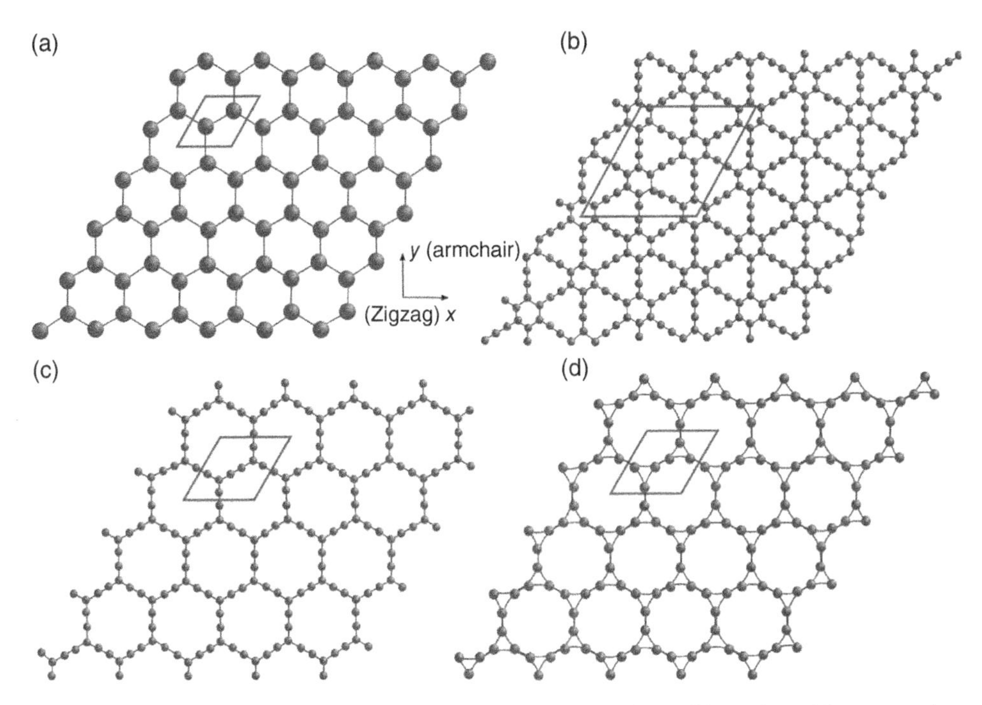

Figure 6.9 Two-dimensional, periodic carbon networks: (a) graphene, (b) graphyne, (c) supergraphene, and (d) structure made from dodecagons.

(e.g. gradient-based ones) have problems with multimodal functions and reveal tendencies to stuck-in local minima. That is why many heuristic and artificial intelligence–based approaches have recently been applied to deal with a similar problem of minimizing the potential energy of the atomic clusters. From the historical point of view, the first group of the methods combined searching for the PES with the simulation of certain physical processes. For example, the random search, the MC, and the simulated annealing methods were applied by [35] to investigate isolated atomic systems, called atomic clusters, while [64] used the basin-hopping MC to study the clusters with atomic interactions described by Morse and Lennard-Jones potentials. The second group of computational intelligence methods is inspired by biological mechanisms present in the natural environment and living organisms. The bioinspired optimization methods of atomic structures have become very popular in the past years. The work in [57] applied, probably for the first time, the GA [40] to the atomic clusters with the Morse potential. In a similar manner, [60] adopted an AIS [13] for optimization of the Lennard-Jones clusters and [70] minimized the potential energy of nickel clusters using the PSO [14].

Bioinspired algorithms, the DEA, AIS, and PSO, were successfully implemented for investigation of small aluminium clusters with various atomic potentials by [43, 47].

Searching for new graphene-like materials can be performed in a similar way; however, a more sophisticated interatomic interaction model is required in the form of the bond-order potential, which is able to handle various hybridization states of carbon atoms, allowing the formation of bondings with proper, neighbourhood-dependent geometry. Additionally, opposite to the perfectly isolated atomic clusters, the new algorithm should impose the periodicity of the created nanostructure.

The Memetic Algorithm (MA), which combines the parallel EA prepared, and the classical conjugated-gradient (CG) minimization of the total potential energy of the optimized atomic system are applied. Since the processed structure is considered as a discrete atomic model, the behaviour and the potential energy of carbon atoms are determined using the adaptive intermolecular reactive empirical bond order (AIREBO) potential [61].

The presented algorithm has a modular structure. Thus, each component can be replaced with a functional equivalent (e.g. EA with AIS, gradient optimization with molecular dynamics) or adapted to be used on new computer architectures: massively parallel supercomputers and GPGPU-based clusters.

This method can be extended to optimize three-dimensional molecular structures and may be considered as an alternative approach to the ab initio–based PSO algorithm called CALYPSO [66].

The algorithm of optimization of carbon-based clusters, presented in this section, is a continuation of investigations and modelling of atomic systems [42] and an expanded version of the approach applied in the minimization of energy of aluminium clusters [43].

The EA implements the mechanisms and rules known from biological evolution of species [40, 67]. In this particular application, the individuals contain only one chromosome with vectors of genes which represent design variables ($g_1, ..., g_m$) in the form of real-valued Cartesian coordinates of each atom in the considered unit cell of the newly created atomic lattice. Thus, the total number of design variables (genes) equals 2 m, where 2 is the dimension of the considered problem and m denotes the number of atoms. The chromosome structure is presented in Figure 6.10.

Each individual represents a certain spatial arrangement of atoms. In the initial population, atoms have randomly generated coordinates and are placed in the unit cell area with periodic boundaries. Dimensions, the rectangular or triclinic type of the unit cell, as well as the number of atoms, are part of a set of parameters of the simulation. Such an approach allows controlling the value of atomic density of the newly created structure. The periodicity of the atomic structure significantly reduces the number of design variables.

The crucial role in creating a new atomic structure is played by the fitness function, which is formulated as the total potential energy of the considered atomic system, i.e. the total sum of all potential energies of particular atomic interactions.

Choosing the proper interaction model is very important for this kind of task. In the presented case, the potential energy, as well as neighbourhood-dependent behaviour of the carbon atoms, is determined on the basis of the AIREBO potential for hydrocarbons [61]. This is an extended REBO model proposed in [2] with additional torsion and long-range terms. This kind of bond-order interaction mode is based on a set of neighbourhood-dependent,

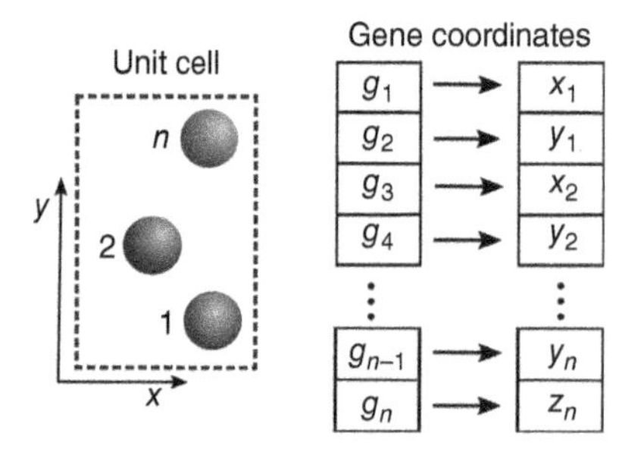

Figure 6.10 Structure of a chromosome.

switched, mathematical formulas, parameterized, in this case, to the properties of hydrocarbons. The AIREBO potential in the following form was used for the research (please refer to Section 2.1.2.4 for more detailed description):

$$FF = \sum_{i}\sum_{j\neq i}\left(E_{ij}^{REBO} + E_{ij}^{LJ} + \sum_{k\neq i,j}\ \sum_{l\neq i,j,k} E_{kijl}^{TORSION}\right). \tag{6.6}$$

The potential is used to compute both: the fitness function for EA and the objective function for the CG minimization where the term denoted by E^{REBO} corresponds to the short-range interactions between covalently bonded pairs of atoms. The long-range interactions in the AIREBO model are computed in a simplified way, using the Lennard-Jones-like function (term E^{LJ}) with additional distance-dependent switching functions, which expand the abilities to form different spatial configurations of carbon atoms. The last, torsional potential ($E^{TORSION}$) depends on the neighbouring atom's dihedral angles. All coefficients in (6.6) depend on design variables in the form of coordinates of atoms.

Other, more accurate approaches of modelling carbon–carbon interactions can be applied, e.g. the ReaxFF model, based on the first-principles calculations, developed in [15] or the second generation long-range carbon bond order potential (LCBOPII) [36]. However, the AIREBO potential is fitted to handle different spatial configurations and hybridization types of carbon atoms properly and is computationally more effective than the ReaxFF approach, which requires additional equilibration of the atomic charge every certain number of iterations [51, 56]. The application of the AIREBO potential to the examination of mechanical properties of various 2D graphene-like materials was performed in [42] and revealed good agreement with the results obtained by other researchers.

An important fact is the way in which generation of the initial positions of atoms may result in very small distances between them, even with overlapping coordinates. Such configurations result in high potential energies and are very unstable. A similar phenomenon may occur after mutation and crossover operations that are performed by the EA. That is why initial and offspring populations have to be equilibrated, i.e. the potential energy has to be minimized by correction of the positions of atoms before the first run of the EA. The CG algorithm is used for this purpose. The minimized functional is formulated by (6.6). This routine is invoked in each iteration of EA for all individuals in the processed population and temporarily pushes solutions into the local minima. Such an approach, combined with the bond-order interaction model, assists in forming of the new, real carbon-based molecular structure, i.e. during the conjugate gradient minimization; each individual, a certain spatial configuration of atoms, starts to form a unique, hybridization-dependent, geometry of flat carbon networks. This step ensures that EA does not process the sets of randomly placed atoms but operates on fragments of properly bonded carbon structures. Additionally, this method ensures that the optimized structure of atoms is properly equilibrated.

The coordinates of atoms are exchanged between EA and the LAMMPS software package, and the equilibration process is performed using the minimization method based on the Polak-Ribiere algorithm [55]. The periodicity of the newly created structure is also achieved in this step by proper boundary conditions imposed on the unit cell. After the CG minimization of the potential energy, the objective function is computed for each individual in the population. This part of the algorithm is also handled by the LAMMPS. The CG optimization is the most time-consuming part of the algorithm. To overcome this problem, the authors decided to parallelize the proposed algorithm and make it suitable for running on multiprocessor computers [30, 31]. Thus, the population is scattered into a certain number of parts using the MPI library. In the next step, each part is further processed in a parallel way using the dedicated instance of LAMMPS running on a separate core or node of the computer. Such minimized atomic structures, along with estimated values of the fitness function and modified design variables, are gathered together and imported to the EA, which performs the selection and invokes evolutionary operators such as the mutation and the crossover. The selection chooses chromosomes for a new parent subpopulation taking into account the values of the fitness function. Evolutionary operators change chromosomes' genes and create new chromosomes for the offspring population. The ranking selection, the simple crossover with the Gaussian mutation, and the uniform mutation are implemented in the presented algorithm. The simple crossover with the Gaussian mutation takes two chromosomes and exchanges some random number of genes between chromosomes, and the offspring chromosome is changed with the use of the Gaussian mutation. The mutation changes some randomly chosen genes' values. The Gaussian mutation changes the genes taking into account the normal distribution – in most cases, a small change of gene value is most likely. The uniform mutation can change the gene value and move an atom in one axis within the considered design space.

The main hybrid evolutionary-gradient algorithm works in an iterative way: the offspring population is scattered into parts again and exported to the CG routine, built-in LAMMPS. In the subsequent step, all the data are gathered and exported to the EA, which performs the selection of individuals and other evolution operations [40, 67]. These steps are repeated until the stop condition is satisfied. The stop condition can be formulated as a given maximum number of iterations or lack of improvement of the fitness functions during a certain time interval. The flowchart of the MA in the form of the hybrid parallel evolutionary algorithm (HPEA) is presented in Figure 6.11. The assumption of constant, predefined dimensions of the unit cell may result in a slightly stressed atomic system. To overcome this problem, the newly obtained structure is relaxed after the main optimization process. This operation is performed as an additional CG energy minimization with free boundaries of the unit cell. Such an approach allows introducing slight changes of the system's dimensions, combined with a further reduction of internal stresses and potential energy.

Validation of the accuracy of the results obtained using the presented approach, certain arrangements of carbon atoms already known from the literature, namely the supergraphene, the graphyne, and the structure made from dodecagons were examined [45]. The selected structures are already presented in Figure 6.9b–d, along with the corresponding unit cells. The configuration of the unit cell is non-unique and represents only one of the possible variants. This is an important fact since the proposed algorithm works on unit cells with prescribed shape and atomic density.

6.1.1.3 New Stable Carbon Networks *X* and *Y*

The described approach to optimization of atomic clusters was applied to the generation of two new stable carbon networks and presented in this section; numerical results of optimization of atomic clusters are based on authors' own results related to carbon atoms [34, 45].

In all of the presented cases, the population consists of 100 individuals. The probabilities of the Gaussian mutation with the simple crossover are equal to 10%, and the probability of the uniform mutation is 90%. The probabilities of operators lead to good exploration of the design space by the algorithm due to the high probability of the uniform mutation. The crossover with the Gaussian mutations also allows exploitation near some basins of attractions. The

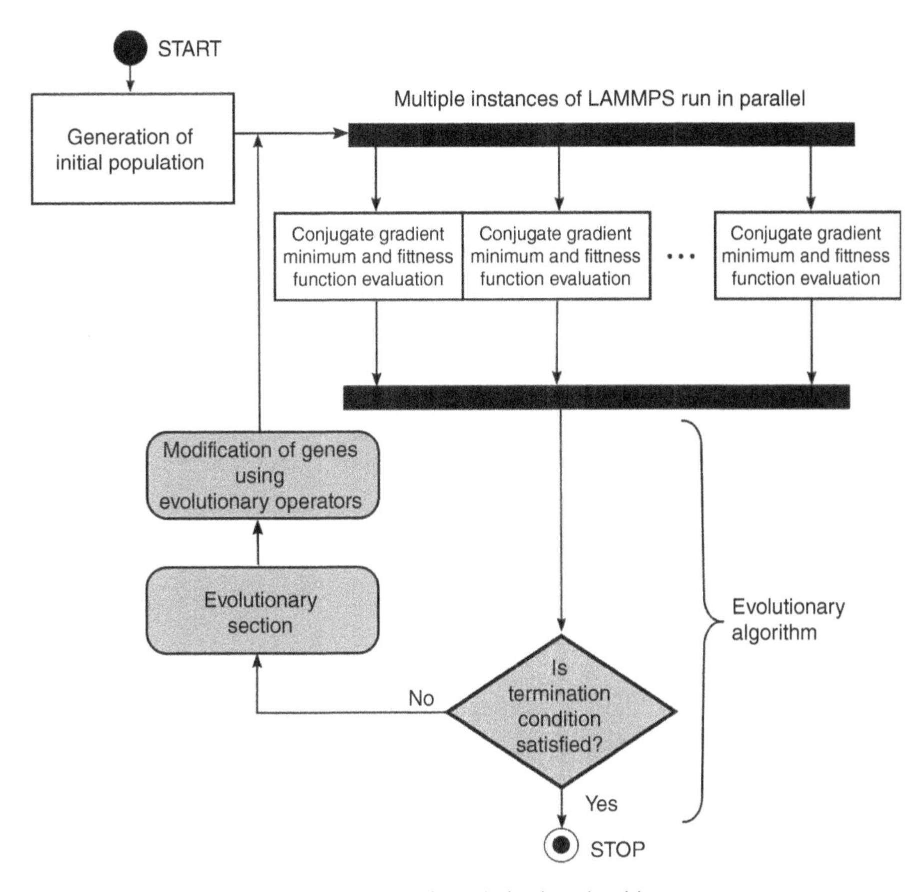

Figure 6.11 Block diagram of the memetic optimization algorithm.

probabilities are chosen on the basis of the authors' experience from previous research. The gradient-based energy minimization and the fitness function evaluation are performed in a parallel way, using four processes.

In the first case, the process of minimization of the total potential energy was carried out for eight carbon atoms placed in the 4 Å × 7 Å rectangular unit cell. The rest of the algorithm's parameters remained unchanged.

The result is a flat network named *X* made of sp^2 hybridized carbon atoms and is presented in Figure 6.12a. A more detailed view of the unit cell (dimensions: 3.94 Å × 6.82 Å – after final relaxing) is shown in Figure 6.12b. The analysis of bond lengths and the potential energy distribution revealed high symmetry in both horizontal and vertical directions. The progress of optimization is shown in Figure 6.13. The final arrangement of the atoms was obtained in the 223rd generation.

In further investigations of the newly obtained structure *X*, two simulations of tensile tests in perpendicular directions were performed. The MD method and the AIREBO potential are applied, and such a simulation can also be considered as an additional test of the structural stability of the atomic lattice. In the first step, the structure is carefully heated from 0 to 10 K with a constant rate of 0.1 K/ps. After that, the atomic system is stretched from 0 to 20% with a constant strain rate corresponding to the velocity equal to 10 m/s. The virtual stress

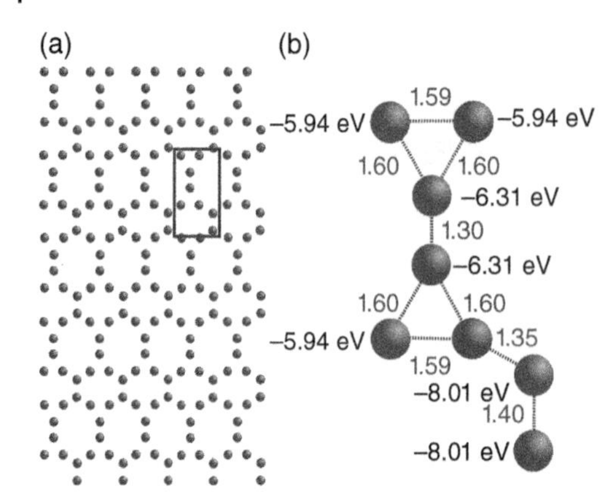

Figure 6.12 New carbon network *X* found by the optimization algorithm: (a) structure and (b) unit cell.

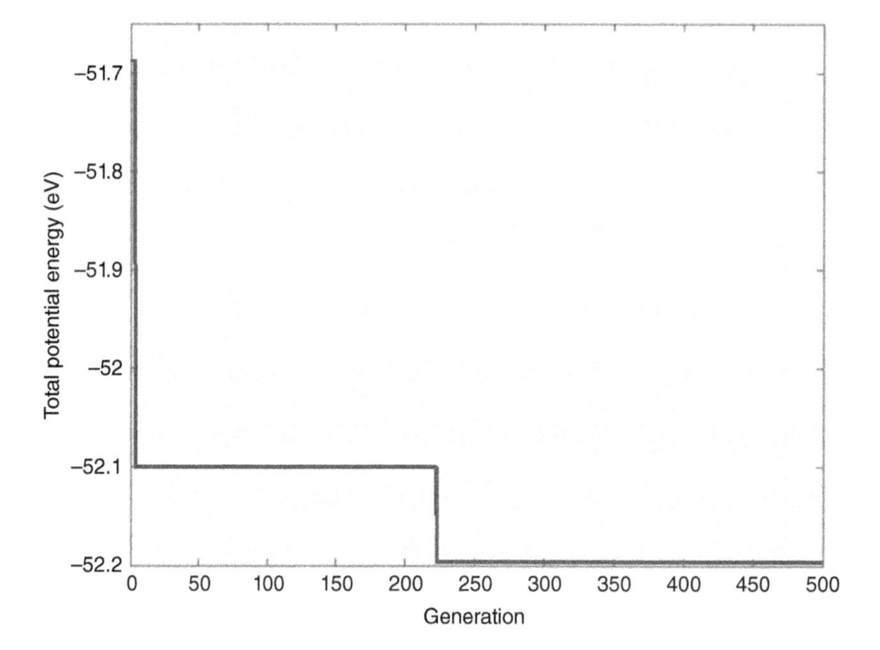

Figure 6.13 Progress of minimization of the total potential energy for the carbon network *X*.

theorem [69] is applied to compute the components of the stress tensor. A detailed description of this procedure is presented in [42].

Since the investigated structures are two-dimensional, all presented stresses and elastic modulus values are provided in the force per unit length (N/m) units. Such values can be expressed in commonly used force per unit area units (GPa) after the evaluation of the effective thickness of each type of the considered structure.

The obtained stress–strain relations in horizontal and vertical directions (according to Figure 6.12a) are presented in Figures 6.14 and 6.15, respectively.

The structure revealed similar behaviour in both directions, and the values of the Young modulus (estimated at 0–5% strain range) are equal to 176 N/m in horizontal and 184 N/m in

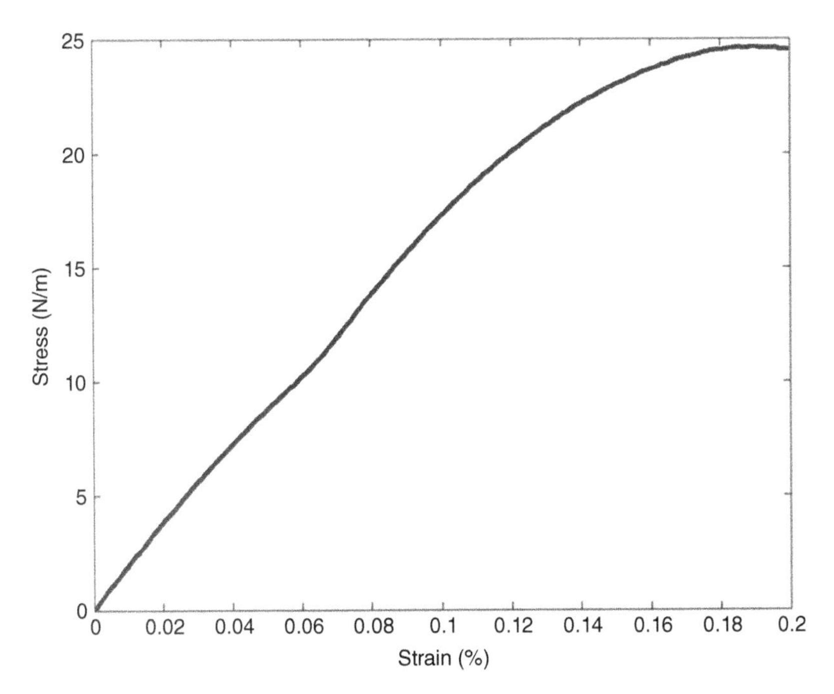

Figure 6.14 Stress–strain relation – horizontal direction for the carbon network *X*.

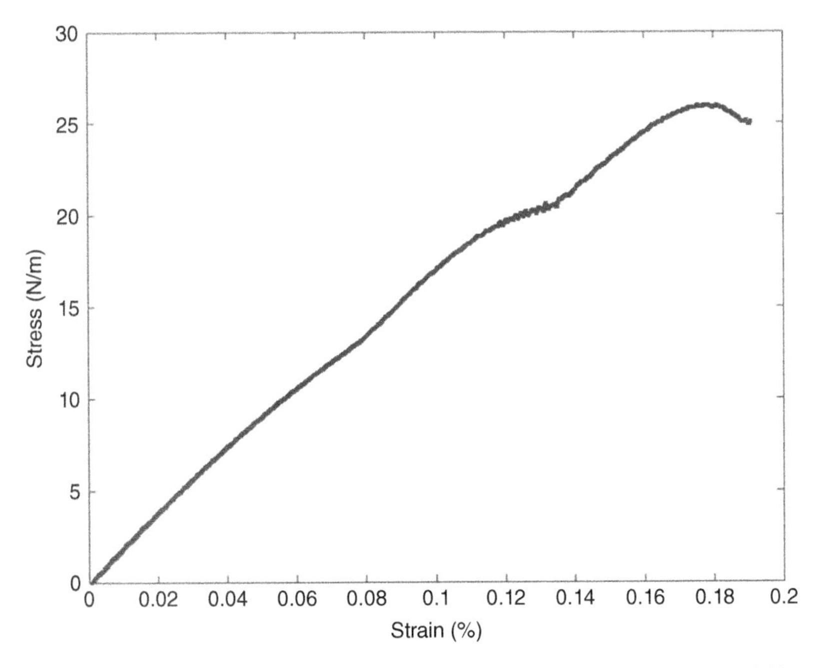

Figure 6.15 Stress–strain relation – vertical direction for the carbon network *X*.

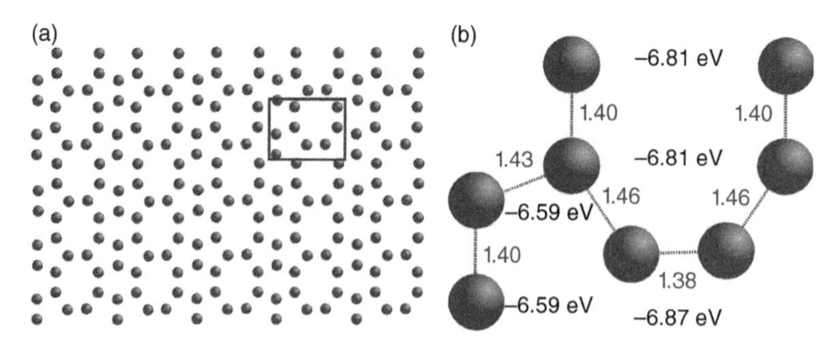

Figure 6.16 New carbon network *Y* found by the optimization algorithm: (a) structure and (b) unit cell.

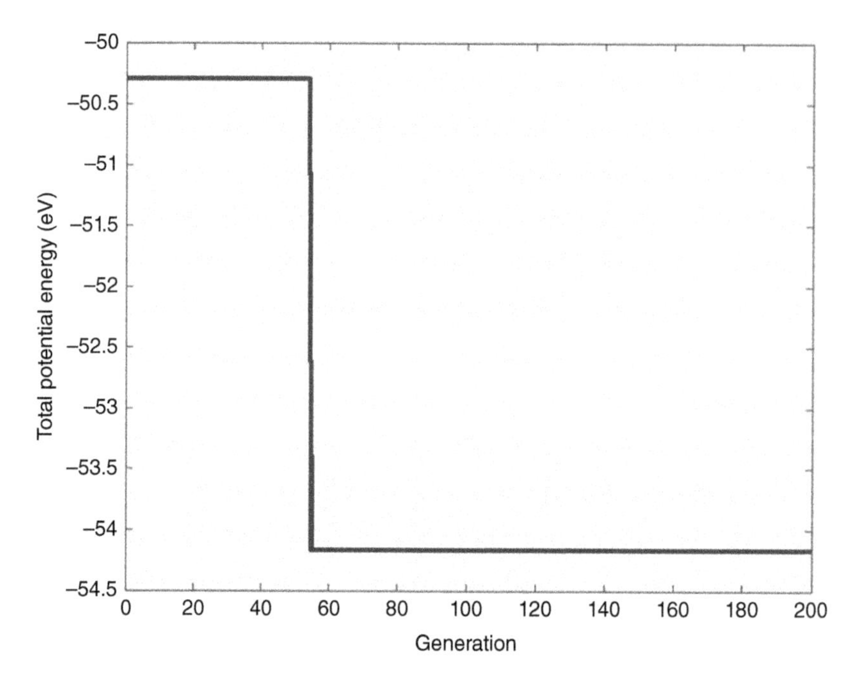

Figure 6.17 Progress of minimization of the total potential energy for carbon network *Y*.

vertical directions. For comparison, graphene, with a higher atomic density than the obtained lattice, has the value of the Young modulus of approx. 300–330 N/m (0.9–1 TPa, assuming that the effective thickness is equal to 3.35 Å), depending on model and computation method.

In the second numerical example [45], the dimensions of the rectangular unit cell are changed to 4 Å × 6 Å. The optimization process again yielded a flat network named *Y* made of sp^2 hybridized carbon atoms, symmetrical in horizontal and vertical directions (Figure 6.16a). A more detailed view of the unit cell (dimensions: 5.76 Å × 3.77 Å) along with bond lengths and potential energy distribution after the final relaxing is shown in Figure 6.16b. The proposed algorithm does not have problems with minimization of the total potential energy of the structure (Figure 6.17), and the final configuration of the atoms is obtained in the 55th generation. The optimization was dominated by the local search algorithm, but it can be observed that without the EA, the optimum could not be found.

However, opposite to the previous example, tensile tests, performed under analogical conditions as in the previous example, revealed anisotropic behaviour of this atomic network. Stress–strain relations in horizontal and vertical directions (according to Figure 6.12a) are presented in Figures 6.18 and 6.19, respectively. The values of the Young modulus (estimated at 0–5% strain range) are equal to 226 N/m in horizontal and 280 N/m in a vertical direction.

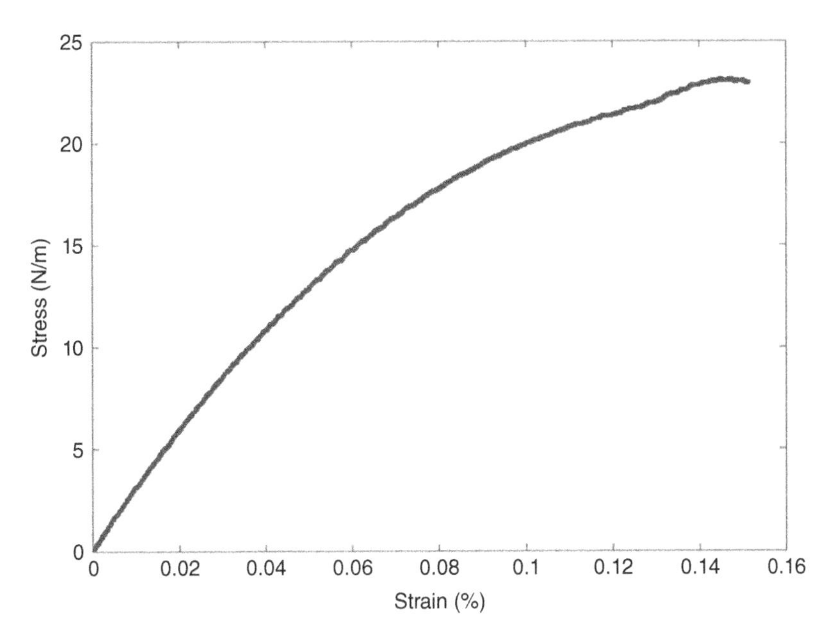

Figure 6.18 Stress–strain relation – horizontal direction for the carbon network *Y*.

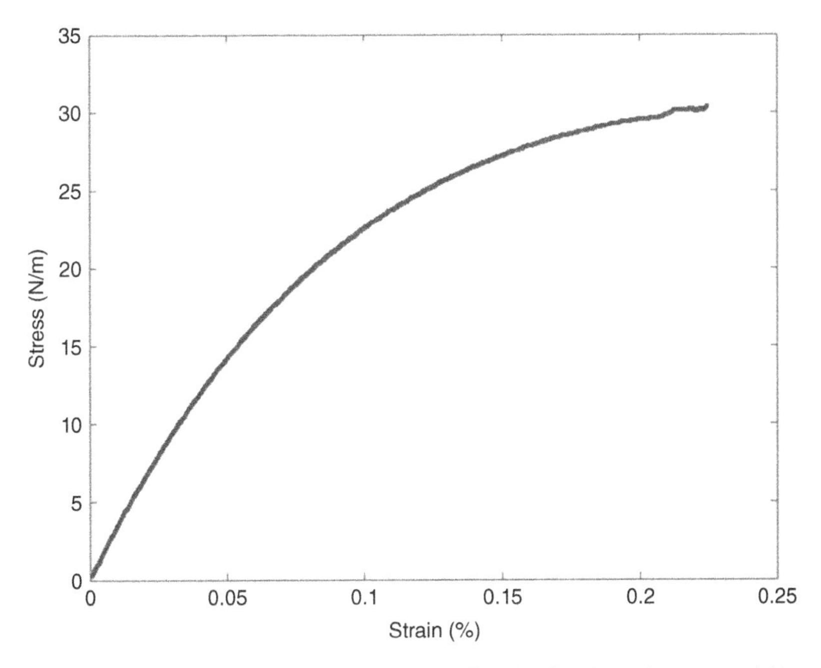

Figure 6.19 Stress–strain relation – vertical direction for the carbon network *Y*.

Such a disproportion is caused by the greater number of double bonds oriented along a vertical direction.

The computational time of the full run of the algorithm (eight atoms in the cell, maximal number of generations) is close to five hours on the four-core i7 2.66 GHz Linux-based personal computer.

The proposed optimization method of cluster atoms is able to find already-known structures like supergraphene and graphyne as well as new stable ones, named X and Y. The last two examples performed for new carbon networks X and Y clearly show that the final form and properties of optimized structures depend on the assumed size, type, and atomic density of the unit cell. Thus, the considered optimization problem can be reformulated, even using multi-criteria approach, and applied to searching for a molecular structure with predefined material properties (e.g. the stiffness tensor).

Every component of the presented algorithm can be replaced with a functional equivalent (e.g. optimization method, atomic potential); additionally, the proposed approach is ready to use in optimization of three-dimensional molecular structures. The parallel approach used in the most time-consuming parts of the algorithm, such as the CG minimization and evaluation of the fitness function, significantly reduces the time of computations.

The proposed 2D-graphene-like materials X and Y were thoroughly analyzed within the framework of the first principles by density functional theory (DFT) from the structural, mechanical, phonon, and electronic properties point of view using Cambridge Serial Total Energy Package (CASTEP). All calculations were performed with ultra-fine quality settings and the modified Perdew–Burke–Ernzerhof generalized gradient approximation for solids exchange-correlation functional being as accurate as the most advanced hybrid functional. The proposed polymorphs of graphene X and Y have been turned out mechanically and dynamically stable [38].

Optimization of atomic carbon clusters was also considered for some constrained conditions imposed on certain mechanical and electronic properties. The new graphene-like material with predefined stiffness parameters was considered in [46]. This proposed 2D-graphene-like material was also analyzed within the framework of the first principles by DFT from the structural, mechanical, phonon, and electronic properties point of view using CASTEP.

The proposed polymorph of graphene named *graphene AC* has been turned out mechanically and dynamically stable to other *cyclicgraphenes*. *Anisotropic-Cyclicgraphene* (AC) can be semiconducting, with a direct bandgap with a value of 0.829 eV [39].

6.1.2 Material, Shape, and Topology Optimization

The optimization problem in the case of multiscale modelling can be generally formulated as minimization of a given objective functional in one scale with respect to design variables in another scale of a structure.

The goal of optimization in multiscale modelling, considered in this section, is to find a vector (chromosome) **ch** of material or geometrical parameters (design variables) on the micro-level representative volume element (RVE), which minimizes an objective function $J_o = J_o(\mathbf{u}, \boldsymbol{\varepsilon}, \boldsymbol{\sigma})$ dependent on state fields of displacements $\mathbf{u}$, strains $\boldsymbol{\varepsilon}$, and stresses $\boldsymbol{\sigma}$ on the macro-level of the structure

$$\min_{ch} J_O, \tag{6.7}$$

where:

$$\mathbf{ch} = [x_1, x_2, ..., x_n], \tag{6.8}$$

is a vector (chromosome) of design variables; x_i are (genes).

For the practical problems, some kind of constraints are imposed in the form

$$\begin{aligned} &x_i^{\min} \le x_i \le x_i^{\max}, \quad i = 1, 2, ..., n \\ &J_\alpha(\mathbf{u}, \varepsilon, \sigma) \le 0, \quad \alpha = 1, 2, ..., A \end{aligned} \tag{6.9}$$

where $x_i^{\min}$ and $x_i^{\max}$ are minimum and maximum values of x_i, respectively, and J_α, $\alpha = 1, 2, ..., A$ are performance functionals.

The chromosome **ch** defines material properties and/or shape/topology of the microstructure components. Genes stored in **ch** can describe the micromodel material as the Young's moduli, Poisson's coefficients, friction coefficients, and parameters of the elastoplastic curve.

The numerical example of evolutionary multiscale optimization for a 2D structure made from the composite is considered (Figure 6.20a).

The optimization criterion is to minimize the maximal reduced displacement of the structure. The objective function is evaluated for each chromosome. The shape of the fibre is coded into chromosomes. The shape of the microstructure fibres in the RVE is described by using NURBS curves [54]. The NURBS are built on the top of a polygon with control points lying on the line ends. The coordinates of control points are used as design variables. The vector of parameters can directly reflect these variables. Coordinates of the NURBS control polygon points play the role of genes (Figure 6.20b).

The RVE is used as a microstructure. The periodic boundary conditions are applied (the periodicity of displacements is used). The own programme based on the computational homogenization utilizing *MSC. Nastran* FEM code [50] was used in the direct problems.

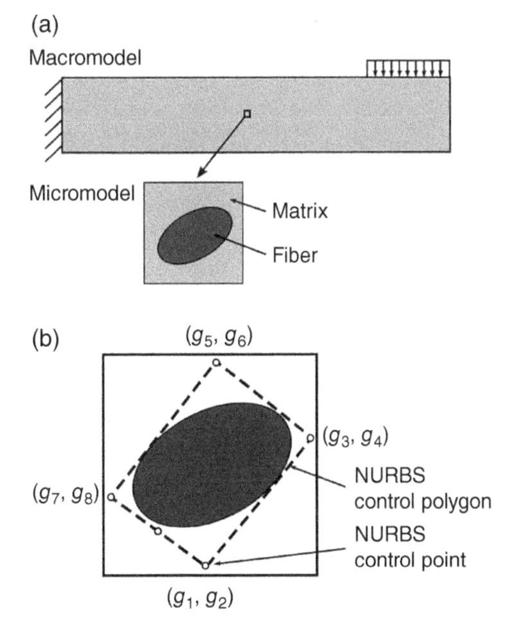

Figure 6.20 Multiscale optimization of a 2D structure: (a) the analyzed 2D beam, macromodel and micromodel and (b) the fibre shape definition using NURBS.

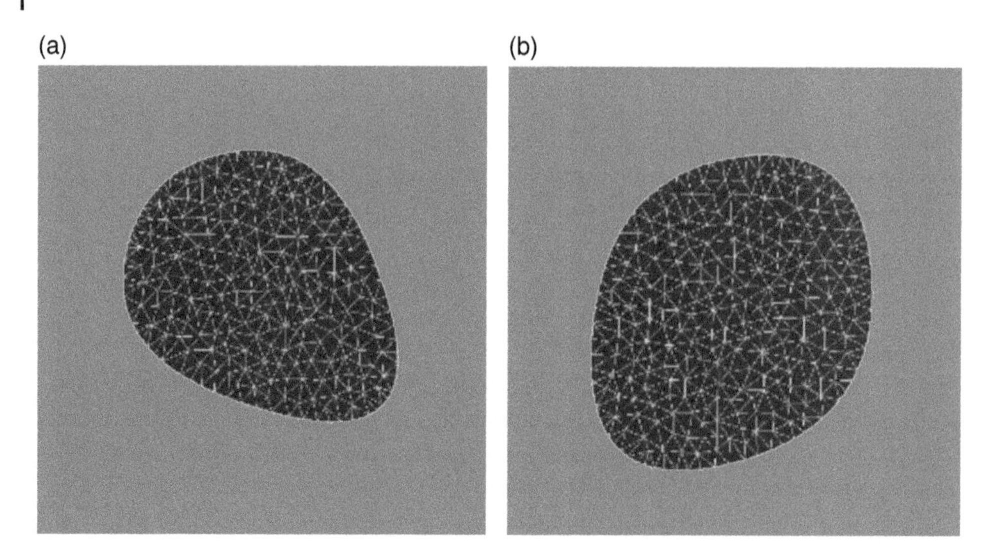

Figure 6.21 The best result of optimization (a) after the first generation and (b) after optimization.

Numerical results of optimization are shown in Figure 6.21. The best microstructures after the first and last generations are shown in Figure 6.21a and b, respectively. The simple crossover and the Gaussian mutation were applied in the optimization process.

Multiscale optimization for the 2D elastic body (Figure 6.22) was also considered [7, 8]. The body consists of reinforcements whose parameters are defined on the basis of microstructures.

The microstructure is modelled by means of the NURBS curve, whose control points play the role of design variables. The analysis problem was solved by means of FEM and computational homogenization.

Microstructure with the reinforcement modelled by means of NURBS with coordinates of control points g_1–g_8 is presented in Figure 6.23. Constraints imposed on design variables are presented in Table 6.3. The objective function represents the minimum displacement of the body in a macroscale. The constraint on the volume V_{max} of the reinforcement was imposed.

Multiscale optimization was performed using three algorithms: EA[29, 40], AIS [13], and swarm algorithm [26]. Parameters of applied algorithms are presented in Table 6.4. The number of chromosomes, the number of B-cells, and the number of particles were the same.

Numerical tests were repeated several times for each algorithm. The best solutions and the reference solution for the circular reinforcement are presented in Figure 6.24. The map of resultant displacements for the macromodel structure is presented in Figure 6.25.

The best results obtained by means of optimization algorithms are presented in Table 6.5.

Gradient and functionally graded materials play an important role in the design of new structures and devices [41]. Such materials have microstructures whose strength and thermal parameters have been designed for the special case. A variable microstructure adapted to the aim which should be fulfilled is the special characteristic of these materials.

The problem of multiscale optimization of the structure whose topology depends on local parameters of microstructure is considered. In the case when microstructures are the same in the whole domain of the body, the situation will be identical as in previous considerations [37]. But when parameters of microstructures are non-identical in different parts of the domain of the body, then one can obtain various topologies in the microscales (Figure 6.26).

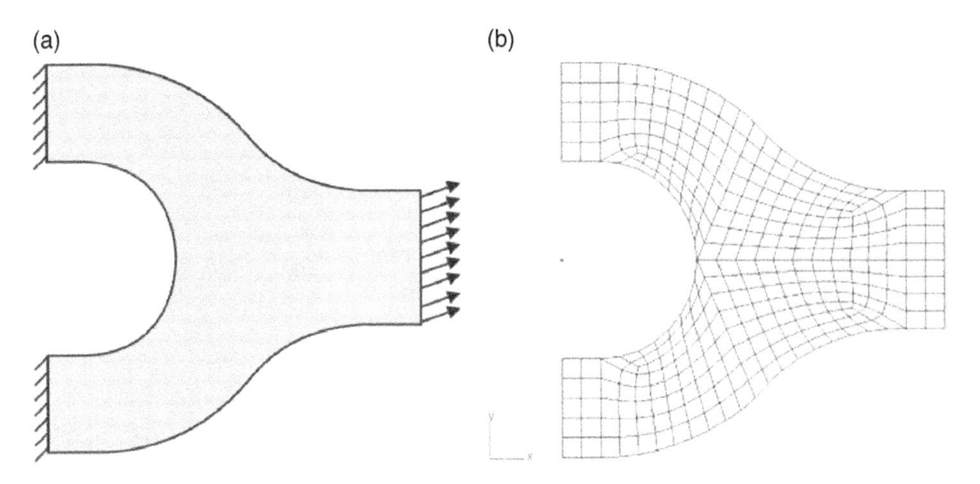

Figure 6.22 Elastic body with microstructure reinforcements: (a) model with boundary conditions and (b) FEM model.

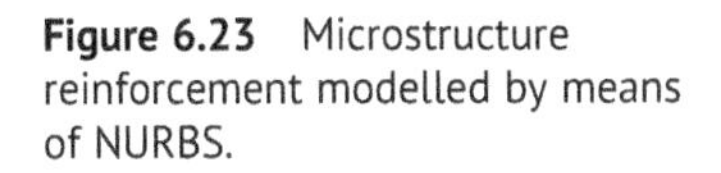

Figure 6.23 Microstructure reinforcement modelled by means of NURBS.

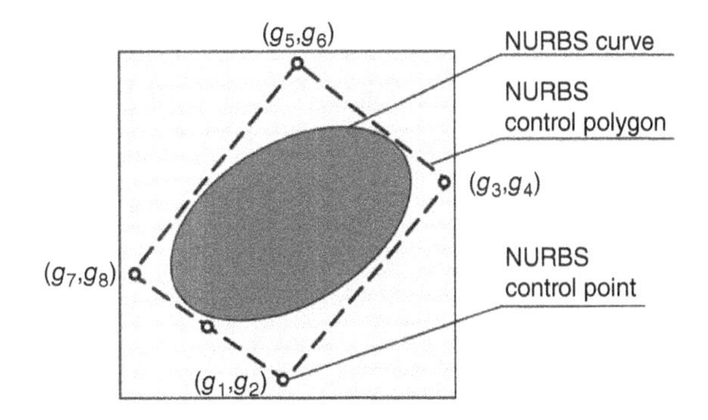

Table 6.3 Constraints imposed on design variables.

Design variables	Minimum	Maximum
g_1	0.10	0.47
g_2	0.10	0.47
g_3	0.50	0.90
g_4	0.10	0.47
g_5	0.50	0.90
g_6	0.50	0.90
g_7	0.10	0.47
g_8	0.50	0.90

The optimization problem is formulated as minimization of an objective function that depends on parameters of microstructures as design variables.

The topology of the body can be variable by changing material or shape parameters of microstructures in selected parts of the body, taking into account also the critical situation relying on void generation.

Table 6.4 Parameters of optimization algorithms.

Parameter	Value
Evolutionary algorithms	
Number of chromosomes	50%
Probability of Gauss mutation	90%
Probability regular mutation	10%
Probability of simple crossover	90%
Selection	Rank
Artificial immune system	
Number of memory cells	5
Number of B cells	50
Probability of Gauss mutation	100%
Swarm algorithm	
Number of particles in swarm	50
Impact parameter of the best particle in swarm	0.33
Impact parameter of the best particle in optimization	0.33
Impact parameter of the best neighbour	0.33

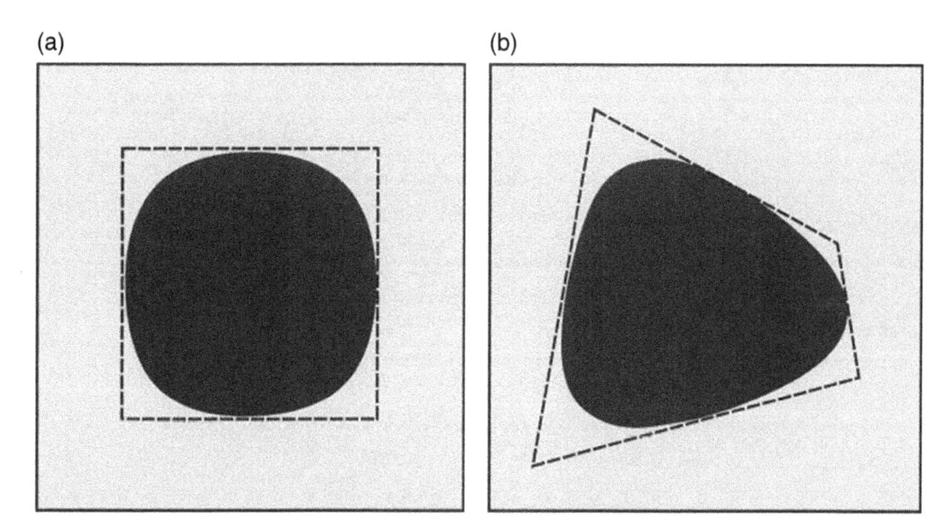

Figure 6.24 Microstructure (a) reference solution and (b) the best-obtained solution. The dashed lines represent the control polygons with control points in corners.

Consider the case of microstructures being composites with variable circular hard inclusions with diameter d as design variables (Figure 6.27). The aim of optimization is minimization of the hard phase in microstructures. The objective function represents aggregate input of hard phase in the body:

$$F = \sum_{z=1}^{n} \int_{A_Z} h \, dA_z, \tag{6.10}$$

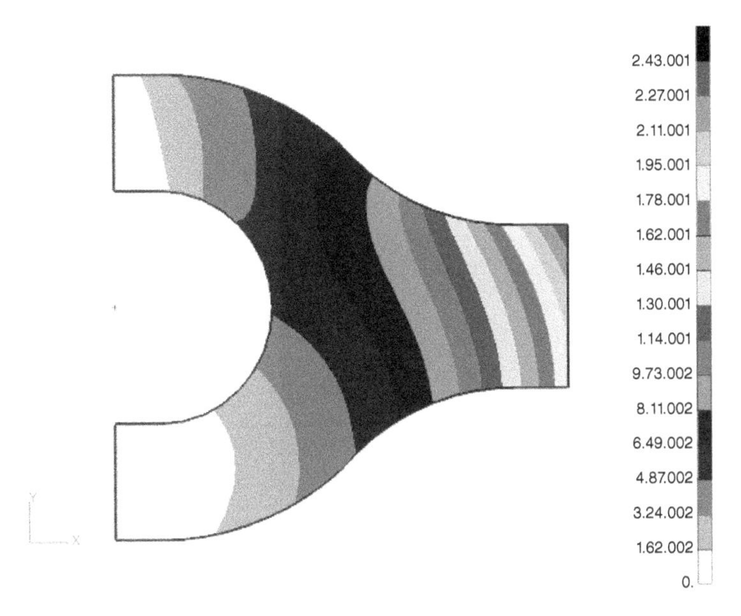

Figure 6.25 Distribution of resultant displacements for the best solution of the structure.

Table 6.5 Results of optimization.

Kind of algorithm	The best value of the objective function
Reference solution	0.2605
Evolutionary algorithm	0.2483
Artificial immune system	0.2433
Swarm algorithm	0.2435

where: $h = 1$ for an inclusion and $h = 0$ for matrix, z – the number of microstructures, A_z – an area of z microstructure, n is a number of microstructures.

The problem of multiscale optimization of the structure presented in Figure 6.28 for the minimization of the objective function (6.10) was performed. A constraint was imposed on maximum deflection:

$$u_i \leq u_{\max}, \tag{6.11}$$

where $u_{\max}$ is maximum admissible deflection.

Constraints imposed on diameters of inclusions equal $[0.1, 0.3]$ (the size of RVE 1×1). Six subdomains with different microstructures were assumed. Diameters of inclusions play the role of design variables in each subdomain. Computational homogenization was applied using the commercial package *MSC. Nastran* [50]. The optimization was performed by means

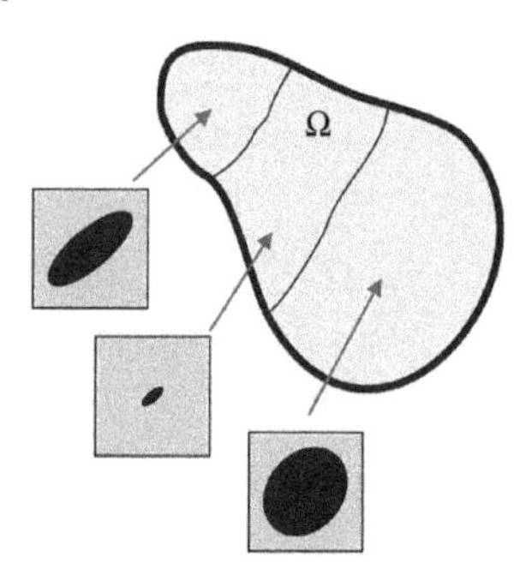

Figure 6.26 The structure with local periodical microstructures.

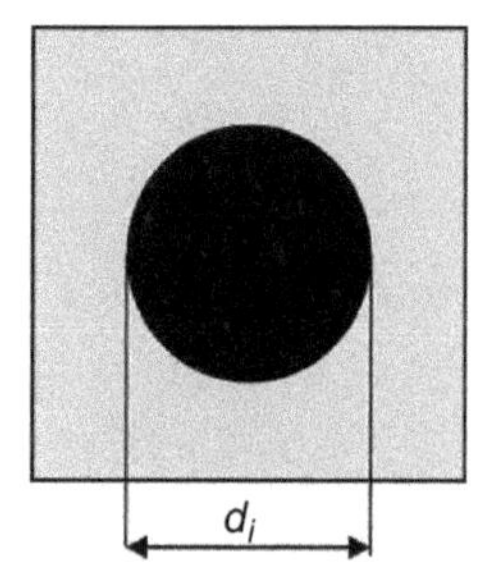

Figure 6.27 Microstructure with a circular inclusion with diameter *d*.

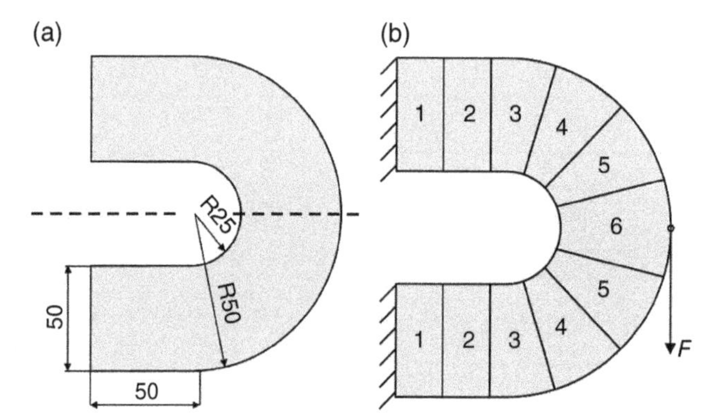

Figure 6.28 Considered structure: (a) geometry with dimensions and (b) boundary conditions.

of an EA [40]. The distribution of equivalent displacements for the optimal solution is presented in Figure 6.29.

Numerical results of diameters of inclusions for each subdomain are presented in Table 6.6.

6.2 Identification in Multiscale Modelling

The general aim of the identification in the case of multiscale description is the evaluation of some geometrical or material parameters of the structures in one scale having measured information in another scale. In this section, the case of two-scale problem of identification is considered. In this case, the goal of the identification is to obtain material properties or shape of the fibres in the microstructure, having measurements of state fields made on

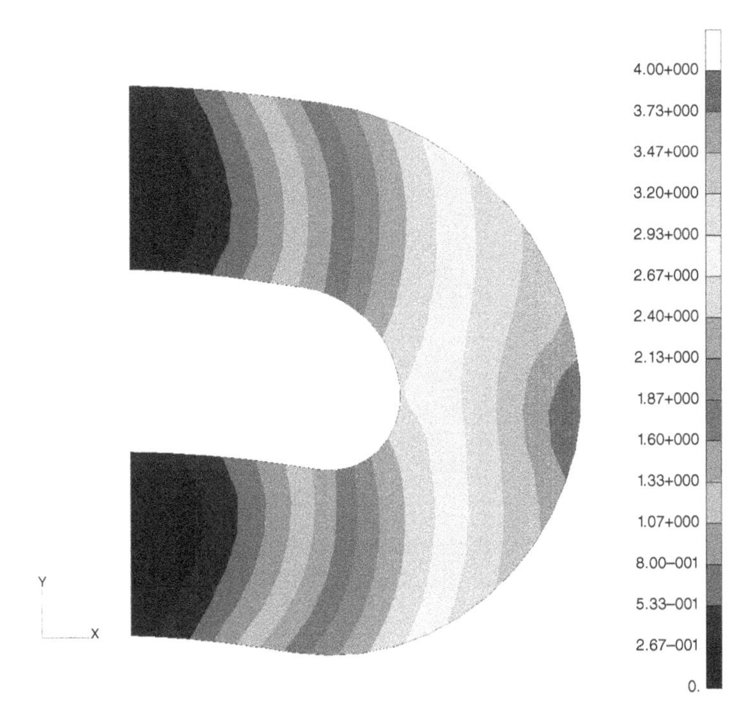

Figure 6.29 Distribution of resultant displacements in the body after optimization.

Table 6.6 Values of design variables.

Number of subdomains	Diameter of an inclusion
1	0.187
2	0.137
3	0.123
4	0.104
5	0.143
6	0.101

the macroscopic object. The measurement data in the macroscale are collected in sensor points. The measured strains or displacements are used in the identification process. The identification is performed iteratively, and the qualities of microscopic material parameters or shape have to be evaluated for each considered solution. The body with sensors in the real object and the computational model are shown in Figure 6.30.

The objective function J_o in identification problems is expressed as an absolute difference between measured and computed values of strains or displacements in sensor points.

$$J_o = a\sum_{i=1}^{m}|\hat{u}_i - u_i| + b\sum_{j=1}^{s}\left|\hat{\varepsilon}_j - \varepsilon_j\right|, \tag{6.12}$$

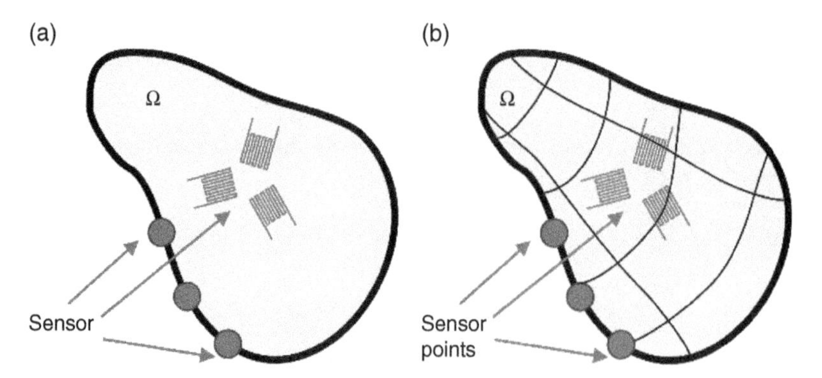

Figure 6.30 Sensors in (a) the real object and (b) the computational model.

where: m – number of displacement sensor points, s – number of strain sensor points, $\hat{u}_i$ – a measured displacement in a sensor point i, u_i – a displacement value in a sensor point i or the vector of parameters **ch** obtained by solving the direct problem, $\hat{\varepsilon}_j$ – a measured strain in a sensor point j, ε_j – a strain value in a sensor point j for the vector of parameters **ch** obtained by solving the direct problem, a and b – scaling coefficients.

Either different strain components or an equivalent strain and either displacements in different directions or a reduced displacement may be used in the objective function (6.12). Scaling coefficients are important when both displacements and strains are measured, and they should be carefully selected to allow the same influence of strains and displacements on the objective function value. The identification can also be performed using only displacements or only strains. The number and distribution of sensor points have an impact on the uniqueness of inverse problem solutions. It is very difficult to give a proper number of sensor points for identification problems because it depends on the shape of the macromodel, shape of the micromodel, material parameters of the micromodel, and boundary conditions in macroscale.

6.2.1 Material Parameters Identification

The identification of material properties of the microstructure is considered. The unknown parameters are obtained on the basis of measured displacements in the macromodel. The rectangular shell, clamped on the left side, and loaded as shown in Figure 6.31a is taken into account. The microstructure RVE model is presented in Figure 6.31b. The dark (red) and white colours show the two different materials. The elastic materials in the microscale are used. The chromosome contains four genes, each connected with one material parameter (two Young's moduli and two Poisson's coefficients). The reference displacement values in sensor points are generated using the numerical model; the glass fibre and the epoxy matrix material parameters were used. The 70 displacements are considered, 35 in x axis and 35 in y axis. The distribution of sensor points can be seen in Figure 6.31a. The macrostructures and microstructures were analyzed using FEM for each chromosome. The programme utilizing *MSC. Nastran* FEM code [50] was used in the direct problems. The material data for the macroscopic model were obtained using computational homogenization. The RVE model was 3D, and the macromodel 2D shell. The DEA was used, while the number of subpopulations was selected to 2, with the number of

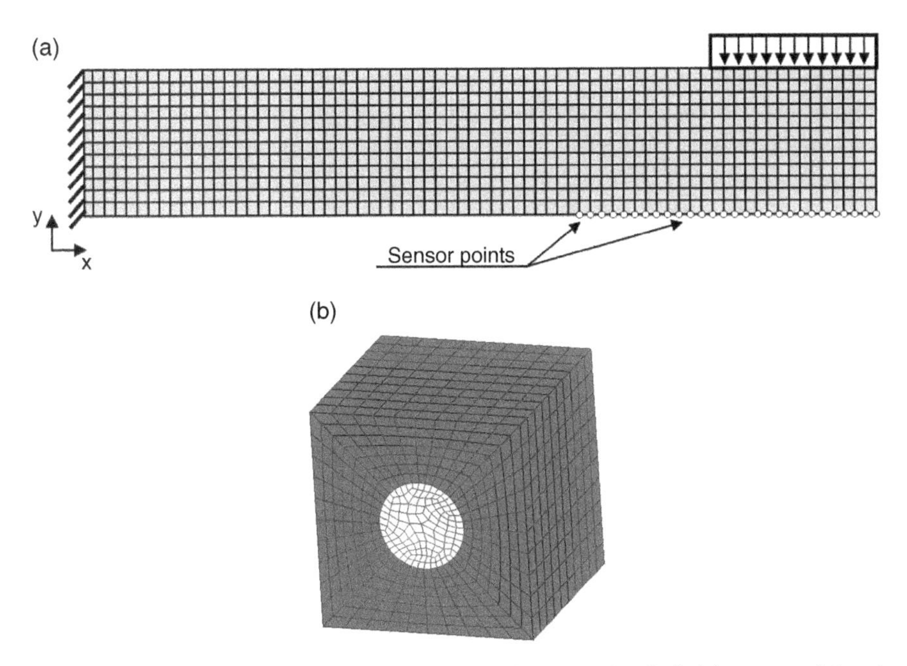

Figure 6.31 The considered model of the clamped rectangular shell: (a) macromodel and (b) microstructure.

Table 6.7 Results of material parameters identification.

Material parameter	Actual value	Found value
Epoxy Young's modulus E (MPa)	1 103	1 123
Epoxy Poisson's ratio	0.30	0.31
Glass Young's modulus E (MPa)	124 105	132 370
Glass Poisson's ratio	0.25	0.25

chromosomes equal to 10. The simple crossover and the Gaussian mutation were applied in the identification process.

The results of identification are shown in Table 6.7. The actual value is a material parameter value used to prepare the numerical experiment, and the found value is the best-found value of the parameter by the EA. The results were obtained after one hour using two processors (each with 6GFLOPs peak performance).

6.2.2 Multiscale Identification Problem in Stochastic Conditions

Very often, material and geometrical parameters of the structure have uncertain nature in the microscale. One of the most important models of uncertainty used in identification problems is based on the theory of probability and stochastic processes [10].

The methodology presented in this section takes into account the stochastic nature of parameters in the microscale, and the identification problem is formulated as the minimization of a certain stochastic objective function. The problem is transformed into a deterministic one in which a new objective functional dependent on mean values and variances is minimized with respect to moments of stochastic parameters. An approach based on evolutionary computing is presented in the minimization problem. Computational homogenization [62] is used for multiscale modelling of the structures. The multiscale analysis is performed for structures with stochastic material or geometry which occur in the RVE.

Suppose that stochastic material and geometrical parameters of the RVE are described by random variables $X_i(\gamma)$, $i = 1, 2, ..., n$, $\gamma \in \Gamma$, where Γ is the space of elementary events which represents all the possible simplest outcomes of a trial associated with the given random phenomenon. In the theoretical model of random phenomena, the basic role is played by the probability space (Γ, F, P), where F is a σ-algebra of subset of Γ. Elements of F are called random events, and P is a probability defined on F [52]. A random variable $X_i = X_i(\gamma)$, $\gamma \in \Gamma$ is defined on a sample space Γ and measurable with respect to P, i.e. for every real number x_i, the set $\{\gamma \ : \ X_i(\gamma) < x_i\}$ is an event in F.

A random vector:

$$\mathbf{X}(\gamma) = [X_1(\gamma), X_2(\gamma), ..., X_i(\gamma), ..., X_n(\gamma)] \tag{6.13}$$

is a function, measurable respect to P, which takes every element $\gamma \in \Gamma$ into a point $\mathbf{x} \in R^n$ and has an n-dimensional Gaussian distribution of the probability density function given as follows:

$$p(x_1, x_2, ..., x_i, ...x_n) = \frac{1}{(2\pi)^{n/2}\sqrt{|\mathbf{K}|}}\left[-\frac{1}{|\mathbf{K}|}\left|K_{ij}\right|(x_i - m_i)\left(x_j - m_j\right)\right]. \tag{6.14}$$

$|\mathbf{K}| \neq 0$ is the determinant of the matrix covariance, $\mathbf{K}[k_{ij}]$, $i, j = 1, 2, ..., n$, where $k_{ij} = \mathbf{E}[(X_i - m_i)(X_j - m_j)]$, $|K_{ij}|$ is the co-factor of the element k_{ij} in the matrix $\mathbf{K}$, and $m_i = \mathbf{E}[X_i(\gamma)]$ is the mean value of $X_i(\gamma)$. $\mathbf{E}[\cdot]$ indicates expectation or ensemble average.

It is assumed that random parameters are independent random variables. The joint probability density function is expressed by the probability density function of single random parameters as follows:

$$p(x_1, x_2, ..., x_i, ...x_n) = p_1(x_1)p_2(x_2)...p_i(x_i)...p_n(x_n). \tag{6.15}$$

Here,

$$p_i(x_i) = N(m_i, \sigma_i) = \frac{1}{\sigma_i\sqrt{2\pi}}\exp\left[-\frac{(x_i - m_i)^2}{2\sigma_i^2}\right] \tag{6.16}$$

is the probability density function of the random parameter $X_i(\gamma)$, where $\sigma_i^2 = \mathbf{E}[(X_i - m_i)]^2$ is the variance and σ_i denotes the standard deviation of $X_i(\gamma)$.

It is seen that if the random parameters $X_i(\gamma)$, $i = 1, 2, ..., n$, are random independent Gaussian variables, two moments – the mean value m_i and the standard deviation σ_i (or the variance σ_i^2) – describe the probability density function of each random variable $X_i(\gamma)$.

The random vector (6.13) now can be replaced by a deterministic vector:

$$Y = \left[(m_1, \sigma_1^2), (m_2, \sigma_2^2), ..., (m_i, \sigma_i^2), ..., (m_n, \sigma_n^2)\right] \tag{6.17}$$

which is described by moments m_i and $\sigma_i^2, i = 1, 2, ..., n$.

The goal of identification in multiscale modelling is to find a vector of material and geometrical parameters $\mathbf{X}$ treated as design variables on the microlevel which minimize an objective function $J_o = J_o(\mathbf{u}, \varepsilon)$ dependent on state fields of displacements $\mathbf{u}$ and strains ε on the macrolevel of the structure [7]. In the considered problem, the vector $\mathbf{X} = \mathbf{X}(\gamma)$ contains random parameters which should be determined from experimental statistical data.

One assumes that available measurement data of random displacements $\hat{\mathbf{u}}(\mathbf{x}_k, \gamma) \equiv \hat{\mathbf{u}}_k(\gamma)$ at sensor points (Figure 6.30) $\mathbf{x}_l$, $l = 1, 2, \ldots, K$, and strains $\hat{\varepsilon}(\mathbf{x}_l, \gamma) \equiv \hat{\varepsilon}_l(\gamma)$ at sensor points $\mathbf{x}_l$, $l = 1, 2, \ldots, L$, are characterized by the mean values $m_{\hat{\mathbf{u}}_k} = \mathbf{E}\hat{\mathbf{u}}_k(\gamma)$ and $m_{\hat{\varepsilon}_l} = \mathbf{E}\hat{\varepsilon}_l(\gamma)$ and variances $\sigma^2_{\mathbf{u}_k} = \mathbf{E}[\hat{\mathbf{u}}_k(\gamma) - m_{\hat{\mathbf{u}}_k}]^2$ and $\sigma^2_{\hat{\varepsilon}_l} = \mathbf{E}[\hat{\varepsilon}_l(\gamma) - m_{\hat{\varepsilon}_l}]^2$.

The identification problem is solved by converting it into a constrained minimization problem. A functional is constructed which represents a distance between measured $\hat{\mathbf{u}}_k(\gamma)$ and $\hat{\varepsilon}_l(\gamma)$ and theoretical values of displacements $\mathbf{u}(\mathbf{x}_k)$ and strains $\varepsilon(\mathbf{x}_l)$:

$$J = a\sum_{k=1}^{K} |\mathbf{u}(\mathbf{x}_k) - \hat{\mathbf{u}}_k(\gamma)| + b\sum_{l=1}^{L} |\varepsilon(\mathbf{x}_l) - \hat{\varepsilon}_l(\gamma)|, \tag{6.18}$$

where a and b are scaling coefficients.

Unknown material and geometrical parameters at the microlevel are described by random variables $\mathbf{X}(\gamma) = [X_i(\gamma)]$, $i = 1, 2, \ldots, N$. It means that $J = J(\mathbf{X})$.

The problem of finding parameters $\mathbf{X}(\gamma)$ is formulated as the nonlinear stochastic programming problem, which is stated as follows:

Find a vector $\mathbf{X}(\gamma) = [X_i(\gamma)]$, $i = 1, 2, \ldots, n$, which minimizes the objective functional $J = J(\mathbf{X})$ with imposed constraints $x_i^- \le X_i \le x_i^+$.

Here, x_i^- and x_i^+ are the limits of X_i.

The stochastic problem stated previously can be converted into an equivalent deterministic task. The objective functional $J = J(\mathbf{X})$ can be approximated as follows:

$$J(\mathbf{X}) \approx J(m_{\mathbf{X}}) + \sum_{i=1}^{N} \left(\frac{\delta J}{\delta X_i}\bigg|_{m_{\mathbf{X}}} \right)^2 (X_i - m_{\mathbf{X}}) = J_o(\mathbf{X}). \tag{6.19}$$

$J_o(\mathbf{X})$ being a linear function of normally distributed variables $X_i(\gamma)$ also has the normal distribution, which is described by the mean value and variance as follows:

$$m_{J_o} = J_o(m_{\mathbf{X}}) \quad \text{and} \quad \sigma^2_{J_o} = \sum_{i=1}^{N} \left(\frac{\delta J}{\delta X_i}\bigg|_{m_{\mathbf{X}}} \right)^2 \sigma^2_{X_i}. \tag{6.20}$$

Now a new deterministic objective functional can be defined as follows:

$$I = c_1 m_{J_o} + c_2 \sigma^2_{J_o}, \tag{6.21}$$

where c_1 and c_2 are non-negative scaling weights indicating the relative importance for minimization of the mean m_{J_o} and the variance $\sigma^2_{J_o}$. Setting $c_2 = 0$ would mean that the mean value of m_{J_o} is to be minimized with no regard to the variance, while the choice $c_1 = 0$ would imply that one is interested in minimizing the dispersion about an arbitrary mean value. The case $c_1 = c_2 = 1$ attaches equal importance to both characteristics.

Now the problem is formulated as follows:

$$\min_{Y} I$$
$$\text{with constraints } m_i^- \le m_i \le m_i^+ \quad \text{and} \quad \sigma_i^{2-} \le \sigma_i^2 \le \sigma_i^{2+}, \tag{6.22}$$

where m_i^-, m_i^+ and σ_i^{2-}, σ_i^{2+} are limits of mean values and variances, respectively.

Thus, the original stochastic programming problem is reduced to the nonlinear deterministic problem, which can be solved using the well-known standard procedure, but the main problem consists in the calculation of the derivatives $\delta J/\delta X_i$, $i = 1, 2, ..., N$. It is caused by the fact that in the general situation, it is impossible to express J in the explicit form with respect to random parameters $X_i(\gamma)$, $i = 1, 2, ..., N$. The problem of finding derivatives $\delta J/\delta X_i$ can be solved using the idea of stochastic shape sensitivity analysis [5].

If one assumes that

$$\begin{aligned} c_1 a &= \frac{a_1}{|m_{\hat{\mathbf{u}}}|} \quad \text{and} \quad c_1 b = \frac{b_1}{|m_{\hat{\varepsilon}}|} \\ c_2 a^2 &= \frac{a_2}{|\sigma^2_{\hat{\mathbf{u}}}|} \quad \text{and} \quad c_2 b^2 = \frac{b_2}{|\sigma^2_{\hat{\varepsilon}}|} \end{aligned} \tag{6.23}$$

where a_1, a_2, b_1 and b_2 are new scaling coefficients, then the deterministic objective function I can be expressed as follows:

$$I = a_1 \sum_{k=1}^{K} \left| \frac{m_{u_k} - m_{\hat{u}_k}}{m_{\hat{u}_k}} \right| + a_2 \sum_{k=1}^{K} \left| \frac{\sigma^2_{u_k} - \sigma^2_{\hat{u}_k}}{\sigma^2_{\hat{u}_k}} \right| + b_1 \sum_{l=1}^{L} \left| \frac{m_{\varepsilon_l} - m_{\hat{\varepsilon}_l}}{m_{\hat{\varepsilon}_l}} \right| + b_2 \sum_{l=1}^{L} \left| \frac{\sigma^2_{\varepsilon_l} - \sigma^2_{\hat{\varepsilon}_l}}{\sigma^2_{\hat{\varepsilon}_l}} \right|. \tag{6.24}$$

The objective functional (6.24) incorporates the differences of mean values and variances of displacements and/or strains obtained from numerical analysis and measurements. The mean values and variances for the numerical analysis can be obtained by solving a direct problem with the use of computational stochastic methods as the stochastic finite element method [27] or the stochastic boundary element method [11].

Averaged material properties for the RVE are obtained with the use of the homogenization method based on the stochastic FEM analysis. The MC based FEM method is also applied. The main disadvantage of the MC is the large number of FEM computations needed to obtain stochastic results. The material properties of the composite are randomly generated with the prescribed mean value and variance. The Muller-Box randomization is used to create random material properties. The homogenization method is used for each set of material properties in the MC method. The results obtained in each run of MC method are collected and used to present stochastic results of the analysis. The presented approach is very time-consuming. The multiscale analysis can be performed in a parallel way. The averaged material properties are obtained on the basis of six independent FEM analyses of RVE (for 3D case). This step can be easily parallelized with speedup close to linear. The MC method also can be parallelized; the multiscale analyses can be performed for each set of material parameters. The stochastic FEM analysis can be also shortened significantly by using other stochastic methods like the perturbation method [25, 58].

To minimize the objective functional I (6.19), the EA is used as an optimization method [6, 32, 33].

As an example of identification of random parameters, the two-scale structure is considered (Figure 6.31). The microstructure is built from two materials with stochastic parameters: Young's moduli $E_1(\gamma)$ and $E_2(\gamma)$ and Poison's ratios $\nu_1(\gamma)$ and $\nu_2(\gamma)$.

The mean values of Young's moduli m_{E_i}, $i = 1,2$ and Poisson's ratios m_{ν_i}, $i = 1,2$ are considered to be known, and the Poisson's variances $\sigma^2_{\nu_i}$, $i = 1,2$ are also known (see

Table 6.8 The material properties for the microscale.

Material parameter	Value
Young's modulus mean value m_{E_1}	3.4 (MPa)
Young's modulus mean value m_{E_2}	72.0 (MPa)
Poisson's ratio mean value m_{ν_1}	0.18
Poisson's ratio mean value m_{ν_2}	0.20
Poisson's ratio variance $\sigma^2_{\nu_1}$	0.000 625
Poisson's ratio variance $\sigma^2_{\nu_2}$	0.000 625

Table 6.8). The goal of the identification is to find variances of Young's moduli $E_1(\gamma)$ and $E_2(\gamma)$:

$$Y = \left[\sigma^2_{E_1}, \sigma^2_{E_2}\right] \tag{6.25}$$

with constraints

$$\sigma^{2-}_{E_i} \leq \sigma^2_{E_i} \leq \sigma^{2+}_{E_i}, \; i = 1, 2$$

where limits of variances are given as follows:

$$\sigma^{2-}_{E_1} = 0.0025 \; [\text{MPa}]^2; \; \sigma^{2+}_{E_1} = 2.2500 \; [\text{MPa}]^2$$

and

$$\sigma^{2-}_{E_2} = 0.25 \; [\text{MPa}]^2; \quad \sigma^{2+}_{E_2} = 225.00 \; [\text{MPa}]^2.$$

The identification is performed on the basis of stochastic values of displacements for the macromodel. The number of displacements sensor point was $K = 70$, and displacements were measured in two directions on the lower part of the macromodel. The objective function described by (6.19) was applied with parameters $a_1 = a_2 = 1, b_1 = b_2 = 0$. The stochastic direct problem was solved by using MC FEM. The *MSC. Nastran* [50] was applied for single FEM analysis in microscales or macroscales. The homogenization procedure was parallelized in the presented approach.

The minimization of the functional I (6.19) with respect to Y (6.25) was performed with the use of the EA. The simple crossover combined with the Gaussian mutation and the uniform mutation and the ranking selection were used. The parameters of the EA are shown in Table 6.9. The operators and their probabilities were chosen on the basis of previous numerical experiments based on mathematical test functions and test engineering problems.

The statistically measured displacements were numerically simulated for testing purposes. The exact results were known before the identification process.

The results of identification after 10 iterations of the EA are presented in Table 6.10. The obtained variances of Young's moduli are close to the actual ones. The change of properties of material number 2 had a bigger influence on the displacements in the macroscale, and the results obtained for this material are closer to the actual one.

The number of identified parameters influences the number of genes and iterations of the EA. The cost of computations will be higher in case of an increasing number of identified parameters.

Table 6.9 The evolutionary algorithm parameters.

Parameter	Value
Number of genes	2
Number of chromosomes	10
Probability of Gaussian mutation and simple crossover	0.9
Probability of uniform mutation	0.1
Ranking selection pressure	0.8

Table 6.10 The results of identification.

Parameter	Actual value	Obtained value
Young's modulus variance $\sigma_{E_1}^2$	0.25 $(MPa)^2$	0.14 $(MPa)^2$
Young's modulus variance $\sigma_{E_2}^2$	25.0 $(MPa)^2$	25.7 $(MPa)^2$

6.2.3 Shape and Topology Identification

The shape identification of the fibre in the microstructure is taken into account. The shape of the fibre is described using the NURBS curve. The design parameters are coordinates of the NURBS curve polygon control points [54]. The macrostructures and microstructures are shown in Figure 6.32.

The identification procedure is examined for displacements measured in the macromodel. The chromosome contains eight coordinates of NURBS control point plying the role genes. The reference displacement values in sensor points are generated using the numerical model. The 70 measured displacements were considered, 35 in x axis and 35 in y axis. The distribution of sensor points can be seen in Figure 6.32a. FEM was used for solving the direct problems. Similar to the previous example, the own programme of computational homogenization utilizing *MSC. Nastran* FEM code [50] was used in the direct problems. The material data for the macroscopic model were obtained using numerical homogenization. The RVE microstructure and the macromodel are modelled as 2D shells. The DEA algorithms were used, and the number of subpopulations was 2, with the number of chromosomes equal to 30. The simple crossover and the Gaussian mutation were applied in the identification process.

The results of identification are presented in Figure 6.33. The actual shape is marked using bright colour. Several evolutionary computing tests were performed. Selected two found shapes are shown using the dashed lines. The identification problem is hard, and the information about displacements in the macroscale is not sufficient for the exact shape of the fibre identification. The found shapes have common features with the actual shape, the location of fibres is close to each other, and the areas are comparable.

The application of the EA in the considered identification problem plays a crucial role because the objective function (6.12) has several local minima. The graph of the objective function J_o versus the design variable g_1 near the global optimum is shown in Figure 6.34. The rest of the seven design variables are fixed and have the optimal values. The found optimum for the design variable g_1 equals 0.1146.

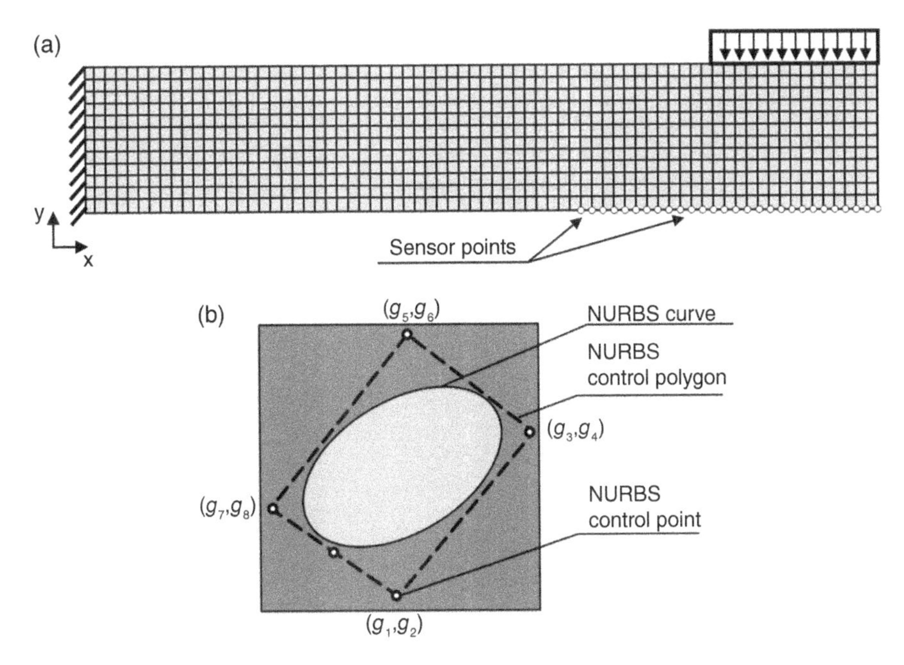

Figure 6.32 The macromodel (a) and micromodel (b) for the fibre's shape identification problem.

Figure 6.33 The results of identification.

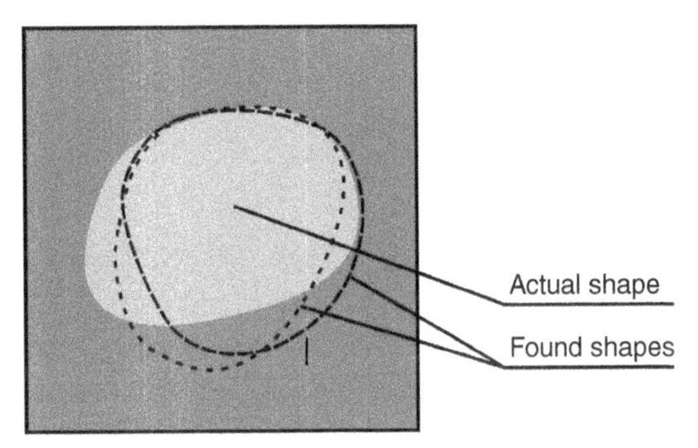

6.2.4 Identification of Shape for Multiscale Thermomechanical Problems

Consider multiscale models of porous materials being under thermomechanical loads [63]. A two-scale model of computational homogenization was applied [7, 12, 28]. The microstructure is represented by RVE with periodical boundary conditions. Computational homogenization is applied to determine equivalent thermomechanical parameters of porous microstructure in the form of elastic constants and coefficients of thermal conductivity [19]. These parameters are used in solving the two-dimensional boundary-value problem in macroscale. The finite element method is applied in computational homogenization in both scales.

The identification problem relied on determining thermomechanical parameters (Variant 1) and shapes of voids (Variants 2–5) in microstructure on the basis of measurement of displacements, and temperatures in sensor points of the macroscale was considered by [19].

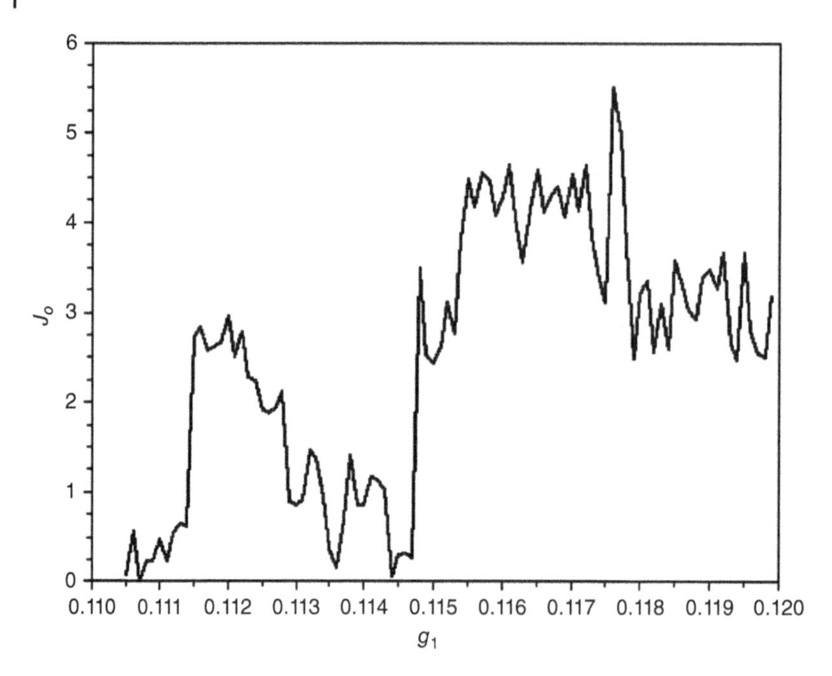

Figure 6.34 The objective function value versus the first design variable (with fixed values for the rest of design variables).

The identification functional is formulated in the form

$$J_O = a\sum_{i=1}^{m}|\hat{u}_i - u_i| + b\sum_{j=1}^{s}|\hat{T}_i - T_i|, \tag{6.26}$$

where: m – number of displacement sensors, s – number of thermal sensors, $\hat{u}_i$ – measured displacements at sensors i (1, m), u_i – displacements at sensors i calculated by solving the boundary-value thermoelastic problem for current design variables included in vector **ch**, $\hat{T}_j$ – measured temperature at sensors j (1, s), T_j – temperatures at sensors j calculated by solving the boundary-value thermoelastic problem for current design variables included in vector **ch**, a and b – scale coefficients.

As an example that illustrates the idea of the multiscale thermomechanical identification problem, consider a square plate of dimensions 50 × 50 × 1 mm supported on the left part of the boundary. Traction force $P_0 = 100$ N/mm is applied on the upper part of the boundary.

Thermal boundary conditions in the form of the first kind are prescribed on two parts of the boundary $T_1 = 10\,°\text{C}$ and $T_2 = 80\,°\text{C}$ (Figure 6.35a). The top and right parts of the boundary are insulated ($q = 0$). The plane state of stress state is assumed. The model of microstructure is presented in Figure 6.35b. Nine sensor points are localized on the right and upper parts of the boundary. Temperatures and displacements are for each sensor. During the work five variants of identification were performed.

The first variant of identification (Variant 1) of microstructure includes a circular void with a diameter equal to half the length of the bottom part of the boundary. Three material parameters in the form of Young's modulus E, Poisson's ration ν, and thermal conductivity λ are

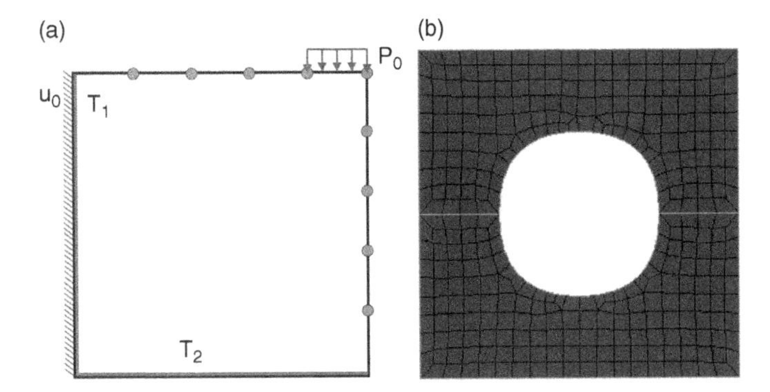

Figure 6.35 The square plate made with a porous material: (a) macromodel under thermoelastic loading and (b) RVE of micromodel with a void.

Table 6.11 Results of identification for Variant 1.

Material parameter	Exact value	Value obtained after identification	Error
E (MPa)	70 000	70 111	0.16%
ν	0.3	0.31	3.4%
λ (W/mK)	200	222	11%

identified by minimization of the functional (6.26) in the Variant 1. The following parameters of the EA were applied:

- The size of the population – 15, number of generations – 100, probability of simple crossover – 0.1, probability of the Gaussian mutation – 0.7, probability of arithmetic crossover – 0.1.
- Design variables are identified material parameters on which the following box constraints are imposed: $E = [40\,000 \div 80\,000]$, $\nu = [0.25 \div 0.35]$, $\lambda = [100 \div 300]$.

Table 6.11 contains results of identification for Variant 1.

Shapes of voids and their size in the porous aluminium microstructure (Figure 6.35b) are also identified on the basis of measured temperatures and displacements in boundary sensors points for the macromodel (Variant 2–5). Elliptical voids and arbitrary shapes of voids modelled by NURBS curves are considered [54]. In the case of elliptical voids, design variables of major and minor axes of an ellipse play the role of design variables, whereas, for NURBS curves, coordinates of control points are considered as design variables. The symmetry of the void is assumed, and the total number of design variables is equal to 5.

The EA with previous parameters is used, but the population size is 10 for elliptical voids and 20 for NURBS voids. Table 6.12 contains numerical results for Variants 2–5, whereas shapes of voids are presented in Figure 6.36. Changes of identification functional values with respect to a number of generations for all Variants 1–5 are presented in Figure 6.37.

All presented variants of identification of voids in RVE of porous materials show very good effectiveness of elaborated multiscale identification methodology.

In the case of identification of material parameters (Variant 1) only the value of coefficient of the heat conductivity is encumbered by the greater error. It is caused by the small

Table 6.12 Numerical results of identification for Variants 2–5.

Variant number – shape of the void design variables	Exact value	Value obtained after identification	Error
Variant 2 – elliptical void			
R1	0.2	0.197	1.6%
R2	0.8	0.802	0.2%
Variant 3 – elliptical void			
R1	0.8	0.801	0.07%
R2	0.2	0.199	0.36%
Variant 4 – void generated by NURBS			
Z1	0.9	0.912	1.4%
Z2	0.1	0.224	124%
Z3	0.4	0.342	14.5%
Z4	0.1	0.099	0.5%
Z5	0.2	0.153	23.4%
Variant 5 – void generated by NURBS			
Z1	0.8	0.787	1.6%
Z2	0.2	0.09	55%
Z3	0.1	0.117	17%
Z4	0.2	0.166	16.9%
Z5	0.2	0.317	58.7%

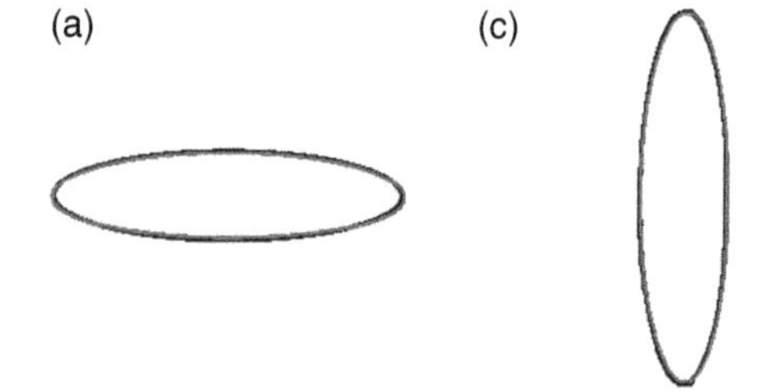

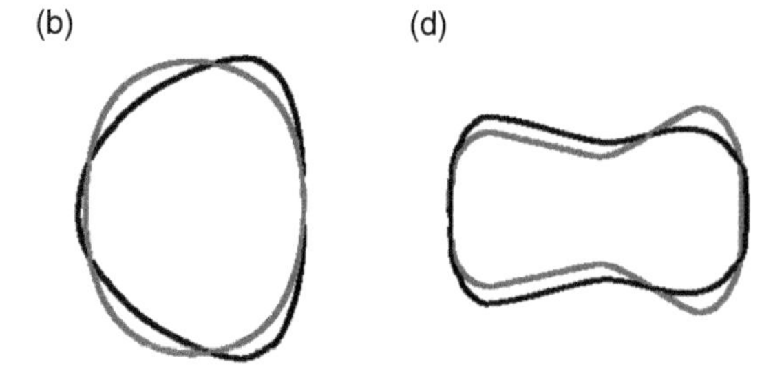

Figure 6.36 Results of identification of voids in microstructures. Reference shapes are blocked in red colour, whereas obtained shapes are blocked in black colour: (a) Variant 2, (b) Variant 3, (c) Variant 4, and (d) Variant 5.

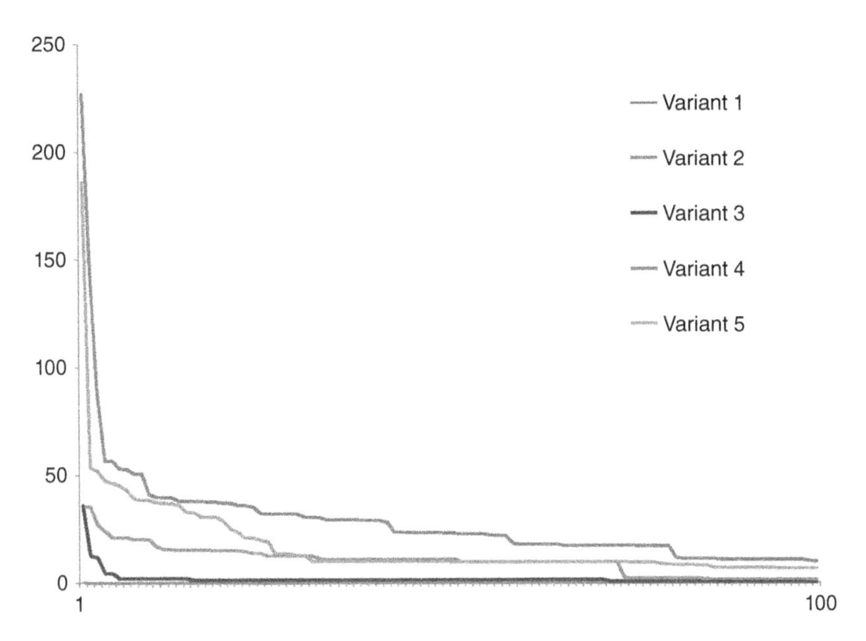

Figure 6.37 Changes of identification functional values with respect to a number of generations for Variants 1–5.

sensitivity of this parameter on the field of displacements and temperature. For the identification of elliptical voids, errors of identification of radius are very small. In the case of voids modelled by NURBS, errors of identification of selected design parameters are substantial, but obtained shapes are very consistent with reference shapes.

As presented, optimization and identification approaches in the area of multiscale modelling are very valuable tools for engineering applications. However, efficient interaction algorithms between the optimization and analysis software have to be often developed by the users from scratch. At the same time, optimization and multiscale calculations, especially in 3D space, often provide a significant amount of data. Issues related to software interactions and visualization of large data sets are discussed in the following chapter.

References

1 Baughman, R.H., Eckhardt, H., and Kertesz, M. (1987). Structure–property predictions for new planar forms of carbon: layered phases containing sp^2 and sp atoms. *Journal of Chemical Physics* 87: 6687–6699.

2 Brenner, D.W., Shenderova, O.A., Harrison, J.A. et al. (2002). A second-generation reactive empirical bond order (REBO) potential energy expression for hydrocarbons. *Journal of Physics: Condensed Matter* 14: 783–802.

3 Bucknum, M.J. and Castro, E.A. (2008). The squarographites: a lesson in the chemical topology of tessellations in 2- and 3-dimensions. *Solid State Sciences* 10: 1245–1251.

4 Burczyński, T. (1995). *The Boundary Element Method*. Warsaw: WNT.

5 Burczyński, T. (1995). Boundary element method in stochastic shape design sensitivity analysis and identification of uncertain elastic solids. *Engineering Analysis with Boundary Elements* 15: 151–160.

6 Burczyński, T., Beluch, W., Długosz, A. et al. (2006). Inteligent computing in inverse problems. *Computer Assisted Mechanics and Engineering Sciences* 13: 161–206.
7 Burczyński, T. and Kuś, W. (2009). Microstructure optimisation and identification in multi-scale modeling. In: *ECCOMAS Multidisciplinary Jubilee Symposium – New Computational Challenges in materials, Structures and Fluids, Computational Methods in Applied Science*, vol. 14 (ed. J. Eberhardsteiner et al.), 169–181. Springer Science.
8 Burczyński, T. and Kuś, W. (2010). *Bioinspired Algorithms in Multiscale Optimization, Advanced Structured Materials, Computer Methods in Mechanics*, 183–192. Springer.
9 Burczyński, T., Kuś, W., Beluch, W. et al. (2020). *Intelligent Computing in Optimal Design*. Springer.
10 Burczyński, T. and Orantek, P. (2009). Uncertain identification problems in the context of granular computing. In: *Human-Centric Information Processing*, vol. 182 (ed. A. Bargiela and W. Pedrycz), 329–350. Springer-Verlag.
11 Burczyński, T. and Skrzypczyk, J. (1999). Theoretical and computational aspects of the stochastic boundary element method. *Computer Methods in Applied Mechanics and Engineering* 168: 321–344.
12 Buryachenko, V. (2007). *Micromechanics of Heterogeneous Materials*. Springer Science.
13 de Castro, L.N. and Timmis, J. (2003). Artificial immune systems as a novel soft computing paradigm. *Soft Computing* 7 (8): 526–544.
14 Chan, F.T.S. and Tiwari, M.K. (2007). *Swarm Intelligence: Focus on Ant and Particle Swarm Optimization*. I-Tech Education and Publishing.
15 Chenoweth, K., van Duin, A.C.T., and Goddard, W.A. (2008). ReaxFF reactive force field for molecular dynamics simulations of hydrocarbon oxidation. *The Journal of Physical Chemistry A* 112: 1040–1053.
16 Chou, M.Y. and Cohen, M.L. (1986). Electronic shell structure in simple metal clusters. *Physics Letters A* 113: 420–424.
17 Cox, H., Johnston, R.L., and Murrell, J.N. (1997). Modelling of surface relaxation and melting of aluminium. *Surface Science* 373: 67–84.
18 Cranford, S.W. and Buehle, M.J. (2011). Mechanical properties of graphyne. *Carbon* 49: 4111–4121.
19 Długosz, A. and Burczyński, T. (2013). Identification in multiscale thermoelastic problems. *Computer Assisted Methods in Engineering and Science* 20: 325–336.
20 Enyashin, A.N. and Ivanovskii, A.L. (2011). Graphene allotropes. *Physica Status Solidi* 248 (8): 1879–1883.
21 Fish, J. (2013). *Practical Multiscaling*. Wiley.
22 Geers, G.D., Kouznetsova, V.G., Matous, K., and Yvonnet, J. (2017). Homogenisation methods and multiscale modelling: nonlinear problems. In: *Encyclopedia of Computational Mechanics* (ed. E. Stein, R. de Borst and T.J.R. Hughes). Wiley.
23 Girifalco, L.A. and Weizer, V.G. (1959). Application of the Morse potential function to cubic metals. *Physical Review* 114: 687.
24 Haley, M.M. (2008). Synthesis and properties of annulenic subunits of graphyne and graphdiyne nanoarchitectures. *Pure and Applied Chemistry* 80 (3): 519–532.
25 Kamiński, M. and Kleiber, M. (2000). Perturbation based stochastic finite element method for homogenisation of two-phase elastic composites. *Computer and Structures* 78: 811–836.
26 Kennedy, J., Eberhart, R.C., and Shi, Y. (2001). *Swarm Intelligence*. Morgan Kaufmann Publishers.
27 Kleiber, M. and Hien, T.D. (1992). *Stochastic Finite Element Method*. Wiley.

28 Kouznetsova, VG. (2002). Computational homogenisation for the multi-scale analysis of multi-phase materials. PhD thesis. TU Eindhoven.

29 Kuś, W. (2007). Grid-enabled evolutionary algorithm application in the mechanical optimisation problems. *Engineering Applications of Artificial Intelligence* 20: 629–636.

30 Kuś, W. and Burczyński, T. (2008). Parallel bioinspired algorithms in optimisation of structures. In: *Parallel Processing and Applied Mathematics PPAM 2007, Lecture Notes in Computer Science*, vol. 4967 (ed. R. Wyrzykowski, J. Dongarra, K. Karczewski and J. Wasniewski), 1285–1292. Springer.

31 Kuś, W. and Burczyński, T. (2008). Optimisation of microstructutre in multiscale problems with use of parallel evolutionary algorithm. In: *8th. World Congress on Computational Mechanics (WCCM8) 5th. European Congress on Computational Methods in Applied Sciences and Engineering (ECCOMAS 2008)*. Venice.

32 Kuś, W. and Burczyński, T. (2010). Bioinspired algorithms in multiscale optimisation. In: *Computer Methods in Mechanics* (ed. M. Kuczma and K. Wilmański), 183–192. Springer.

33 Kuś, W., Długosz, A., and Burczyński, T. (2011). OPTIM - library of bioinspired optimisation algorithms in engineering applications. *Computer Methods in Material Science* 11 (1): 9–15.

34 Kuś, W., Mrozek, A., and Burczyński, T. (2016). Memetic optimisation of graphene-like materials on Intel PHI Coprocessor. In: *Lecture Notes in Artificial Intelligence*, vol. 9692, 401–410. Lammps Software http://lammps.sandia.gov (Apr 2015).

35 Lloyd, L.D. and Johnston, R.L. (1998). Modelling aluminum clusters with an empirical many-body potential. *Chemical Physics* 236: 107–121.

36 Los, J.H., Ghiringhelli, L.M., Meijer, E.J., and Fasolino, A. (2005). Improved long range reactive bond-order potential for carbon I. *Construction, Physical Review B* 72: 214102.

37 Madej, L., Mrozek, A., Kuś, W. et al. (2008). Concurrent and upscaling methods in multi scale modeling – case studies. *Computer Methods in Materials Science* 8 (1): 1–15.

38 Maździarz, M., Mrozek, A., Kuś, W., and Burczyński, T. (2017). First-principles study of new *X*-graphene and *Y*-graphene polymorphs generated by the two stage strategy. *Materials Chemistry And Physics* 202: 7–14.

39 Maździarz, M., Mrozek, A., Kuś, W., and Burczyński, T. (2018). Anisotropic-cyclicgraphene: a new two-dimensional semiconducting carbon allotrope. *Materials* 11 (3): 432.

40 Michalewicz, Z. (1996). *Genetic Algorithms + Data Structures = Eolutionary Algorithms*. Berlin: Springer-Verlag.

41 Miyamto, Y., Niino, M., and Koizumi, M. (1997). FGM research programs in Japan – from structural to functional uses. In: *Functionally Graded Materials* (ed. I. Shiota and Y. Miyamoto). Elsevier Science.

42 Mrozek, A. and Burczyński, T. (2013). Examination of mechanical properties of graphene allotropes by means of computer simulation. *Computer Assisted Methods in Engineering and Science* 20 (4): 309–323.

43 Mrozek, A., Kuś, W., and Burczyński, T. (2010). Searching of stable configurations of nanostructures using computational intelligence methods. *Technical Science* 20 (107): 85–97.

44 Mrozek, A., Kuś, W., and Burczyński, T. (2011). Computational intelligence methods in searching of stable configurations of nanostructures, Computer Methods in Materials Science 11: 46–52.

45 Mrozek, A., Kuś, W., and Burczyński, T. (2015). Nano level optimisation of graphene allotropes by means of a hybrid parallel evolutionary algorithm. *Computational Materials Science* 106: 161–169.

46 Mrozek, A., Kuś, W., and Burczyński, T. (2017). Method for determining structures of new carbon-based 2D materials with predefined mechanical properties. *International Journal For Multiscale Computational Engineering* 15 (5): 379–394.

47 Mrozek, A., Kuś, W., Orantek, P., and Burczyński, T. (2005). Prediction of the aluminium atoms distribution using evolutionary algorithm. In: *Recent Developments in Artificial Intelligence Methods* (ed. T. Burczyński, W. Cholewa and W. Moczulski), 127–130. Gliwice.

48 Murrell, J.N. and Mottram, R.E. (1990). Potential energy functions for atomic solids. *Molecular Physics* 69 (3): 571–585.

49 Murrell, J.N. and Rodriguez-Ruiz, J.A. (1990). Potential energy functions for atomic solids. II. Potential functions for diamond-like structures. *Molecular Physics* 71 (4): 823–834.

50 MSC. 2006. Nastran User Guide.

51 Nakano, A. (1997). Parallel multilevel preconditioned conjugate-gradient approach to variable-charge molecular dynamics. *Computer Physics Communications* 104: 59–69.

52 Papoulis, A. (1991). *Probability, Random Variables and Stochastic Processes.* New York: McGrawHill.

53 Peng, Q., Ji, W., and De, S. (2012). Mechanical properties of graphyne monolayers: a first-principles study. *Physical Chemistry Chemical Physics* 14 (38): 13385–13391.

54 Piegl, L. and Tiller, W. (1997). *The NURBS Book*, 2e. Berlin: Springer.

55 Press, W.H., Teukolsky, S.A., Vetterling, W.T., and Flanery, B.P. (2007). *Numerical Recipes: The Art of Scientific Computing.* Cambridge: Cambridge University Press.

56 Rappe, A.K. and Goddard, W.A. (1991). Charge equilibration for molecular dynamics simulations. *Journal of Chemical Physics* 95 (8): 3358–3363.

57 Roberts, C., Johnston, R.L., and Wilson, N.T. (2000). A genetic algorithm for the structural optimisation of Morse clusters. *Theoretical Chemistry Accounts* 104: 123–130.

58 Sakata, S. and Ashida, F. (2011). Hierarchical stochastic homogenisation analysis of a particle reinforced composite material considering non-uniform distribution of microscopic random quantities. *Computational Mechanics* 48: 529–540.

59 Scarpa, F., Adhikari, S., and Phani, A.S. (2009). Effective elastic mechanical properties of single layer graphene sheets. *Nanotechnology* 20: 065709.

60 Shao, X., Cheng, L., and Cai, W. (2004). An adaptive immune optimisation algorithm for energy minimisation problems. *Journal of Chemical Physics* 120 (24): 11401–11406.

61 Stuart, S.J., Tutein, A.B., and Harrison, J.A. (2000). A reactive potential for hydrocarbons with intermolecular interactions. *Journal of Chemical Physics* 112 (14): 6472–6486.

62 Terada, K. and Kikuchi, N. (2001). A class of general algorithms for multi-scale analyses for heterogeneous media. *Computer Methods in Applied Mechanics and Engineering* 190: 5427–5464.

63 Terada, K., Kurumatani, M., Ushida, T., and Kikuchi, N. (2012). A method of two-scale thermo-mechanical analysis for porous solids with micro-scale heat transfer. *Comput Mechanics* 46: 69–285.

64 Wales, D.J. and Doye, J.P.K. (1997). Global optimisation by basin-hopping and the lowest energy structures of Lennard-Jones clusters containing up to 110 atoms. *Journal of Physical Chemistry A* 101: 5111–5116.

65 Wales, D.J. and Scheraga, H.A. (1999). Global optimisation of clusters, crystals and biomolecules. *Science* 285: 1368–1372.

66 Wang, Y., Lv, J., Zhu, L., and Ma, Y. (2010). Crystal structure prediction via particle-swarm optimisation. *Physical Review B* 82 (9): 094116–0941123.

67 Yo, X. and Gen, M. (2010). *Introduction to Evolutionary Algorithms.* London: Springer-Verlag.

68 Zhodi, T.I. and Wriggers, P. (2005). *Introduction to Computational Micromechanics*. Berlin: Springer.

69 Zhou, Z. (2003). A new look at the atomic level virial stress: on continuum-molecular system equivalence. *Proceedings of the Royal Society of London Series A—Mathematical Physical and Engineering Sciences* 459 (2037): 2347–2392.

70 Zhou, J., Li, W., and Zhu, J. (2008). Particle swarm optimisation computer simulation of Ni clusters. *Transactions of Nonferrous Metals Society of China* 18: 410–415.

71 Zienkiewicz, O.C. and Taylor, R.L. (2000). *The Finite Element Method*. Oxford: Butterworth-Heinemann.

72 Zienkiewicz, O.C., Taylor, R.L., and Zhu, J.Z. (2005). *The Finite Element Method: Its Basis and Fundamentals*, 6e. Butterworth-Heinemann.

7

Computer Implementation Issues

7.1 Interactions Between the Analysis and Optimization Solutions

The multiscale analysis as well as optimization of structures and materials uses well-known direct problem-solving methods like FEM, BEM, MD, CA, or MC. However, most of the time, the new methods of analysis and optimization lead to the need for modification of available software. In many cases, the software is not created from scratch, but open-source or commercial programs are used. Some of the programs are distributed with API (Application Programming Interface), allowing modifications of part of the algorithms, but in some cases, only access to input and output files is provided.

Some of the FEM software available on the market, like Comsol [5] and Ansys [1], contain specialized modules allowing optimization with well-known gradient-based or bioinspired methods. The simplicity of use of integrated optimization algorithm with FEM/BEM/MD/CA/MC software is very important, but in some cases, when a new optimization algorithm is present as an external software package, the user has to decide how to connect together optimization and analysis solutions. First, the software to be treated as a master algorithm should be chosen. Both the optimization and analysis programs can play this function. However, sometimes it is necessary to introduce another solution like shell script to play the role of a master program. The choice of the plan depends on possible modification, parametrization of optimization, and analysis software. Figure 7.1 presents a variant of the master–slave approach where the optimization algorithm is leading the process. The optimization is performed in an iterative way, where the first stage in most optimization algorithms generates an initial set of vectors containing design variables (in the case of the gradient-based algorithm, it starts just from one point in design space, but more starting points may improve results when the unimodality of objective function is not guaranteed). The design vectors describing geometry and materials parameters have to be transferred to the input file formatted in a native way for a direct problem solver. The solver is executed for each design vector, but many vectors can be analyzed in parallel if enough computational resources are available. On the basis of results in the form of displacements, strains, stress, and temperature fields, the objective function is computed for each design vector. The data is transferred to the optimization algorithm where is used for modifications of design vectors to be analyzed in the next iteration. The algorithm stops when the stop condition is fulfilled.

Multiscale Modelling and Optimisation of Materials and Structures, First Edition. Tadeusz Burczyński, Maciej Pietrzyk, Wacław Kuś, Łukasz Madej, Adam Mrozek, and Łukasz Rauch.

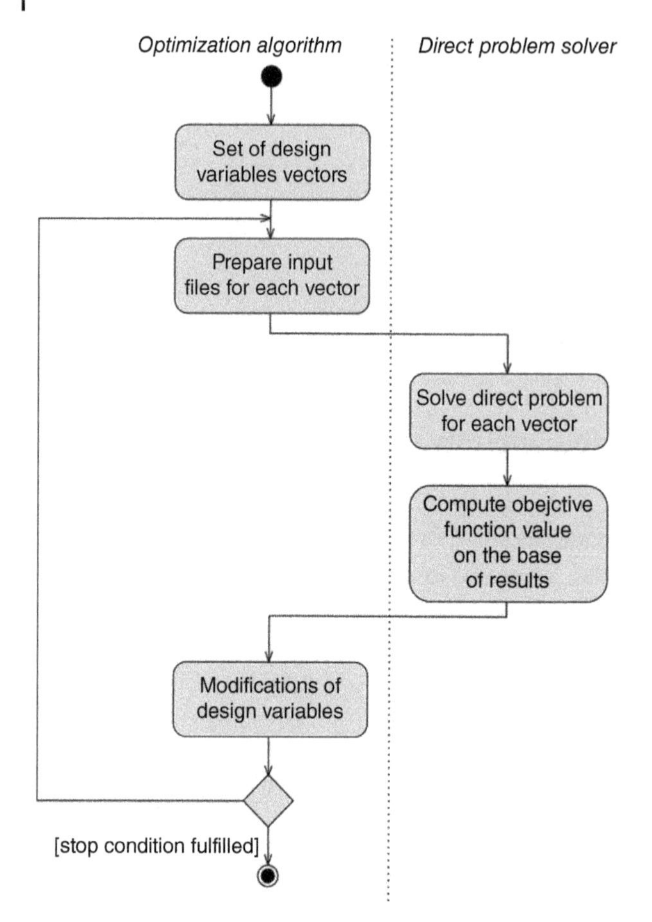

Figure 7.1 The master optimization algorithm and slave direct problem solver.

The optimization algorithm as a slave can be used if the program allows to execute one iteration, store the design vectors and states, e.g. in files, database, and later execute another iteration retrieving the previous state of the algorithm. Such an approach is shown in Figure 7.2. The capabilities of direct solver scripting must be used to invoke external optimization algorithm.

The external script coordinating optimization and objective function analysis algorithms can be efficiently useful when the parallel environment is present, and the optimization algorithms need an evaluation of many objective function values in each iteration.

The initialization of the optimization algorithm is executed at the beginning, and the design vectors are returned to the direct problem solver. Next the direct problem and objective function values are computed, and again the optimization algorithm is invoked. The algorithm modifies the design variables and returns their values to the direct problem solver.

The third variant is external script invoking optimization algorithm (just for one iteration) and later direct solver, the stop condition and the loop are realized by script. The external script coordinating optimization and objective function analysis algorithms can be efficiently

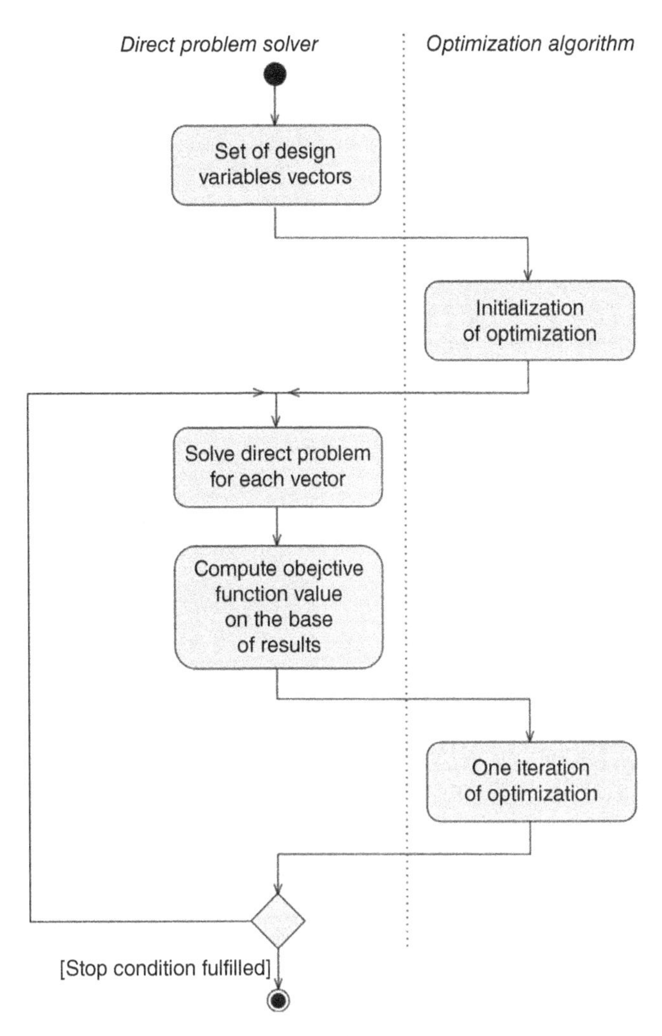

Figure 7.2 The master direct problem and slave optimization algorithm.

useful when the parallel environment is present, and the optimization algorithms need an evaluation of many objective function values in each iteration (e.g. when the optimization is performed with the use of cluster or supercomputer).

7.1.1 Example of Direct Problem Solver File Access

The direct problem solver MSC.Nastran [9] is used as an example of access to input and output files. Let us assume the problem where design variables are material properties of the structure, and the objective function depends on maximum displacement. The input files contain a description of boundary conditions, mesh definition (nodes and elements), and also material properties. The information about structure and analysis properties are stored in cards, that contain name and parameters. Part of the input file used in multiscale optimization in [4] is shown next.

```
SOL 101
CEND
(...)
BEGIN BULK
(...)
PARAM     POST     0
PARAM    PRTMAXIM YES
PSOLID          1       1       0
GRID     1                 .0830661-.1625    0.
GRID     2                 .0830662-.142643 0.
GRID     3                 .0830663-.122786 0.
(...)
CHEXA    1       1       9       8       3       2       297     300
         299     298
CHEXA    2       1       8       7       4       3       300     302
         301     299
CHEXA    3       1       7       6       5       4       302     304
         303     301
(...)
MAT1     1       12560.  7692.31 .35
(...)
```

The material properties for linear elastic material are defined with the use of MAT1 card. The execution of MSC.Nastran creates the output files, and depending on version and parameters, they can be just simple text, native binary, or Hierarchical Data Format (HDF5) files. For the clarity of the example, part of the output text file (.f06), which can be used to obtain maximum displacements of the structure, is presented.

```
0
0                              MAXIMUM  DISPLACEMENTS
  SUBCASE/
  DAREA ID      T1             T2             T3             R1             R2             R3
0        1 1.7449645E+00  1.2232061E+01  0.0000000E+00  0.0000000E+00  0.0000000E+00  1.7694198E-01
```

Many software can perform the one-scale analysis, but multiscale approaches are particularly interesting for the current book. The direct solvers can be used in multiscale analysis, e.g. when simple linear two-scale upscaling is performed. In this case, a few separate runs of direct problems for the microscale have to be followed by macroanalysis. There are already many examples of more complicated multiscale problems packages like Digimat [8] or DAMASK [20] that the user can select for the research.

7.1.2 Examples of an Internal Script in Direct Problem Solver

The possibility to use scripts in some of the direct problem solvers allows use of external optimization software or even build optimization as a script. The Comsol Multiphysics [5] FEM software is used as an example of scripting possibilities in direct problem solvers. The Comsol allows to create scripts based on Java language and interact with Comsol models. The code can be invoked from forms built by the user using internal tools. The loops, functions, and tables allow building complex scripts. The example of part of the code used for optimization and multiscale analysis in [19] is shown next.

```
(...)
for (int j = 0; j < 50; j++)
{

  String[][] dataTable = new String[][]{};   // define table
  dataTable = readCSVFile(inputFileName);    // read data from CSV file into
                                                table
  Integer numberOfRows = dataTable.length;   // define number of rows in table
  (...)
  with(model.param());                       // modify parameters of the FEM
                                                model
       set("m1", dataTable[0][0]);
       set("m2", dataTable[0][1]);
  endwith();

  model.sol("sol9").runAll();                // solve one of the load cases in
                                                FEM model

  (...)
  fileOpen("optimizationOneStep.exe");       // execute external program
  (...)
}
```

The code executes loop 50 times, reads CSV files, transfers data as finite element model parameters, runs geometry recreation on a base of models parameters, mesh creation, FEM analysis, and execution of an external program (e.g. one step of optimization algorithm) in each iteration.

7.2 Visualization of Large Data Sets

Visualization of huge data sets was always a challenge. Even now, when high-performance computing assets are available on demand or when cloud computing and other XaaS (X –Anything as a Service, where X may be replaced with various letters, e.g. S – Software, I – Infrastructure, or P – Platform) computing architectures are highly available, fluent visualization procedure through large volumes of variate data is very difficult. This may be observed in the case of the presentation of DMR results described in Chapter 4. Such results are composed of many scales containing data of multi-physics calculations. Therefore, large and diverse sets of information must be grappled with during visualization. Such a process often requires not only a large amount of memory resources inside the client's computer but also strong processors or other computing accelerators supporting visualization.

In the software market, there are many applications for visualization of data resulting from computational procedures. Most of them are general programs, able to visualize all kinds of results stored in the form of standard file formats. In practice, these representations of data are very inefficient and do not help in the deep analysis of a large number of datasets, which nowadays exceed gigabytes. This is usually restricted by capabilities of computers, i.e. central processing unit (CPU) and general-purpose graphics processing unit (GPGPU) efficiency, random access memory (RAM) capacity, bus bandwidth, and many other hardware parameters. Data representing structures of real materials are one of such data types, which due to many details in each scale, e.g. grains, dislocations, boundaries, and nanoparticles, require significant space in memory and specific dedicated algorithms to be visualized efficiently. Available visualization tools, however, do not provide algorithms for this kind of data. Therefore, the process of visualization is often very time-consuming for ordinary programs and sometimes even impossible for large datasets.

The first part of this chapter reviews programming libraries and available software, which allow visualizing various datasets. The second part shows an idea and implementation details of a computer application that supports visualization of datasets exceeding the capabilities of the hardware. This will be presented by using examples of multiscale material structures visualization in an efficient and interactive way. The following modules of the proposed software will be presented in detail:

- preprocessing of datasets stored in large files exceeding gigabytes,
- engine for visualization of large dataset implemented in C++ and Open Graphics Library (OpenGL [17]),
- mechanisms of interactive transition between different length scales, and
- multithreaded loading of data organized in neighbouring sections.

The proposed methodology is interesting for many users, which do not possess access to highly developed innovative hardware architectures like power walls. Hence, methods described in this chapter are performed on average personal computers or even laptops, offering fluent visualization of datasets exceeding RAM capacity. The chapter presents not only theoretical aspects of this novel approach to visualization of large datasets but also its practical usage. The results obtained for efficiency analysis of created software are also described in the chapter.

7.2.1 Implementation Aspects and Tools

This chapter presents a review of graphical libraries allowing a flexible implementation of high-performance visualization procedures and off-the-shelf solutions available on the market, including commercial and open-source software. The most current versions of the software are described in this chapter, while most of the reviewed papers from the scientific literature are published between the years 2000 and 2021.

7.2.1.1 Graphical Libraries

The main criteria for the selection of the most suitable graphical libraries for scientific visualization are their efficiency, wide API, and possibilities of 3D scene modelling. The great majority of graphical libraries, which can be found in the Internet, are implemented in OpenGL, OpenCL, or WebGL. From the scientific point of view, these choices are the best because of the high efficiency of these libraries as well as their portability between many operation systems. They easily operate on the most popular systems like Microsoft Windows or various versions of Linux, but there are also distributions dedicated to mobile devices, using Android or IOS, e.g. OpenGL ES. These are the most popular libraries applied in visualization, physics, and game development, allowing usage of graphics card acceleration.

7.2.1.1.1 OpenGL (Open Graphics Language)

The main advantage of this library as mentioned earlier is the portability between different operating systems, including Microsoft Windows, Linux, BSD, Solaris, Android, iOS, and even Playstation or Nintendo Wii. Moreover, there exist many programming frameworks, which through their API allow using OpenGL functionality, e.g. Qt, WinAPI, Tao, OpenTK, and many others. Some of these frameworks also facilitate graphical user interface (GUI) creation, sound management, or data processing and offer efficient mechanisms like GL_POINT_SPRITE, which allows treating graphical objects as one point to reduce memory usage.

According to the latest statistics, OpenGL library is much more popular than DirectX. It was implemented much earlier and is very well documented in many tutorials and samples available on the Internet. In comparison to competitive solutions, OpenGL library is much more often used in scientific applications dedicated to numerical simulations. The main reason lying beyond this state is the possibility of low-level programming fostered by its high performance and portability [3].

7.2.1.1.2 OpenCL (Open Computing Language)

The framework is dedicated to writing programs that execute across heterogeneous platforms consisting of various CPUs, GPUs, digital signal processors (DSPs), field-programmable gate arrays (FPGAs), and other accelerators. This technology specifies programming languages (based on C99, C++14, and C++17) used for programming devices and APIs to control the platform and execute programs on the computing devices. OpenCL provides a standard interface for parallel computing using task- and data-based parallelism. In visualization, it is used for the fast processing of data on the server-side. This is an open standard maintained by Khronos Group, which is a non-profit organization. Conformant implementations are available from many suppliers, e.g. AMD, Apple (implementation still available, but planned to be deprecated in favour of Metal 2), ARM, IBM, Intel, Nvidia, Qualcomm, and Samsung [11].

7.2.1.1.3 WebGL (Web Graphics Language)

In the era of dominating web applications composed of a separated server and client sides, the development of a library dedicated to visualization was a matter of time. The first WebGL version was released in 2011 as a JavaScript implementation for rendering interactive 2D and 3D graphics with any compatible web browser. The most important advantage lies in the independence of the browser, while no other plugin is required for the execution of WebGL-based applications. The library is fully integrated with current web standards and allows GPU acceleration of physics and image processing. WebGL can be integrated with other technologies like HTML and CSS. The number of available visualization frameworks based on WebGL library exceeds fifteen in 2021.

7.2.1.2 Software

Currently, the lack of software dedicated to the visualization of multiscale material microstructures is observable. However, in the area of scientific visualization programs, a lot of computer applications exist, which can be adapted to visualize different scales of material. Nevertheless, in all of these cases, such visualization is performed separately. The software listed in this section belongs to the group of the most popular and highly downloaded solutions.

7.2.1.2.1 ParaView

ParaView [18] is open-source, multi-platform software, which supports server as well as local side actions aiming at data processing and rendering. It allows to build visualizations quickly and to analyze data using qualitative and quantitative techniques. The data exploration can be done interactively in 3D or programmatically using ParaView's batch processing capabilities. The module of 3D visualization is implemented by using visualization tool kit (VTK) [22]. It allows for visualization of a large number of experiments, including data of different types and different sizes. ParaView was developed to analyze extremely large datasets using distributed memory computing resources. It can be run on supercomputers to analyze datasets of exascale size as well as on laptops for smaller data.

File with the microstructure data is loaded into the program and visualized as the set of particles. Figure 7.3 presents example of microstructure visualized by ParaView engine. It should be emphasized that no input format is submitted to the program, and only the text file is loaded. In the case of larger files, the application is slow, and a number of visible points during visualization is constrained. Beyond this, the digital microstructures cannot be

Figure 7.3 Visualization of the microstructure using ParaView software.

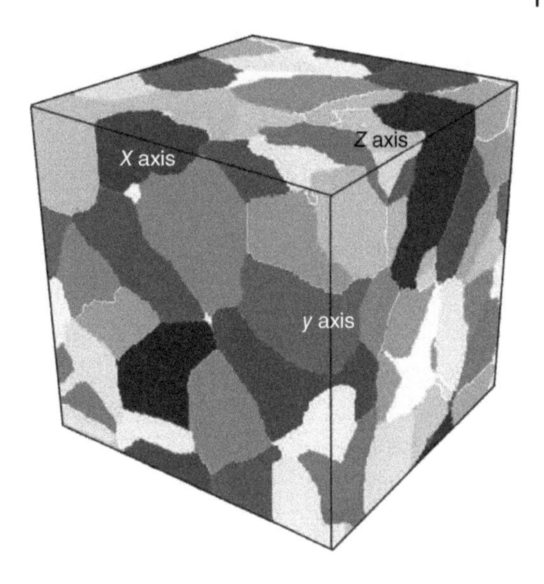

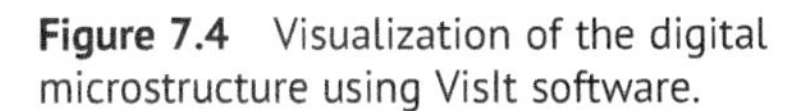
Figure 7.4 Visualization of the digital microstructure using VisIt software.

analyzed from the inside. The program is not adapted to do this. It is possible, but not very likely that some options of the support of the efficient visualization of multiscale materials exist.

7.2.1.2.2 VisIt

Before loading the file, the format of this file has to be selected among a number of available formats. After loading the file, there are several possibilities of visualization of the data. When images of microstructures are considered, the available colours and shapes of grains are important. The former is presented in Figure 7.4.

Lack of the possibility of getting an inside view is a weak point of this program. Although an object in Figure 7.4 is composed of only about 130 thousand particles, visualization in the form of particles was slow, and it did not meet expectations. Recapitulating, VisIt software guarantees a good quality visualization of the digital microstructure. On the other hand, it is fairly slow, and it lacks some important functionalities for visualization of multiscale materials.

7.2.1.3 Frameworks

There are several frameworks dedicated to visualization. They are described in several publications and on the Internet. VTK is one of the popular systems, and it was mentioned in the presentation of the ParaView application. Analogously to similar solutions, VTK shares classes and libraries in C++. VTK is a universal tool and, therefore, cannot be classified as an optimal solution for the visualization of multiscale data. Schemes described in the literature are mostly created by scientists or developers; therefore, they are not widely known, and they probably do not provide sufficient quality and efficiency. Some solutions, which are better known, like system VTK of the Kitware Company, guarantee good performance, but they are not dedicated to the visualization of materials models.

7.2.1.4 Data Storing

Data storing is crucial for the selection of the implementation technology for the system. Both software and hardware solutions are important. Before discussing this problem, information on the potential content of the files is needed. When visualization of material structures is considered, the files have to contain coordinates of the particles (e.g. X, Y, Z) and certain variable, which allows distinguishing groups of the particles, for example grains, stresses, strains, and temperatures. It can be an identifier or a colour (e.g. R, G, B). If the latter is the case, there is no need to change an identifier into colour in real time. Although such a data set is a base for the presentation of material structure, other variables can be eventually added to describe some selected physical features.

7.2.1.4.1 Text Files

The text file is the simplest method to store the data. This is the most popular solution in a majority of publications. It does not require special architectures of the system. It needs only an optimal library for operations on files. These libraries may differ in efficiency depending on the available system and its architecture. The main disadvantage of such a solution is a large size of a text file, which is usually accessed linearly, causing some delays in data reading. However, clever indexing of subsequent file lines and the mapping between memory addresses and line numbers can solve this problem in most cases.

7.2.1.4.2 Binary Files

In recent years, the HDF became the most popular format for storing data for scientific visualization. HDF5, the representative of this group of formats, is a standard for gathering data sets, which are large, complex, and very often heterogeneous. This is often the case for the digital material representation data. Moreover, if fluent visualization is required, e.g. dynamic moving through a material with the presentation of all its layers, parallel input/output operations, and random access are also needed. These features are supported by HDF5 as well. HDF5 is an open-source standard organized inside in the form of folders and subfolders similar to a typical file system. This facilitates the usage of such data and is structured in groups and datasets, which are additionally described by metadata. The latter information makes HDF5 a self-describing format.

7.2.1.4.3 Databases

The solutions which assume using the database to store information about data for scientific visualization can be encountered in the scientific literature [2, 12]. First of the cited positions presents a wide review of methods, numerical procedures, and software used for scientific visualization, while the data is really huge and there is a need for compromises between aggregation or preserving details. This usually requires databases to store such large data sets. Authors in [2] proposed building of the system based on Java and interactive data language (IDL) technologies. During visualization, the data are taken from an SQL database and are rendered on the high-class server. The authors encountered several problems connected with this solution, such as for example a high number of inquiries or a large dimension of the database. Figure 7.5 shows how this system was implemented and how various technologies were connected. This application was implemented using Java language. It sends queries to the

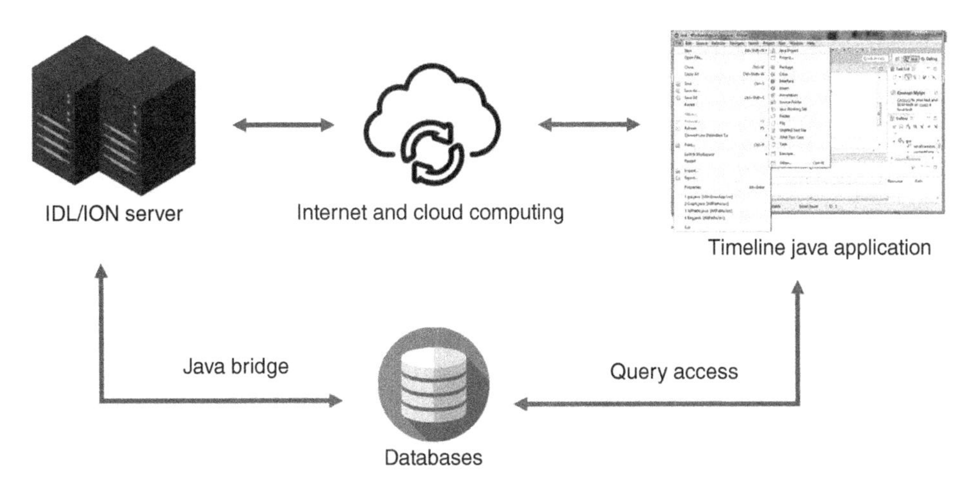

Figure 7.5 The architecture of the proposed solution.

database with the objective of gaining representation of selected data. Java Bridge framework provides communication between database and server, which renders the requested scene. Exchange of the data between server and computer with the visual application is performed through the network, which also causes some delays and influences smoothness of the visualization process.

The presented solution is technologically interesting. Possible expansion of this idea enables improvement of the results. Since rendering is performed on the server, the application of the Java language, which is not very efficient, is not a problem. Using so many technologies (database, server, computer, and connections) makes this solution costly and involves the necessity of fulfilment of many conditions, like the necessity of staying online.

7.2.1.4.4 Text Files Compressed Into Binaries

There are also solutions that focus on decreasing data files using their compression [7]. The size of the data files is decreased in three steps: wavelet transform, quantization of the wavelet coefficients, and their compression. Before visualization such data are decompressed in real time. The presented solution is based on a costly and time-consuming process of data preparation, which is composed of a few steps and includes decompression. This usually requires high-performance CPUs, and using this method to carry out many calculations for large data sets is time-consuming. It is often used for the data, which are represented in the form of animation. Each simulation stem may require a large data set; therefore, compression is a good solution.

7.2.2 High Efficiency of Visualization

The concept of high efficiency of visualization depends on the objective of the visualization. Analysis of the graphical representation of the static model, such as for example, set of the medical data [25], one has to focus on the data rendering. When dynamic visualization is considered, in which visualized objects may move (e.g. fluid dynamics [14, 15]), beyond data management, there are also required operations responsible for changes of the visualized points. The first of the cited papers presents a very interesting and most recent solution to the mentioned problems by using VTK and Python-based library PyQT5.

7.2.2.1 Dedicated Algorithms

To achieve high efficiency on individual computers existing algorithms can be used, or a new algorithm has to be developed, which will enable full usage of all available resources. To avoid losing an available computational power, an application of concurrent programming seems to be necessary. Scrolling of the 3D medical data is an example of this solution [23]. Authors of this solution point out problems with obtaining satisfactory effect without an application of distributed computing. The proposed solution, which can be executed on a standard PC computer, assumes data management before visualization composed of conversion to the compressed and hierarchical wavelet representation. During the running of this application, the data are decompressed in real time and displayed using texture mapping. Details of the displayed scene depend on a local spectrum of data frequencies and their locations with respect to the observer. More importantly, this solution assures satisfactory frames per second (FPS) coefficient.

Visualization of the particles of carbohydrates is another example of the algorithm dedicated to efficient visualization [13]. Displaying large and complex particles requires a more advanced approach. Two algorithms were proposed in [13]. The first is called PaperChain, and it defines the type and structure of the carbohydrate rings. The second is called Twister, and it accounts for relative orientation between mentioned rings. Both these algorithms offer better efficiency compared to the classical ball-and-stick algorithm, which assumes visualization of each particle and connection between particles. PaperChain and Twister algorithms assume recognition of one particle of the carbohydrate as one object.

7.2.2.2 Hardware Parallelism

Application of the parallel computing can be considered in different aspects. Building a cluster and using computers as rendering nodes is one of the possible solutions [21]. This solution assumes the division of the visualized data into blocks, displaying each block on a separate computer and connecting individual displays into one final picture. This approach assures visualization with high resolution and interactivity while satisfactory efficiency is maintained. As it is shown in Figure 7.6, it is very important that the fragments of the object are properly connected. Good quality of the rendered details is needed. When this quality is poor, differences at the boundaries between connected objects can appear. An example shown in the left part of Figure 7.6 missing points at the border can be seen, which is an important weak point of this solution.

Building a cluster of computers is another approach to hardware parallelism. In this solution, the nodes not only render but also carry out parallel computations. System WireGL [10] is an example of such a solution. Due to the execution of 16 displaying nodes and 16 computing nodes, it was possible to obtain animation for 70 million triangles. It was an excellent result in the year 2001 when this work was published. This method is scalable for modern visualization devices as well.

Using parallel operations on the text files with data for visualization is another interesting solution [24]. This method eliminates the delay connected with the management of the input data. Consequently, scrolling the data in good quality is possible. Besides that, the data are loaded very fast, a large amount of data can be displayed. This solution allows building a low-cost computer system, which performs data management in a distributed way while all operations connected with rendering are realized on a dedicated external machine.

Presented methods of hardware parallelism make the system very efficient, assuming that it is properly designed. This method, however, needs extensive hardware facilities, which is

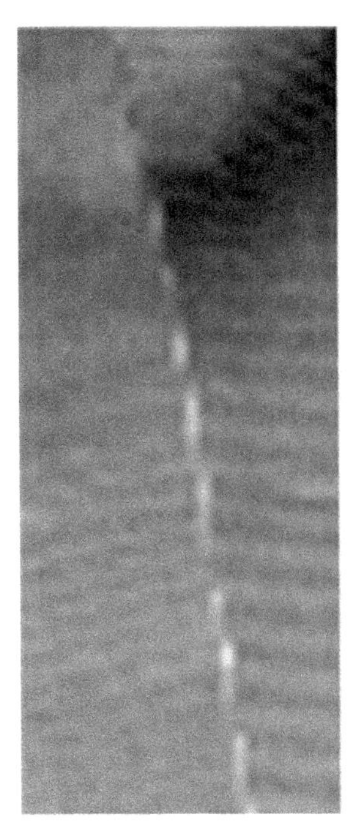

Figure 7.6 Results of the visualization on the GPU cluster with different levels of accuracy.

an obvious weak point. In some cases, displaying all data simultaneously is necessary and presented methods are helpful in reaching this goal. In contrast, when displaying all data simultaneously is not required, using a single computer with dedicated software for efficient data analysis seems to be a better solution.

7.2.2.3 Quality Improvement

The high quality of the visualization is a very important issue for the implementation of the visualization software. It concerns the quality of the displayed objects. Unfortunately, good quality and software efficiency contradict each other. To obtain high-quality visualization, algorithmic approaches can be applied or adjusted to the needs of technologies available in the graphical environment.

7.2.2.3.1 Algorithmic Approaches

Application of the Yin-Yang mesh to visualization of the spherical data is a solution in the search for better efficiency [16]. Spherical data are common in visualization of scientific data, which is demonstrated by previously described algorithm. Due to convergent mesh close to poles, the spherical coordinate system is not perfect. Suggested Yin-Yang mesh guarantees balanced data distribution and effective visualization. Discussed mesh is shown in Figure 7.7. It is seen that visualized sphere has properly distributed points in the whole volume, and there is no deforming influence of the poles. This solution assures very good quality, but the management of the resources for visualization is computationally intensive. Created Yin-Yang mesh contains a large number of points, which are useful only in the case of very

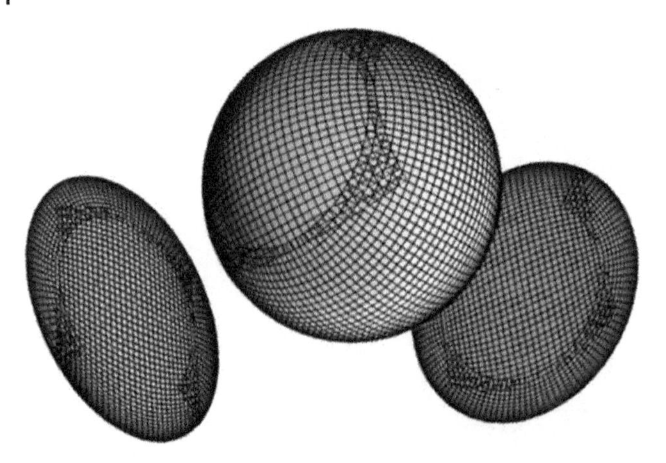

Figure 7.7 View of the Yin-Yang mesh, which is composed of two Yin and Yang constructions, connected with each other and partly superimposed, which create a spherical surface.

close analysis of the particle. This feature is also quite useful in the visualization of materials structures.

7.2.2.3.2 Technological Capabilities

OpenGL library allows using only such figures as spheres. If drawing such a figure involves calling one procedure, its rendering is more complicated. The newest version of OpenGL contains a number of extensions, which are useful in the improvement of quality and efficiency. GL_POINT_SPRITE is one of the solutions, which allows attaching texture to one point in the 3D space. By application of this attachment in the larger amount of particles, one obtains a system of particles, which is an efficient solution for both static and dynamic problems. This extension is used, among the others, in the creation of fog or smoke or visualization of snow [6]. As it is shown in Figure 7.8, when separate particles with attached textures (Figure 7.8a) are subjected to the engine for display, a final effect is obtained (Figure 7.8b). The extensive capabilities of this method and how efficient it is can be seen.

The presented extension of the OpenGL library is important for the present work. Individual particles in microstructure or nanostructure have to be visualized as spheres. As it is shown in Figure 7.9 and in Table 7.1, the creation of a sphere from points requires at least 100 points and the creation of flat figures between these points. Substituting so many points by only one would be an exceptional data gain. Moreover, the introduction of a proper texture should improve the quality of the displayed grains.

The spheres in Figure 7.9 are built by the generation of a mesh of points and the connection of these points. The input parameters for mesh generation compose the number of planes with points and number of parallels along which the points are located (number of points in a plane). These parameters are gathered in Table 7.1. A comparison of the quality of both solutions is presented in Figure 7.10. Beyond noticeable gain in the memory, an improvement of the quality of the visualization was obtained.

7.2.2.4 Material Data for Visualization Purposes

In the present work, data for visualization of the digital material microstructures were generated using the cellular automata method from Chapter 4. Therefore, they are presented as

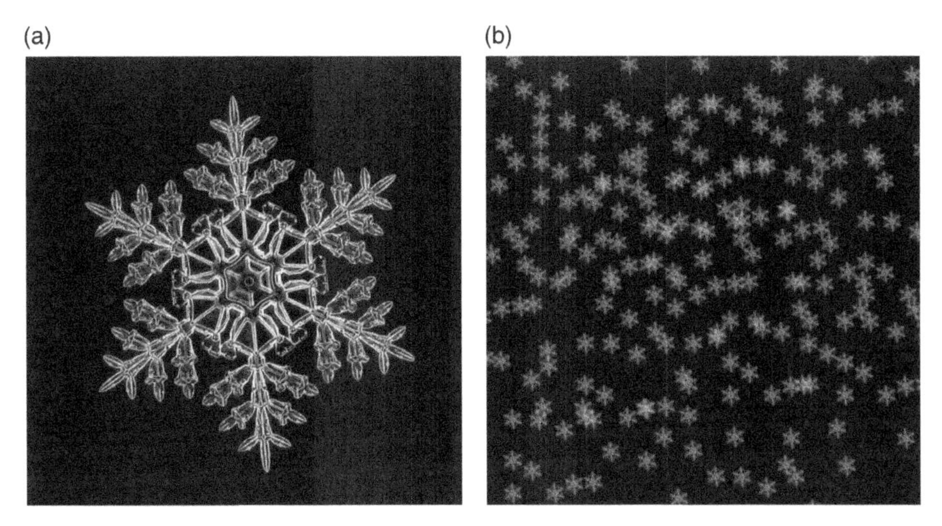

Figure 7.8 Visualization of snow: the texture of a single particle (a) and system of particles subjected to the visualization engine (b).

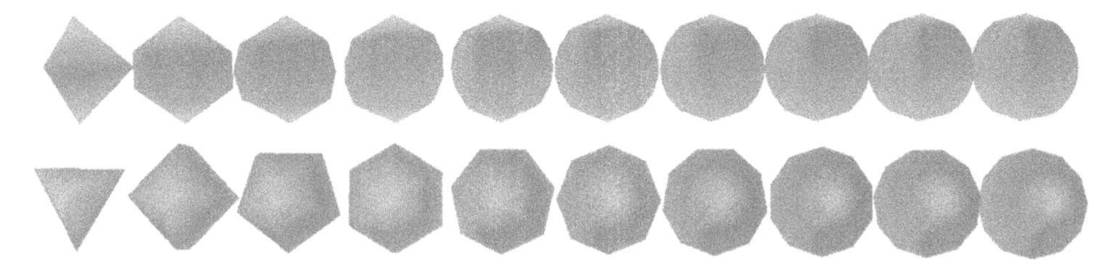

Figure 7.9 Views of spheres built from a various number of points.

Table 7.1 A number of points for spheres in Figure 7.9.

Number of planes with points	3	4	5	6	7	8	9	10	11	12
Number of points in a plane	3	4	5	6	7	8	9	10	11	12
Total number of points	5	10	17	26	37	50	65	82	101	122

the CA space. Coordinates of the particles and their colours are the data necessary for visualization. Additional parameters, such as for example, label/number of the grain, can be added to the input data. The method of presentation of these data is shown in Figure 7.11. As it is seen, the visualization reassembles cubic CA space. Different colours represent individual grains in the microstructure.

Visualization in a lower scale (nanoscale) is based on the visualization of individual particles. Thus, this solution reveals all defects like dislocations, which are not seen in the CA generated microstructures. One has to realize how much data are needed to visualize the microstructure. Figure 7.12 shows the cycle of presentation of the sample after the compression test. The current diameter of the sample is about 11 mm. Analysis of this figure shows that a cubic fragment with a side of 300 μm is visualized to show a small part of the whole sample. The further cycle of the display for a lower scale is represented by a cube with a side of

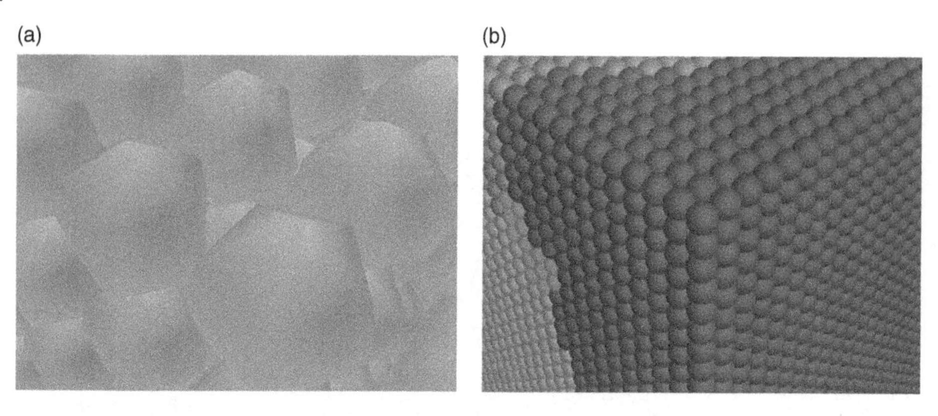

Figure 7.10 Visualization of the particles. Building the sphere from points by their connection (a) and displaying the sphere as a texture attached in a single point (b).

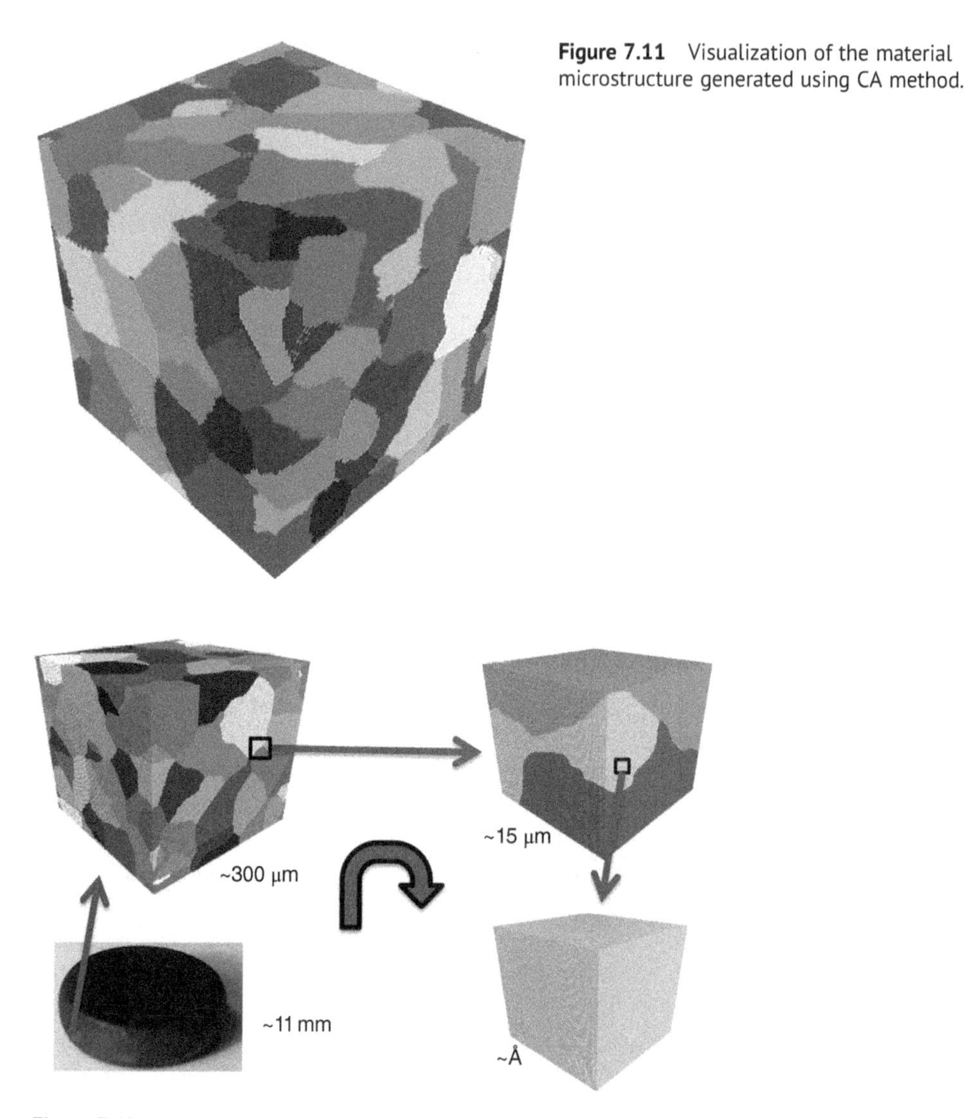

Figure 7.11 Visualization of the material microstructure generated using CA method.

Figure 7.12 Cycle of visualization from coarse to fine scale.

15 μm. This cube contains a triple point. Zooming inside towards lower scales involves an analysis of areas measured in Angstroms. Today, computers deal with visualization of the first 300 μm cube without any problems. Visualization of the whole object, composed of a number of such fragments, would require a dedicated solution. The method has to find a balance between unsatisfactory computer power and a large number of particles.

7.2.3 Visualization Based on Sectioning

According to explanations presented in the previous section, the data describing material structures are very large, and usually, it is not possible to display and scroll all these data simultaneously. The minimum data for the presentation of the digital microstructure contains coordinates and colours of particles. However, these data are not sufficient for interactive visualization. Data for the real-size objects may occupy hundreds of gigabyte memory that shows how difficult their visualization is. To reach this goal, a dedicated algorithm has to be created. This algorithm should allow displaying as much data as is needed for the analysis of the selected part of the microstructure. The algorithm dedicated to interactive visualization of the material structures is described in this chapter.

7.2.3.1 Algorithm Idea

The main idea of the algorithm was a division of the large data file into sections following geometrical criteria. The number of such sections depends on the dimensions of the object. The size of the section depends on the capabilities of the available graphical card, and this size can be configured individually. The division into sections in 2D is presented in Figure 7.13. The central square (section number 5) is the area in which an observer is located. The surrounding area contains all neighbouring sections (numbers 1, 2, 3, 4, 6, 7, 8, 9). The data, which are displayed during the stay in section 5, are external edges of the neighbouring sections marked in yellow. Thus, one or two sides of each neighbouring section are displayed which finally gives a square. Due to this, for a side of the section equal N, analysis of the cube with the side $3N$ is possible. Of course, displayed sections are empty which allows a more accurate interpretation of the material structure. This solution assures continuity of the interactive visualization of large material structures. When a user moves to the next section, it is not necessary to enter into the wall of the displayed cube. Better results can be obtained by a decrease in the size of the section and application of the method presented in Figure 7.10.

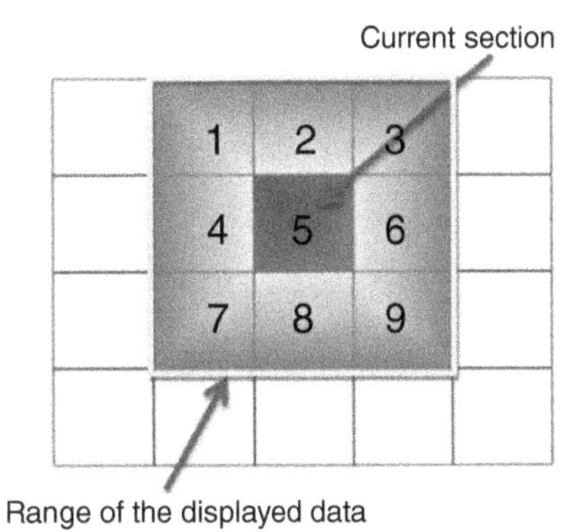

Figure 7.13 The idea of the division into geometrical sections.

The solution with displaying sections is initiated when the observer enters inside the object. The first loading of the section begins at the time of the transfer, and it involves a break in the interaction for less than 0.5 seconds. This break is caused by the fact that it is not possible to store all the external sections while waiting for the transfer to one of them. During rendering of a part of the data, neighbouring sections are loaded in the background.

7.2.3.2 Background Buffering

The data for the neighbouring sections are loaded in the background during the whole time of visualization and dependent on the user's interaction (especially directions of moving through the digital material structure), which allows for smooth transfer between them. It should be mentioned that the loading of the sections is simplified. Data for a specific part are loaded line by line, and an address of the beginning of the section is placed in the metafile. This format is reached after the procedure described next. Figure 7.14 explains the idea of data management. The black arrow presents the operation of the visualization algorithm in time. Red rectangles represent events, which occurred during interactive scrolling. The event presented in the figure is a *change of the section* event and transition to another area. Of course, the new data are immediately displayed, but the diagram in Figure 7.14 presents only processes of loading the data to subsequent neighbouring sections. After displaying new data, the data management thread is run. This master thread runs slave threads to load the data for the preferred section. A possibility of doing all this in one thread exists, but if the visualization procedure has to wait for the end of this thread, the display would have to be stopped. If the application did not wait for the end of the thread, it would not be clear which sections have been loaded. The management thread can wait for the actions of other threads and send messages to the user about the state of the loading.

The moment when a change of the sections occurs is very important. If this change appears after the end of the action of the management thread, there will be no problem for the application. In contrast, if this change appears before the end of the action of the management thread or the preferred section is selected before the first message was sent, the new data will not be loaded. The algorithm is prepared for such a case. If it appears all threads are stopped, and the data are loaded directly. Visualization has to be stopped, but the waiting time does not exceed one second. The time needed to load the section has an influence on the rate of scrolling of the object. Numerical tests allowed to match the rate of the process in a way that smooth transfer to neighbouring sections was possible. It should be mentioned that this rate

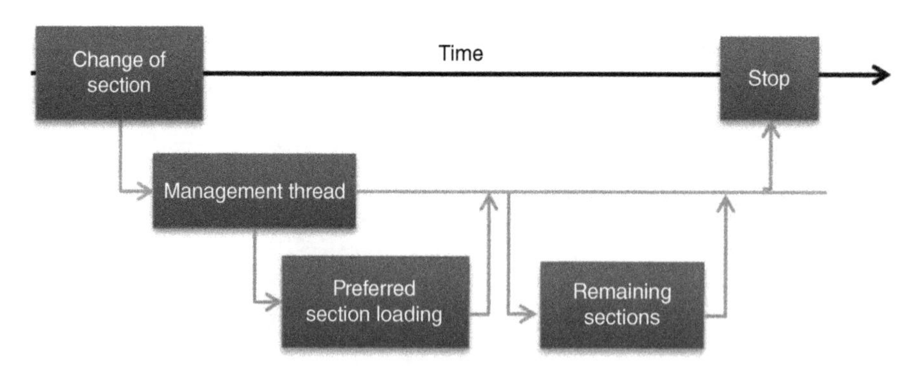

Figure 7.14 Management of the data in the background.

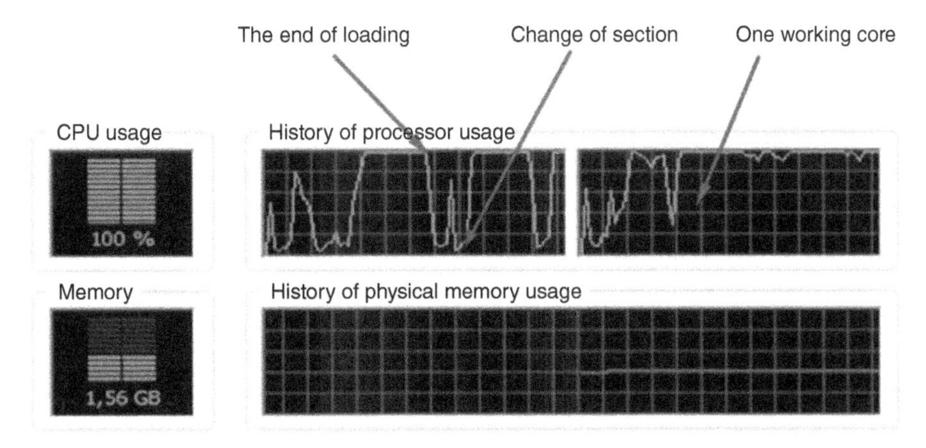

Figure 7.15 Loading of the processors during visualization with casual data loading in the background.

does not impede an analysis, and the user does not feel the application is slowing down. The behaviour of the computer during such data loading is presented in Figure 7.15. As it is seen in this figure, the core used for operations other than rendering is under maximal loading only when the sections are loaded. After loading the sections, the load on the particular CPU core decreases.

7.2.3.3 Preferred Sections

Preferred section is the one transfer from the current section that is the easiest. Selection of this section is difficult, and it has to be kept in mind that the number of operations during rendering has to be as small as possible. The simplest solution is fixing this section as the one in front (number 2 in Figure 7.13, under the condition that number 8 was the previous section). This is the most logical approach to the selection of the preferred section, assuming that direction of the user's interactive visualization will not change. However, in the real process of visualization it cannot be predicted where the user will go during the analysis of the material structures. It depends, to a large extent, on the characteristic of the structures and their various defects. Simultaneous loading of all the sections would be the best solution. It will be possible when parallel operations on files are used.

7.2.3.3.1 Simplifications

Large gain of the computer memory is possible by simplification of the view of the polycrystal structures. Instead of rendering of each individual microstructure, as for example, it is shown in Figure 7.16, displaying only one particle is possible. This simplification is commonly used to obtain better efficiency [13]. Since often relative locations of crystal lattices are analyzed, not locations of single atoms, using this simplification in the implementation of visualization algorithm is fully justified.

7.2.3.3.2 Section Presentation

To enable analysis of the interior of the microstructure, displayed sections can be visualized as empty inside. This idea has already been presented in Figure 7.13. In the presented algorithm, 3D sections are visualized differently (Figure 7.17).

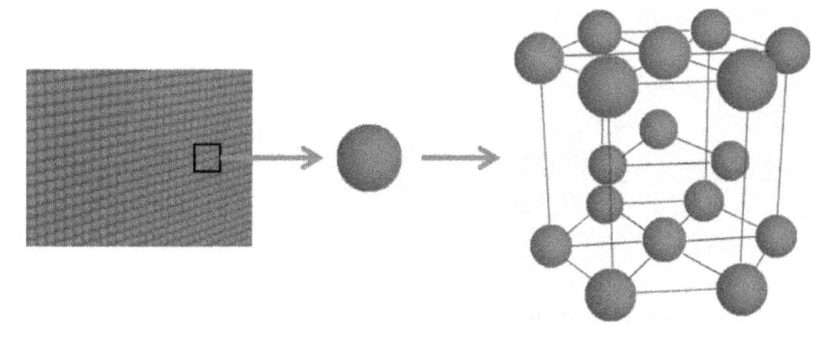

Figure 7.16 Simplification of the visualization of the particles.

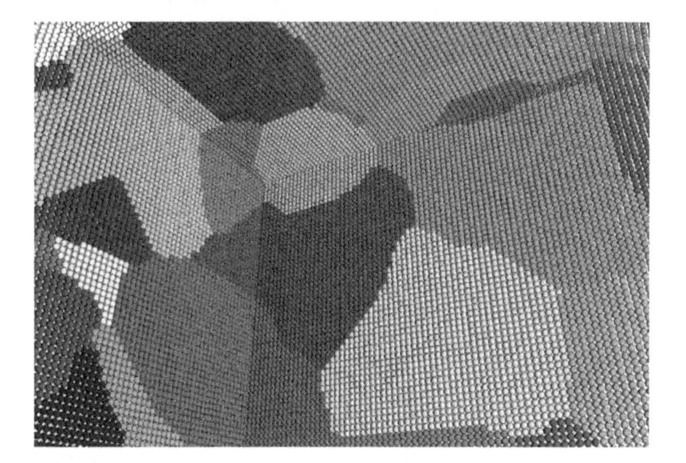

Figure 7.17 View of the section from inside.

The selection of the section, which has to be currently displayed, depends on the location of the observer. On the basis of the coordinates of this location, the section number is calculated from the equation:

$$SN=\left(\frac{Z+(Z_{\dim}/2)}{S_{\dim}}\right)\left(\frac{X_{\dim}Y_{\dim}}{S_{\dim}^{2}}\right)+\left(\frac{Y+(Y_{\dim}/2)}{S_{\dim}}\right)\left(\frac{X_{\dim}}{S_{\dim}}\right)+\left(\frac{X+(X_{\dim}/2)}{S_{\dim}}\right)\left(\frac{Y_{\dim}}{S_{\dim}}\right), \tag{7.1}$$

where: SN – number of the section, X, Y, Z – coordinates of the current location of the observer, $X_{\dim}$, $Y_{\dim}$, $Z_{\dim}$ – dimensions of the object along the coordinates, $S_{\dim}$ – dimension of the individual section.

7.2.3.3.3 Sections Numbering

The number of sections depends on the size of the object and on the length of the side of the section, according to the following formula:

$$SC = \left(\frac{X_{\dim}}{S_{\dim}}\right)\left(\frac{Y_{\dim}}{S_{\dim}}\right)\left(\frac{Z_{\dim}}{S_{\dim}}\right), \tag{7.2}$$

where: SC – sections quantity.

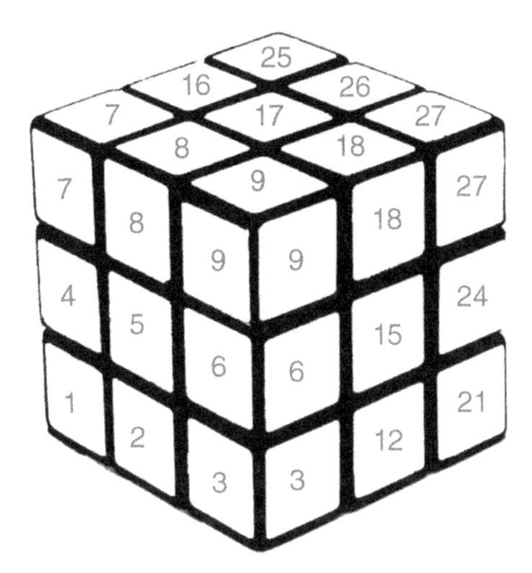

Figure 7.18 Numbering of the sections.

The correctness of this equation is confirmed by Figure 7.18. Sections are numbered, beginning from one of the corners of the cube. Numbering proceeds subsequently in planes perpendicular to X, Y, and Z, as shown in Figure 7.18. This numbering is in agreement with Eq. (7.1).

7.2.4 Functional Assumptions

7.2.4.1 Data Preprocessing

Preprocessing is the first operation before the visualization of the data. This task includes the preparation of the data in such a way that the computing time for displaying of the microstructure and loading of new information in the background is minimal. The process of data preparation is divided into stages dedicated to the preparation of the data for the visualization subsequent scales of material structures in descending order, e.g. at first macroscale, then microscale and nanoscale. The finest scale is usually more complicated; thus it will be described in more detail.

As a result of preprocessing for the data file, three new files are obtained (Figure 7.19). Edge is a file, which contains information about the surface of the object. The remaining two files describe the interior. Sections are subsequently located in the inside file, and metadata file contains information where these sections are located.

7.2.4.1.1 Nanostructures – External Part

Preprocessing is composed of two stages. The first is dedicated to obtain the surface of the object, which will be available for displaying when the interior is not analyzed. In the current version of the application, the preprocessor is adjusted to cubic objects. The file with the data for the microstructure (locations of points with their features) is scrolled line by line. When the point is located on the edge of the object, the coordinates with relevant colour are saved to the edge file. In this case, there is no need to use metadata file because all data are displayed at the same time. The data, which are not visible for the observer, are not displayed. This functionality is enabled by the graphical library.

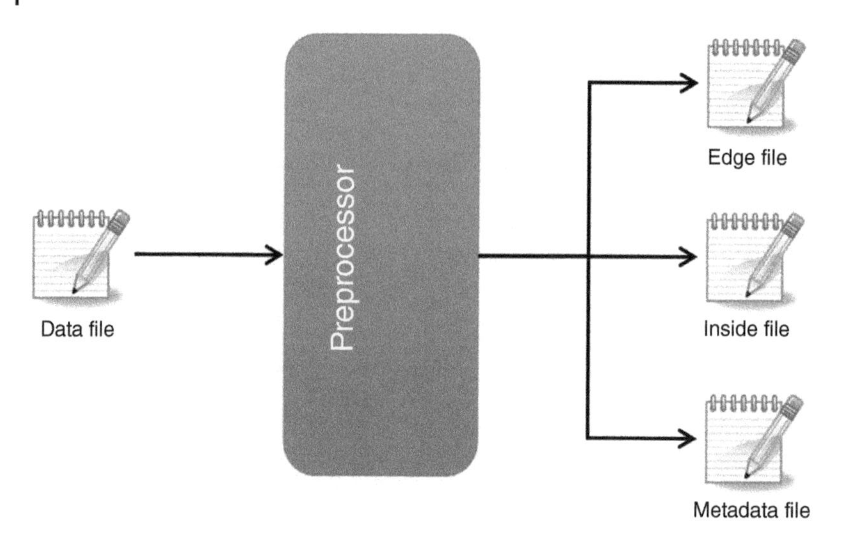

Figure 7.19 The idea of the data preprocessing for visualization of the material structures.

7.2.4.1.2 Nanostructures – Internal Part

The inside file is created in the same way as presented as the edge file. Subsequent sections in the object are considered, and the edges of the object are searched for. When the number of the section is known, it is easy to define its walls. The inside file contains subsequent sections. Simultaneously with this file, the metadata file with locations of the sections is created. The metadata file contains information on how many bytes to move the indicator from the beginning of the inside file in order to load the requested section. The numbers of bytes are saved when the file is created by the application responsible for the data preparation. After these numbers are saved, the size of the file is checked. The received number of bytes is an address for the next section. The address for the first section is zero. The metadata file has exactly as many lines as there are sections in the object. An example of this application is shown in Figure 7.20. It is seen in this figure that the inside file (left column) contains sections, which have an identical quantity of the data. On the other hand, the real data are not identical, and it is difficult to predict what the distances between the beginnings of the sections are. This information is located in the metadata file in which each subsequent line is an address of the beginning of the next section. Three lines in Figure 7.20 show addresses that are stored.

7.2.4.1.3 Digital Microstructures

Data preprocessing for the visualization of the microstructure is similar to the preparation of the edge file. Since another technology of displaying is used, it is necessary to generate different data in a different order. It should be mentioned that visualized microstructure is only the outer layer of the object. Since only the nanoscale allows to analyze details inside the object, displaying the interior is an option for the nanostructure only.

Microstructures are displayed as the cellular automata space using squares (Figure 7.21). To avoid repeating the data, the squares are not defined as points. The mesh composed of all corners of future squares is created and information about connections between corners is stored in a table with indicators. The result of this approach is presented in Figure 7.21b. This method gives gradient transition between grains. In case if real grains do not penetrate each other, this approach does not introduce an error. A comparison of the two methods is

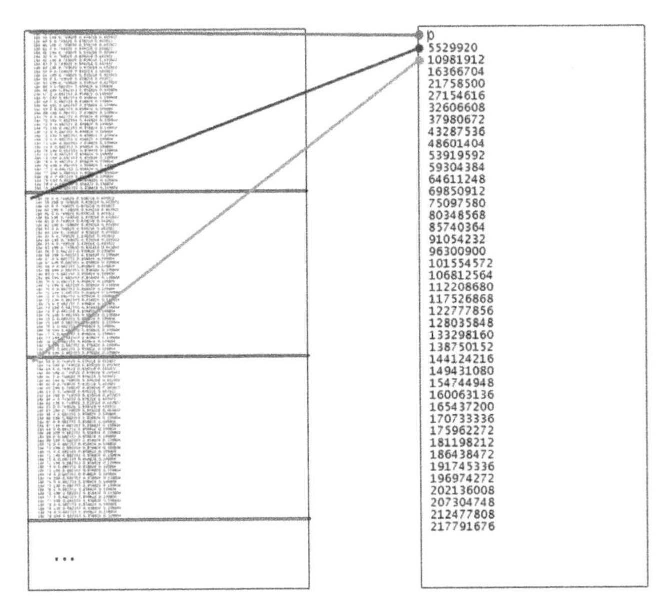

Figure 7.20 The role of the metadata file (right column) for the inside file (left column).

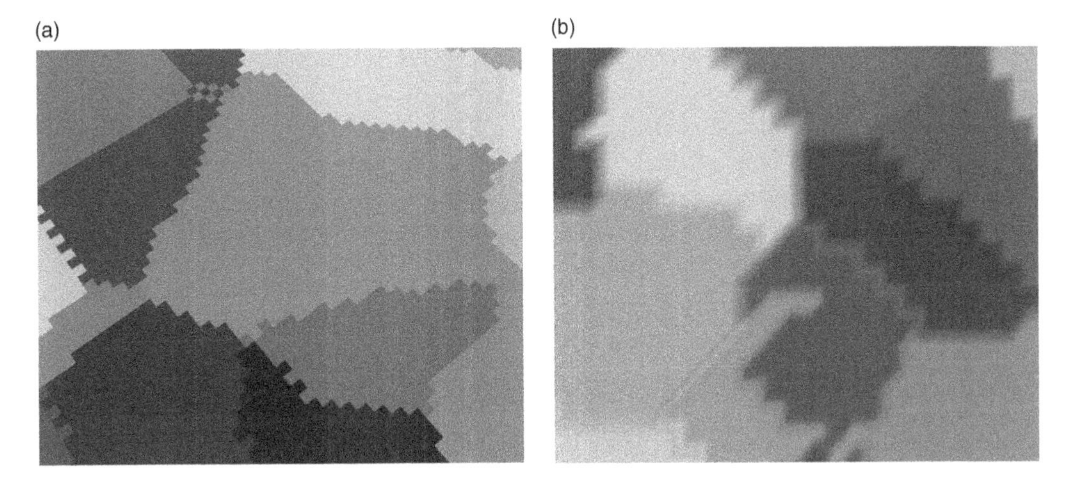

Figure 7.21 Visualization of the microstructure using the method of (a) drawing the squares and (b) corners with indicators.

presented in Figure 7.21. To visualize the data using this method, additional actions are needed. Beyond coordinates of nodes of the mesh and colours, the table with pointers has to be generated. The coordinates are the numbers of four subsequent points which create the square. When the microstructure is to be drawn, the corner which is not located on the edge will be introduced four times for four neighbouring squares. When we use pointers, such a point is defined only once with four different pointers. Each point is represented by its coordinates (x, y, z) and by the colour (r, g, b). To simplify the procedure of generation of components, the structure of the microstructural data has to be defined in a specific way.

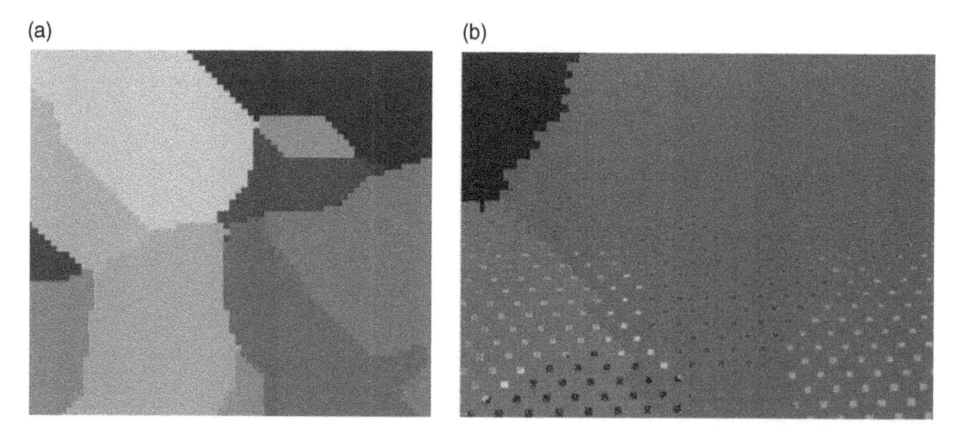

Figure 7.22 Visualization of the microstructure using GL_POINT_SPRITE (a) and effect of clearance (b).

The data are grouped along walls on which they are located. When the number of points on one wall are known, further processing is facilitated.

A method similar to visualization of particles is the second possible approach to visualization of the microstructure. The square is inserted in each defined point. This solution does not need any texture, and the figure defined in the code is satisfactory. This method is more efficient compared to those previously discussed. It requires less memory, and the number of displayed FPS can be larger. The effect of the visualization using this method is identical to displaying ordinary squares (Figure 7.22a).

This approach has some weak points. The geometric figure attached to each point is always facing the observer. During rotation, a clearance between squares can occur for certain viewing angles, and the back wall of the object can be seen through this clearance (Figure 7.22b). The microstructural data are displayed at some distance from the observer. Due to this, the clearance of the next layer is not likely. The efficiency of the method is so high that, in spite of possible breaks in the microstructure, the proposed solution then displays the squares, even with pointers. Therefore, the OpenGL library (GL_POINT_SPRITE) was implemented in our application.

7.2.4.2 Visualization

The visualization procedure is designed to present the digital form of the material in an interactive manner. When displaying data, the application uses files created in preprocessing operation described in the previous section. The purpose of this application is to allow in-depth analysis of the displayed object. The visualization has several mode options to display different length scales. All of these modes are created to allow a highly accurate evaluation of the features of a material.

7.2.4.2.1 First View

When the application is turned on, the outline of the material is displayed, and its surfaces are covered with a texture. It is possible to rotate and view the object. This is an option for a general analysis of the displayed object and possibly selecting a place for more detailed observation. This display can be compared to the visualization tools of modelling applications such as Solid Works.

7.2.4.2.2 Microstructures Presentation

After sufficient zoom into the wall of the component, the microstructure of the material is displayed. Instead of a metallic structure, a cellular automaton space appears, which contains information about the grains. Microstructure data is created using cellular automata, as mentioned before, so its visualization consists of drawing squares. These squares are pinned to points in the space (one square, one point), which allows for a large memory gain.

7.2.4.2.3 Nanostructures Presentation

This part of the visualization implements entirely the dedicated algorithm described in the previous section. After sufficient zooming, the transition from microscale to nanoscale occurs. Due to the need to load the data and load it to the graphics card, the visualization pauses for a couple of seconds. After loading the data, the structure can be analyzed at the molecular level. In this scale, all defects and gaps between grains are already clearly visible. Once the observer comes into an element, a section view of the object is started. Depending on the coordinates at which the observer is located, a specific section is displayed. The data of neighbouring sections are loaded in the background, which ensures continuity of the algorithm and true interaction.

7.2.4.2.4 Transition Between Scales

A key role for the interactivity of the browsing is played by the transition between the displayed scales at runtime. Due to the large difference in the amount of data between successive length scales, it is not possible for the transition to be smooth. Considering the fact that it is not known where exactly the element will be approximated, or where the entrance to the inside of the object will take place, it is necessary to focus on loading the required data as quickly as possible. The process from request to display must include the stage of loading the data and then loading them into the GPU. The data that was previously displayed must be erased. The only possible step that can be optimized is the data loading by choosing an optimal method. The way the data is stored also plays a very important role here. File operations can take a very long time, so it is desirable to review the best methods, apply them to the specific case of loading for material structures, and choose the best one. Due to the fact that the software is created for PC class computers, it is better to focus on using the maximum speed of the hard drive rather than compressing and decompressing data.

7.2.4.2.5 Data Reduction

A way to get better visualization performance when considering large datasets is to reduce the amount of data. However, this reduction must not consist of reducing the object resolution, which would worsen the analysis, but in some simplification. The simplification proposed for visualization, which is described in the previous subsection, allows a significant reduction of the data, without losing resolution. More specifically, in the case of the HCP structure, fifteen times less data remains to be displayed. A big gain in memory is also the fact that the sections in the middle are empty and that the displayed object in the first two scales is also empty. This is possible because the preprocessor prepares the data beforehand. In this case, the gain is huge, and it does not change anything when analyzing the external structure, which is exactly what is assumed when visualizing these scales.

7.2.5 Case Studies

Analysis of the whole system for visualization, many results of different types were obtained. As a result of the application displaying data on the screen, graphical effects and a certain FPS depending on the rendered scene were obtained. During the operation of the preprocessor, the main focus was on the time in which computational operations are performed and how it depends on the number of threads and execution technology.

7.2.5.1 Digital Microstructures

The results of the visualization algorithms are views of the digital microstructures that were generated as a result of numerical simulations. Figure 7.23 shows how the visualization of subsequent scale lengths may look like – from the object outline to the analysis of its internal part. First, the outline of the object is visible (Figure 7.23a), where a texture from the file has been applied to its walls. If the macrostructure is zoomed in enough, a view of the material microstructure is displayed on the screen (Figure 7.23b). Here you can see a visualization of the cellular automaton space – successive squares represent cells. The next step is the presentation of the nanostructure (Figure 7.23c), called in the same way as the previous microstructure view. The molecules of the entire object are visible. The final step is a sectional view

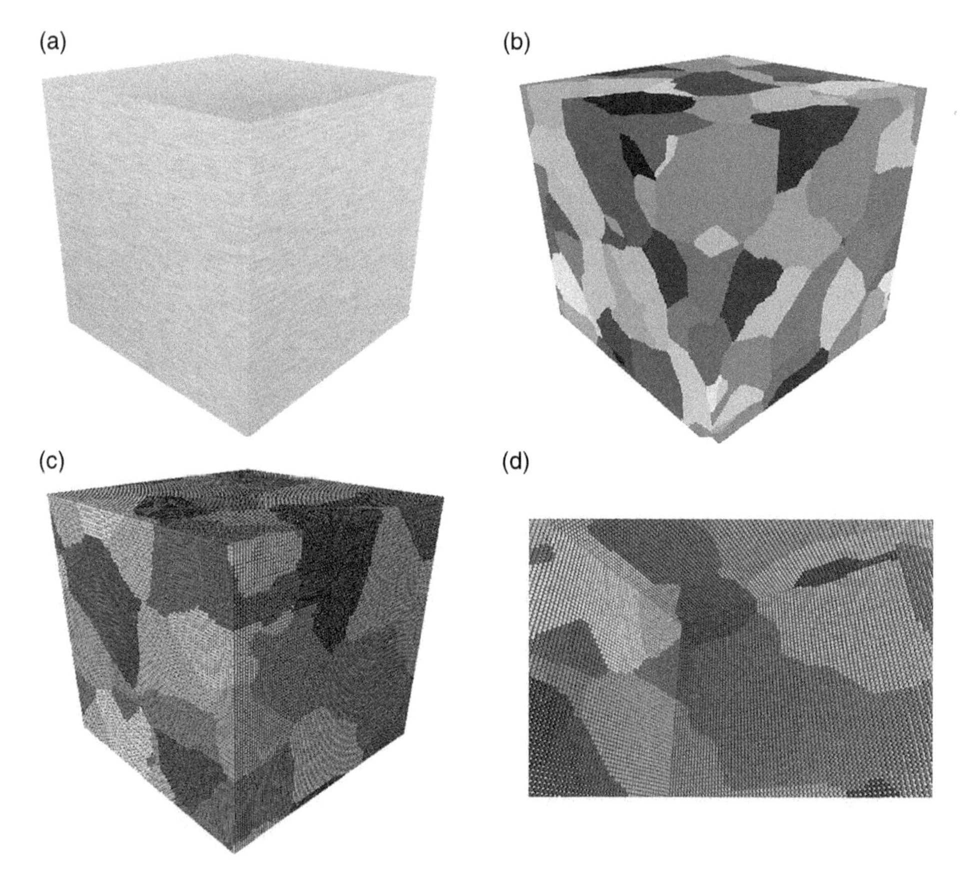

Figure 7.23 Visualization of subsequent length scales of material structures.

of the nanostructures (Figure 7.23d). Details of the sectional viewing algorithm are described in the previous section. It is worth mentioning that the described visualization is completely interactive, with transitions between scales taking less than seconds.

7.2.5.2 Performance Tests

7.2.5.2.1 Data Size Reduction

The visualization algorithm was subjected to performance testing to assess its capabilities. The first test was focused on data reduction during material structures visualization. This basic type of analysis was performed for both micro and nanostructured data where the dimension of the displayed cellular automata and the number of displayed points with an attached texture were changed. The data displayed during visualization were located on the graphics card. Their amount was therefore dependent on the memory of this card. Table 7.2 shows the ball count tests for the best of the implemented methods based on the attachment of the ball textures to single points in space.

As seen in Table 7.2, the data gain is huge. To build a good quality ball of points, it is required to use 99.26% more data than assumed by the GL_POINT_SPRITE extension. The dimensions of the cubes are not random; the last two values are approximate sizes of objects, which can be displayed by using graphics cards with 256 MB and 512 MB memory. The obtained results confirm the tremendous data gain that was predicted during the implementation using the GL_ POINT_ SPRITE extension. Figure 7.24 shows the increase in data as a function of object size. It can be seen that the increase is exponential in nature. This is due to the fact that the visualized object is a cube – a double increase in each side of the wall results in eight times increase in volume.

7.2.5.2.2 Efficiency

The aim of the test is to verify the fluency of visualization in case of increasing data. The results of this test is a number of FPS displayed by the implemented software. This test was performed for a nanoscale visualization implementation. While the microscale uses the same technique, it differs in that no texture is displayed, only a square, drawn using OpenGL library. Tests done with texture display are certainly more reliable.

Tables 7.3 and 7.4 present the results of both performed tests. They were divided into two tables due to the different characteristics of the tests. The measurements were conducted according to two criteria – the number of frames displayed when the object is viewed from a distance, and after a significant zoom and view of only a piece of the wall with the

Table 7.2 Result of tests for a gain of memory during nanostructure visualization.

	Size of macrocube				
	200	500	1000	2200	3200
Number of particles in external layer	79 204	498 004	1 996 004	9 671 204	20 467 204
GL_POINT_SPRITE [MB]	190	1195	4790	23 211	49 121
Data size [MB]	256	1610	6451	31 257	66 150
Data reduction during visualization [%]	9926				

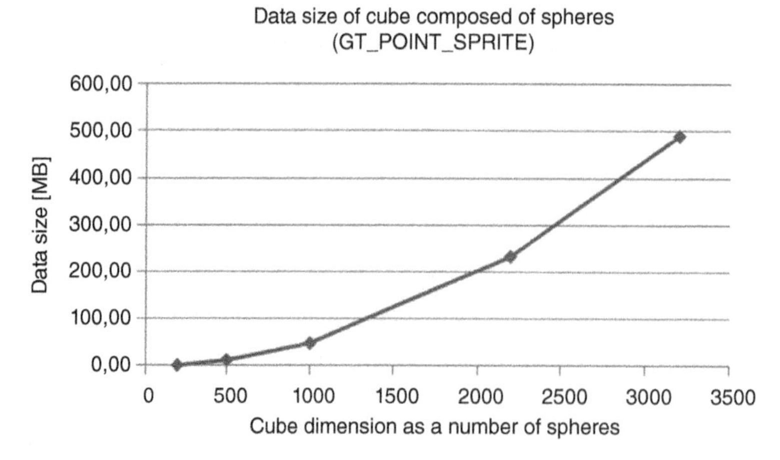

Figure 7.24 Data size used to visualize cube, an object composed of spheres in 3D by using GL_POINT_SPRITE extension.

Table 7.3 FPS obtained for various data sizes (less than two million particles).

	FPS	
Number of particles	**View from distance**	**Zoom in**
200 000	60	60
400 000	60	59
800 000	59	47
1 200 000	45	33
1 500 000	37	27
1 800 000	31	24

Table 7.4 FPS obtained for various data sizes (large sets with more than two million particles).

	FPS	
Number of particles	**View from distance**	**Zoom in**
3 000 000	20	15
4 000 000	16	12
6 000 000	11	8
8 000 000	8	6
10 000 000	7	5
18 000 000	4	3

microstructure. After analyzing the results, it is immediately apparent that the number of frames displayed decreases when the object is zoomed in. Table 7.3 tells how the visualization algorithm behaves with smaller amounts of data. Using the sectional display algorithm described in the previous section, it is not necessary to display a very large amount of data. Based on these results, it is possible to choose an appropriate section size. The application performs very well in displaying data up to two million particles. However, it is worth choosing a much smaller section size, taking into account the amount of data to load in the background of the program during visualization. The results ensure complete interactivity of data viewing. Over 25 FPS is a result satisfactory for human senses. Table 7.4 shows the result of the test for a much larger amount of data. With several system programs running, the largest number of particles that was displayed by implemented software was eighteen million particles by using 256 MB of graphics card memory. Each particle has three geometric coordinates and three colour components.

7.2.5.2.3 Preprocessor

The preprocessor is the most important part of the whole visualization algorithm when it comes to the interactive display of data itself. Its task is to arrange all the points in such a way that they can be viewed in successive sections or the outer layer. For large objects, the application can take a very long time to execute. This is due to the necessity of browsing the whole data files in search of specific points (geometric conditions). Sometimes the whole files have to be browsed, because of the lack of standard order in the analyzed data. In case of the need to optimize the execution time, parallelization with simultaneously running threads on a single machine has to be implemented. The preprocessor for nanostructures is usually composed of one thread, which is reserved for the outer edge, and the other threads, which work on the sections inside the object. For microstructures, only the outer layer is extracted. Finally, the number of threads is hardcoded in the implementation as a function of the size of the visualized object and pointer to the single section. Table 7.5 shows the results of the preprocessing tests for a cubic object built of two hundred particles in each direction. As can be seen, the computation time is reduced the most when five threads are executed.

Table 7.6 shows the distribution of program execution time per task. As can be seen, the vast majority of the time was taken by the I/O (input/output) operations that were performed on the files. The calculations themselves took only a small part of the total time. The presented calculations were performed for a cubic area of a linear dimension of two hundred particles. The row with inside file operations contains four values in the brackets, which are related

Table 7.5 Computing time depending on the number of threads.

Number of threads	Time [s]
1	61.1
2	60.2
3	53.4
5	50.8

Table 7.6 Profiling of the whole algorithm for the subsequent tasks of the visualization.

Task	Time [s]
Data read	24.6
Edge file operations	0.05
Inside file operations	2.8 (2.1 I 2.8 I 2.4 I 2.3)
Data write	23.4
SUM	50.8

to the performance time of each thread. The time of the longest-running thread is the time of the whole task, while the writing operation can start when all threads finish the job.

Possibilities of material structure visualization and the idea of dedicated multithreaded algorithm for efficient interactive visualization of material structures using the OpenGL library were presented in this chapter. An algorithm for sectional processing and viewing was created and used at the nanolevel and microlevel. Vertex buffer object and GL_POINT_-SPRITE were used in the implementation, allowing for very efficient use of the graphics card and maximum reduction of the amount of data to be displayed. Subsequent visualization stages of object outline, microstructure, and nanostructure allow for a very detailed analysis of the object. While the object outline and microlevel view of the structure are indicative, the nanoscale structure has the specific arrangement of particles, the relationships between them, and the relationships between individual grains. Full interactivity was achieved during visualization. The amount of data required for visualization was significantly reduced. It was possible to achieve a high reduction rate by using the GL_POINT_SPRITE extension, reducing visualized object to one point per displayed cell or particle. From the point of view of the OpenGL library, a greater reduction is no longer possible. The displayed data is stored in the hyper-memory of the graphics card, which ensures high performance.

Solutions looking for maximum performance also account for the main idea of creating this visualization to run on personal computers instead of using high-performance servers or dedicated visualization walls. The task of adapting the performance to the appropriate amount of data has been accomplished – by using an average personal computer it is possible to display the structure of real sophisticated objects. Implemented numerical procedures facilitate scientific visualization to study the multiscale structure of materials. The proposed software can still be developed with further functional modules. First of all, it is necessary to create an engine for visualization of non-cubic objects. Often only a part of the analyzed object is considered, but extending the software capability to include any geometric object will provide better and more realistic visualization. At this stage, the application appears to be a tool for visualizing computer-simulated data of the limited shape. Hardware parallelization of the background data loading procedure is one of the most interesting directions of development of the proposed solution. This idea would certainly improve the efficiency of visualization. With such an approach, the section change would always take place at the time of transition. Finally, the GUI is also a very useful module, which has to be developed in the future. The preprocessor is already equipped with a sufficient user interface. It would also be worth adding a database to the system in which various objects for visualization would be stored. Such a solution would consume very large amounts of memory, and therefore file compression techniques would have to be used.

References

1 Ansys Engineering Simulation Software (2021). www.ansys.com (accessed 25 October 2021).

2 Aoyama, D.A., Hsiao, J.T., Cárdenas, A.F., and Pon, R.K. (2007). TimeLine and visualization of multiple-data sets and the visualization querying challenge. *Journal of Visual Languages and Computing* 18: 1–21.

3 Brill, F., Erukhimov, V., Giduthuri, R., and Ramm, S. (2020). *OpenVX Programming Guide.* Academic Press.

4 Burczyński, T. and Kuś, W. (2009) Microstructure optimization and identification in multiscale modeling, in New computational challenges in materials, structures and fluids. ECCOMAS Multidisciplinary Jubilee Symposium, Springer, pp. 169–181.

5 Comsol Software for Multyphysics Simulation (2021). www.comsol.com (accessed 25 October 2021).

6 Ganczarski, J. (2015). *OpenGL Fundamentals of 3D Graphics Programming.* Helion.

7 Guthe, S. and Straßer, W. (2001). Real-time decompression and visualization of animated volume data. IEEE Visualization Conference, San Diego, 349–356.

8 HEXAGON, E-XStream, Digimat (2021). https://www.e-xstream.com/products/digimat/about-digimat (accessed 25 October 2021).

9 HEXAGON, MSC Software, MSC Nastran (2021). https://www.mscsoftware.com/product/msc-nastran (accessed 25 October 2021).

10 Humphreys, G., Eldridge, M., and Buck, I. et al. (2001). WireGL: a scalable graphics system for clusters. 28th Annual Conference on Computer Graphics and Interactive Techniques SIGGRAPH, Los Angeles, 129–140.

11 Kaeli, D., Mistry, P., Schaa, D., and Zhang, D.P. (2015). *Heterogeneous Computing with OpenCL 2.0.* Morgan Kaufmann.

12 Kellehera, C. and Braswell, A. (2021). Introductory overview: recommendations for approaching scientific visualization with large environmental datasets. *Environmental Modelling and Software* 143: 105113.

13 Kuttel, M., Gain, J., Burger, A., and Eborn, I. (2006). Techniques for visualization of carbohydrate molecules. *Journal of Molecular Graphics and Modelling* 25: 380–388.

14 Liang, K., Monger, P., and Couchman, H. (2005). Interactive parallel visualization of large particle datasets. *Parallel Computing* 31: 243–260.

15 Liu, Q., Qiao, Z., and Lv, Y. (2021). PyVT: a python-based open-source software for visualization and graphic analysis of fluid dynamics datasets. *Aerospace Science and Technology* 117: 106961.

16 Ohno, N. and Kageyama, A. (2009). Visualization of spherical data by Yin–Yang grid. *Computer Physics Communications* 180: 1534–1538.

17 OpenGL (2021). The Industry's Foundation for High Performance Graphics. www.opengl.org (accessed 25 October 2021).

18 Paraview (2021). www.paraview.org (accessed 25 October 2021).

19 Poręba-Sebastjan, M. and Kuś, W. (2020). Decoupled homogenization of hyperelastic composite with carbon black inclusion. *Computer Methods in Materials Science* 20: 14–23.

20 Roters, F., Diehl, M., Shanthraj, P. et al. (2019). DAMASK – The Düsseldorf Advanced Material Simulation Kit for modeling multi-physics crystal plasticity, thermal, and damage phenomena

from the single crystal up to the component scale. *Computational Materials Science* 158: 420–478.

21 Strengert, M., Magallon, M., Weiskopf, D. et al. (2005). Large volume visualization of compressed time-dependent datasets on GPU clusters. *Parallel Computing* 31: 205–219.

22 VTK(2021). The visualisation toolkit. vtk.org (accessed 25 October 2021).

23 Xie, K., Yang, J., and Zhu, Y.M. (2005). Real-time rendering of 3D medical data sets. *Future Generation Computer Systems* 21: 573–581.

24 Yu, H. and Ma, K.-L. (2005). A study of I/O methods for parallel visualization of large-scale data. *Parallel Computing* 31: 167–183.

25 Zudilova-Seinstra, E., Yang, N., Axner, L. et al. (2008). Service-oriented visualization applied to medical data analysis. *Service Oriented Computing and Applications* 2 (4): 187–201.

8

Concluding Remarks

Providing advice and guidelines on how to combine the problem of multiscale modelling with the optimization tasks was the main theme of the book. Our main goal was to provide a wide range of possibilities for practical applications and make the design of materials processing more accurate and efficient. The book presents methods that can break through the barriers that today hinder the development and broad application of computational materials design in industrial practise. Beyond the description of various modelling methods allowing to increase the accuracy of predictions, we have tried to emphasize the importance of selecting an appropriate optimization technique for a particular computing task.

As mentioned in Chapter 1, the development of new materials with unusual in-use properties requires advanced modelling methods combined with optimization techniques. The numerical models should support control of phenomena occurring in mesoscale, microscale and nanoscale during manufacturing. It is clearly emphasized that multiscale modelling will be more often used in practical applications both due to the fact that the complexity of models can be reduced within a reasonable manner as well as the computational power of today's computers will continue to increase. However, when working with multiscale modelling techniques, the computational complexity should always be considered and cannot be underestimated.

The driving force for the fast development of multiscale modelling techniques is also exposed in Chapter 1. It is shown that accurate and reliable prediction of the correlation between processing parameters and product properties requires an investigation of the macroscopic material behaviour accounting for phenomena occurring at lower dimensional scales, at grain level or even at atomistic levels. Main problems connected with multiscale modelling techniques are reviewed in that chapter. Finally, the role of optimization in the design of materials is emphasized.

The major physical phenomena responsible for material behaviour under manufacturing and exploitations stages are described in Chapter 2. Simulation methods in the nanoscale are presented first. The laws of quantum mechanics, which underlie the basis of the description of the physical phenomena, are discussed. To lower the high computational complexity of ab initio calculations, various simplified models of interatomic interactions, called atomic potentials, were introduced and presented in that chapter. It is concluded that the majority of the molecular dynamics (MD) solvers has already built-in software routines for embedded atom method (EAM), reactive force fields (ReaxFF), and reactive empirical bond order (REBO) potentials. The EAM, and a more sophisticated modified embedded atom method (MEAM), which overcomes the limitations of the former, is one of the most popular

Multiscale Modelling and Optimisation of Materials and Structures, First Edition. Tadeusz Burczyński, Maciej Pietrzyk, Wacław Kuś, Łukasz Madej, Adam Mrozek, and Łukasz Rauch.

approaches in modelling monometallic materials and alloys. On the other hand, the REBO potential as well as the ReaxFF formulation are suitable for modelling hydrocarbons, carbon, its allotropes, and whole family of two-dimensional materials with surprisingly good accuracy. The MD simulations and optimization techniques using such potentials are presented in Chapters 4 and 6.

The next part of Chapter 2 is dedicated to the microscale material models. A short literature review of the progress in numerical modelling is presented, starting from simple external variable models, internal variable models, and finally to complex discrete solutions. Our goal was to show that the researchers should do their best to avoid uncritical growth of the developed models only due to the growth of processing capabilities and computer memory. It was shown in several instances in the book that the application of a simple model may supply satisfactory data for the process designers and may allow very efficient optimization of the considered process. The importance of the identification of the models is emphasized in Chapter 2, as well. Several examples of identification based on the inverse analysis are also presented. These examples are constrained to the conventional models, as the optimization and identification issues in the multiscale models are described later in the book in Chapter 6.

Chapter 3 is dedicated to the methods which describe thermomechanical processes at the macroscale. The classic computations methods like the finite element method (FEM) and boundary element method (BEM) have been for decades successfully used in the multiscale computations of various physical phenomena. Since these are well-known methods, only brief information is given in this chapter. Additionally, more advanced methods like XFEM and coupled BEM/FEM are also described to highlight the variety of available numerical methods. The computational homogenization as a tool for coupling different analyses scales is also discussed in Chapter 3. A series of crucial methods for modelling phenomena at nanoscale and microscale, that have a stochastic and discontinuous character, is presented in Chapter 3. Molecular statics (MS), MD, Monte Carlo (MC), as well as cellular automata (CA) are described in detail. At the end of this chapter, the most popular classical and modern computational intelligence optimization methods are presented. Firstly, the gradient and nongradient-based classical optimization algorithms are presented, and advantages, as well as disadvantages, are discussed. Secondly, the computational intelligence global optimization algorithms like evolutionary algorithms, artificial immune systems, and particle swarm optimization are described in this section. The pros and cons of using different types of methods are explained to the reader.

Then various approaches for the creation of an explicit representation of the material structure that is of importance in multiscale models are described in Chapter 4. The importance of taking into account such explicit representation in numerical simulations of material behaviour under manufacturing and exploitation stages is emphasized. The methods used for digital material representation (DMR) of nanostructures are presented first. Possibilities of incorporation of such DMR into the nanoscale simulations based on MS or MD methods are addressed. Following this, the methods used for the generation of DMR of microstructure with various features, e.g. grains, grain boundaries, inclusions, and phases, are presented. The two approaches based on image processing techniques and statistically equivalent approaches are discussed. The advantages and limitations of both methodologies are clearly distinguished. Possibilities of incorporation of such DMRs as a basis of the microscale simulations using MC, CA, or FE methods are also addressed.

State of the art in the multiscale simulation is briefly presented in Chapter 5. Based on the wide range of examples of multiscale simulations, a classification of these techniques into two major upscaling and concurrent approaches is presented. Examples of multiscale analyses in the two- and three-dimensional scales are also described.

As mentioned, the combination of the optimization techniques with the nanoscale and microscale simulation is the main added value of the book. The multiscale optimization allows to obtain the optimal performance of structure at one level by changing parameters of structures at other levels, e.g. the modification of material parameter in microstructure allows to obtain the macrostructure with minimal displacements. Chapter 6 describes optimization techniques in practical applications to optimization of atomic clusters, obtaining the best material properties, shape, and topology optimization in the multiscale models.

The identification process can be formulated as a special case optimization problem. The identification can be performed with the use of optimization algorithms. The chapter presents examples of identification in multiscale approaches. The material parameters and shape of structures are also considered in Chapter 6.

The coupling options of optimization algorithms with commercial FEM and BEM software is finally presented in the first part of Chapter 7. Particular attention is put on the formulation of an objective function and related problems in practical applications. This chapter also presents a description of visualization methods used in the pre- and post-processing stages of calculations. The algorithms used for material visualization, slicing, or animation in different length scales are discussed together with their advantages and drawback. The solutions to problems related to these mentioned applications are proposed. The last subsection of the chapter contains a presentation of complex computer methodology dedicated to interactive visualization of multiscale materials. Additionally, the functionality of the most popular visualization programming libraries is presented.

Finally, it can be concluded that benefitting from capabilities provided by modern numerical models combined with optimization methods and sophisticated hardware is the future of computational materials science. We hope that the book will enrich relations between Computational Mechanics and Materials Science communities working on the topic of new material development. This may result in many new opportunities for engineering areas that are currently blocked by the complexity and limitations involved in computational materials design.

Index

Multiscale Modelling and Optimisation of Materials and Structures, First Edition. Tadeusz Burczyński, Maciej Pietrzyk, Wacław Kuś, Łukasz Madej, Adam Mrozek, and Łukasz Rauch.